Ionospheric
Radio Communications

Ionospheric Radio Communications

Edited by Kristen Folkestad

Norwegian Defence Research Establishment
Division of Electronics
Kjeller, Norway

Proceedings of a NATO Institute on Ionospheric
Radio Communications in the Arctic, Organized by the
Canadian Defence Research Telecommunications
Establishment and the Norwegian Defence Research
Establishment, at Finse, Norway, April 13-19, 1967.

℗ SPRINGER SCIENCE+BUSINESS MEDIA, LLC 1968

Library of Congress Catalog Card Number 68-20271

© **1968** Springer Science+Business Media New York
Originally published by **Plenum Press** in 1968
Softcover reprint of the hardcover 1st edition 1968

ISBN 978-1-4899-5513-5   ISBN 978-1-4899-5511-1 (eBook)
DOI 10.1007/978-1-4899-5511-1

PREFACE

Essentially three groups of research workers are concerned with ionospheric radio communications. Radio physicists engaged in a study of the propagation medium, equipment and system designers, and technical personnel responsible for operating terminal facilities. The success of endeavours to solve the numerous challenging problems which still remain depends upon the ability of the groups to coordinate their efforts and functions. The present situation leaves much to be desired. Physicists in particular, in treating the experimental material at hand have been inclined to restrict their research in pursuit of purely scientific goals. Their formulation of problems and terms of analysis have often been poorly matched to the requirements of the users and the design engineers.

A record of a NATO Advanced Study Institute this volume brings together representatives of the above groups to exchange experiences and viewpoints.

This volume emphasizes the phenomena and problems associated with the arctic environment. However, contributions representing novel techniques and advanced system refinements have been included even if they are not particularly related to the high latitude domain. Nearly all the papers are based on experimental research, either in the form of measurements of ionospheric characteristics and properctices of propagated signals or as laboratory investigations to optimize system performances.

Features of the physics of the arctic ionosphere, required as a basis for meaningful discussions, are reviewed in the introductory chapter. Outstanding phenomena bound to the complex mechanisms in the sun-earth relationship are described and their significance for practical communication stressed. Several contributions are devoted to transmissions in the spectrum ranging from the VLF- to the VHF-band. Of particular interest among the multitude of subjects are navigational applications of VLF, aspects of ionospheric and meteor scatter propagation, temporal and spatial characteristics of HF-transmissions, description of modulation techniques, air-to-ground communication, prediction of HF working frequencies and use of anomalous ionozation structures, such as E_s and field-aligned irregularities for practical communication purposes.

Practical engineers will find the section dealing with system techniques particularly important. Features of modern design philosophy are considered, including adaptive transmitting and receiving equipment, methods for avoiding or reducing the effects of frequency and time dispersion, system applications based on electronic computers and employment of artificial ionization in establishing useable channels in operating high priority projects.

Other papers survey the communication facilities in the Canadian and Norwegian Arctic.

A summary of the sessions is given in the final chapter, together with a stimulating panel discussion on future developments.

As a whole, the book provides a comprehensive picture of the present state of affairs in the field of ionospheric communication. Throughout the volume the need for extended cooperation among all groups of communication workers is repeatedly stressed.

Progress presupposes a desire to establish a common language, understood and accepted by all people involved: physicists, designers, and operators. If the book helps to further this view to a broader audience, it will have fulfilled a worthwhile purpose.

CONTRIBUTORS

Allen, R.S., Morphology of fading of radio waves traversing the auroral ionosphere

Bartholomé, P.J., Survey of ionospheric and meteor scatter communications

Blake, K.W. and N.H. Knudtzon, Comparison of propagation predictions and measured performance for two paths over the Norwegian Sea

Burgess, B., Some aspects of the VLF OMEGA navigation system as appropriate to the arctic environment

Coll, D.C., Adaptive receivers for multipath channels

Collins, C. and L.A. Maynard, Simultaneous VHF riometer and forward-scatter observations of the disturbed lower ionosphere

Daly, R.F., see Vincent

Davies, K., Physicist's contribution to panel discussion

Egeland, A., Discussion of VLF records obtained during polar cap absorption and auroral disturbances

Folkestad, K., Results from stepped frequency oblique soundings at high latitudes

Hartz, T.R., The general pattern of auroral particle precipitation and its implications for high latitude communication systems

Hartz, T.R., Summary of sessions

Hatton, W.L., Survey of the effects of ionospheric propagation on HF communication systems

Hatton, W.L., Relationship between the user, system designer and radio physicist

Holt, O., Characteristics of polar cap absorption

Jespersen, M. and B. Landmark, Electron density observations during auroral absorption related to radio wave communication problems

Jull, G.W., HF spatial and temporal propagation characteristics and sounding assisted communications

Knudtzon, N.H., Summary of techniques for overcoming physical problems of high-latitude communications

Kulinyi, R.A., Notes on a multi-mode communication system

Landmark, B., see Jespersen

Lange-Hesse, G., VHF bistatic auroral backscatter communication

Maynard, L.A., Meteor burst communications in the Arctic

Meek, J.H., Physical problems of particular relevance to arctic communications

Meek, J.H., Bridging the gap between physicist, engineer and the user and the needs for the future

Petrie, L.E. and E. Warren, The propagation of high frequency waves on the Winnipeg-Resolute Bay oblique sounder circuit

Petrie, L.E., Developments of HF predictions for the Arctic

CONTENTS

Survey of Existing Communication Facilities and User
Problems in the Arctic

Panel Discussion of Future Developments

OPENING SPEECH

Finn Lied

Director, Norwegian Defence Research

Establishment

1. GENERAL

The noticeable trend in our days to exploit the arctic for industrial, military and scientific purposes has led to an increasing demand for extended and improved communications in that part of the world.

Unfortunately the communication technique most extensively employed hitherto, that of ionospheric HF-propagation, has proved to be rather susceptible to disruption and impairments caused by dynamic processes frequently occurring in the polar ionosphere. The recognized unreliability of propagation at high frequencies in the polar regions naturally has stimulated the interest for transmission systems less critically dependent upon the state of the intervening medium, such as tropospheric scatter circuits, microwave links and the novel technique based on use of artificial satellites.

In spite of the difficulties associated with HF-operation at high latitudes and the persistent efforts to establish more reliable ways of communicating, it is reasonable to believe that this range of the spectrum still for many years will carry a substantial part of the radio-traffic in these regions. Whenever primary importance is placed on simplicity of installations, moderate power consumption and economy, none of the other modes of RF-operation really can compete with an HF-arrangement.

The LF- and VLF-bands have nowadays become firmly established as working domains for several navigational networks, among them the most advanced global system termed OMEGA.

In these frequency ranges it is often convenient to consider a signal as a combination of modes propagated in a spherical wave-guide formed by the earth's surface and the low boundary of the ionosphere. The characteristics of reception are largely determined by the properties of the ionospheric strata involved.

At this NATO Institute we shall confine our interest to a study of ionospheric transmissions only. We are convinced that the part of arctic communication concerned with ionospheric radio propagation is of sufficient future importance to justify such a restriction of the subject.

The arrangement of the Institute should be regarded as an attempt to establish a forum where the significant physical and practical aspects of ionospheric radio communication in the Arctic will receive due attention.

It may be appropriate at the outset, for the sake of perspective to briefly outline some basic concepts pertinent to the polar ionosphere and its relation to solar and interplanetary processes.

2. MAIN PROBLEMS

Generally the electron distribution N(h) in the lower part of the ionosphere, the D- and the E-regions, say between 50 and 95 km, has a profound influence on the propagation of electromagnetic energy at radiofrequencies. Considering waves in the LF- and VLF-ranges the primary effect of an increase in the ionization in the D-layer is a downward displacement of the reflection points with an alteration of the effective signal paths. The resulting changes in the phase patterns on the ground may completely mask the regular diurnal variations. Usually, departures from the normal signal strength levels, detected in conjunction with a disturbance in the lower ionosphere, do not seriously affect the circuit behaviour.

Ionospherically propagated rays in the HF band suffer energy loss in traversing the D-layer on the way to or from the reflection regions in the higher strata. In this part of the spectrum the attenuation is essentially determined by the product of electron density and electron collision frequency, $N \times \nu$. As the collision frequency increases downward in approximate correspondence with an exponential law, one might infer that a given change ΔN at low altitudes leads to a more severe attenuation than a similar deviation in the ionization at larger heights.

Since the propagational difficulties encountered are inherently associated with spatial and temporal changes in the content

of free electrons in the ionosphere, a broad understanding of the
nature of the disturbances implies a consideration of the mechanisms responsible for the production and maintenance of the ionization involved.

At low and middle latitudes N is predominantly caused by
electromagnetic radiation from the sun. In the polar ionosphere
incoming charged particles constitute production agencies equally
or even more important than the ultra-violet source. The particles are known to be of solar or galactic origin. They cover a
broad spectrum of energies and exhibit intensities which depend
upon the conditions on the sun. During periods succeeding solar
eruptions the particle flux may increase considerably.

As the particles approach the earth's surroundings, they are
acted upon by the geomagnetic field and guided into the polar
areas. The resulting spatial distribution pattern is determined
by the energy of the impinging particles. The most serious category of disturbances which may occur, termed Polar Cap Absorption,
is accompanied by an influx of rather energetic particles distributed across the entire polar cap. In the majority of events,
though, the impact tends to concentrate in restricted areas, in
geophysical parlance denoted Auroral Zones. Here they give rise
to a multitude of phenomena, such as magnetic storms, auroral
displays and to the formation of certain types of Es-ionization.

3. HISTORICAL BACKGROUND AND DEVELOPMENT

The pioneering experimental work conducted by Birkeland, and
the subsequent extensive theoretical analyses by Størmer of trajectories of charged particles in a dipole field, mark the starting point of later extensive research of polar disturbances.
Størmer's theories are still found valuable in the study of the
behaviour of cosmic rays. To explain the dynamics of auroral impacts associated with auroral disturbances, more refined models
have to be invoked. It is interesting to note in passing that
Størmer more than 50 years ago showed that charged particles actually could remain trapped in the earth's magnetic field. As
many will know, the existence of regions with geomagnetically
trapped particles was drastically demonstrated in 1958 through the
measurements of Van Allen in Explorer I.

For several decades the lack of experimental data served to
give theories concerned with the solar-terrestrial relationship a
conspicuous character of speculation. The situation is now in
the process of being radically changed, first of all through the
use of rockets and exploring satellites. There are still a variety of unsolved questions which undoubtedly will challenge the
scientists in years to come. However, from the available elements

of information obtained through active research in recent years, a
fairly coherent pattern for the structure of the terrestrial environment is slowly beginning to manifest itself.

Soundings of the ionosphere conducted from the top-side by
the Canadian satellites Alouette I and II and the American space
vehicle Explorer XX have yielded a wealth of information on the
spatial, diurnal and annual variations of the electron distribution, on field aligned irregularities, resonance phenomena and
guided propagation at low frequencies. The output from numerous
experiments designed to measure angular spectra and energy distributions of charged particles, has greatly advanced our knowledge of the dynamics of particle movements in the earth's field.
Our present picture of the geomagnetic field, its boundaries, intensity and directional properties, is in large based on measurements of magnetic sensors in orbiting satellites.

4. PATTERN FOR THE TERRRESTRIAL SURROUNDINGS

The main features of our present-day concept of the earth's
environment may conveniently be displayed in connection with a
representation given by Ness (1), based on measurements of the
IMP satellite.

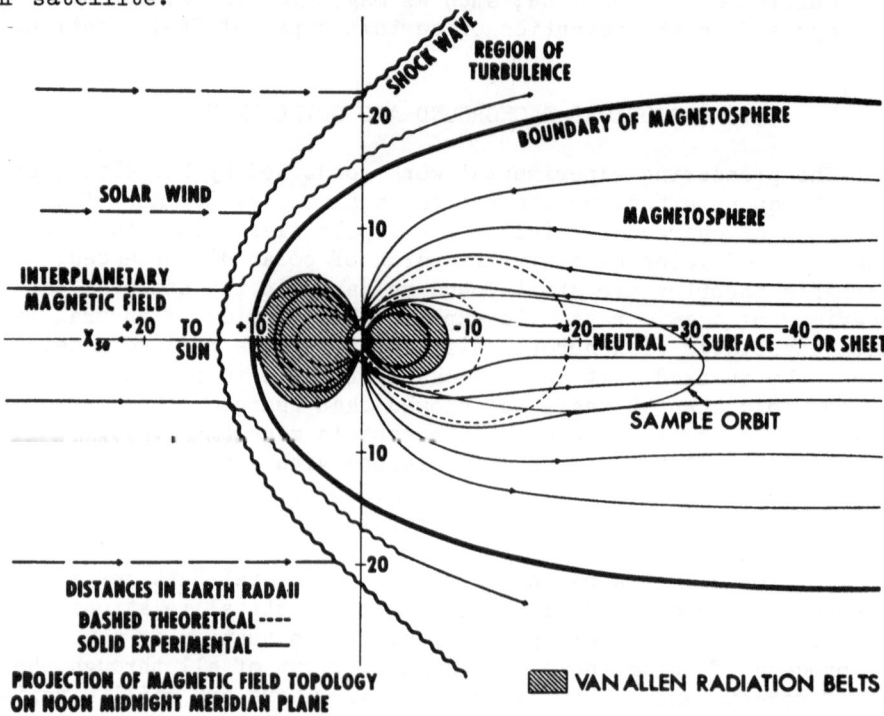

Figure 1. Earth's environment

A highly tenuous plasma blowing radially outward from the sun at a speed of 3-500 km/s at 1 AU constitutes a "solar wind". Embedded in the plasma, and moving with it, is a magnetic field whose order of magnitude is about $5 \cdot 10^{-5}$ Gauss. Due to the interaction of the earth's magnetic field and the streaming plasma flux, a cavity is created inside which the earth's field is confined. Ahead of the cavity boundary a detached shock wave is produced. Between the shock wave and the magnetosphere is a transition region characterized by a turbulent, disordered field configuration. The dimensions of the cavity depend upon the intensity of the streaming jet. The distance to the magnetopause in the solar direction is typically about 10 earth radii. In the direction away from the sun the cavity has been found to stretch out to very great distances, probably to more than 60 earth radii.

The central part of the magnetic tail is made up of a neutral sheet. It is highly probably that this restricted part of the tail plays an important role as a coupling domain where the solar particles are injected into the geomagnetic confinements. From the sheet the particles are constrained to move, under the influence of electrodynamical forces, into the high latitude regions.

The dashed parts in the figure represent radiation belts consisting of magnetically trapped particles. The displayed unsymmetry is easily explained in terms of invariant properties governing this type of particle drift. There are experimental indications that particles in the radiation belts, at times of disturbances, may be released from their confined orbits and penetrate into the lower polar ionosphere. Exact details of the nature of the mechanisms initiating such a dumping of particles remain unknown.

5. CONCLUDING REMARKS

With the rapid expansion in space research in our days, and the resulting steady accumulation of new data, it is unlikely that each detail in the model just described will remain unchanged. It is thought, however, that essential features in the solar-terrestrial relationship are adequately represented. The model does explain the strong linkage of the polar ionosphere to corpuscular interplanetary processes and the observed latitudinal cutoff in precipitation. This is a property which any model pretending to describe the earth's surroundings has to possess.

In the coming week we will hear lecturers supplementing and extending this somewhat sketchy presentation. Several papers will deal with particular forms of particle influx and their implication for arctic communication, the morphology of magnetic disturbances and theoretical aspects of wave-propagation. Other lectures will be devoted to a description of communication tech-

niques, practical results from radio-operation in the Arctic,
users problems and prediction methods.

Participants engaged in practical work, as operators or technical administrators, will have an opportunity of broadening
their concepts and clarifying their notions of the physical processes causing disruption of their circuits. The physicists on
their side may obtain a clearer understanding of the problems and
demands of practical communication services. It is our ambitious
hope that the interchange of information and experience conveyed
through lectures and discussions at this symposium may serve to
focus and concentrate our interests and engagements in the field
of arctic ionospheric communication on those problems where future
efforts may be of greatest benefit.

With these words I would like to open this NATO Study
Institute.

6. REFERENCE

1. N.F. Ness, J. Geophys. Res. 70, 2989, 1965.

Morphology of the Arctic Ionosphere and Ionospheric Absorption

THE GENERAL PATTERN OF AURORAL PARTICLE PRECIPITATION AND ITS IMPLICATIONS FOR HIGH LATITUDE COMMUNICATION SYSTEMS

T. R. Hartz

Defence Research Telecommunications Establishment

Ottawa, Canada

Abstract: Ionospheric radio communications in the Arctic are known to suffer adverse effects during intervals when the auroral activity is high. This paper considers the aurorally associated disturbances of the upper atmosphere, and particularly the anomalous ionization, from a gross statistical point of view. The geographic regions in which the phenomena are most prevalent are indicated, and the latitudinal and diurnal variations of the various ionospheric parameters are summarized. Two dissimilar zones are identified, the characteristics of the particle precipitation in each are reviewed, and the importance of these zones for radio communications is discussed.

1. INTRODUCTION

At high latitudes there are two general types of ionospheric disturbance that adversely affect radio communications, the polar cap phenomena (better known as polar cap absorption, or PCA) and the auroral phenomena. Both of these result from the influx into the upper atmosphere of charged particles, but the basic distinction between them hinges on the energy and number density of the particles. In the main the detailed morphology of the polar cap phenomena is fairly well established, whereas the auroral morphology is much more complex with many uncertain details that still require clarification. The PCA events are associated with incoming protons having energies of the order of 10 MeV or so, but whose number density is sufficiently low that particle interactions are not significant; as a consequence the motions of these particles in the geomagnetic field can be fairly well approximated by Størmer's theory for individual particles in a dipole field.

The auroral phenomena, on the other hand, are associated with a substantially greater flux of lesser energy particles -- electrons and some protons -- and their behaviour in the geomagnetic field is rather more complicated because of particle interaction effects. As a result, these particles do not precipitate in the polar caps but in a characteristic annular pattern that is known as the Auroral Zone. Yet a further complication that is of greater importance for the auroral than the polar cap particles is the fact that the geomagnetic field departs markedly from a dipole field at great heights. This paper will describe the average diurnal precipitation pattern for the auroral particles and, as well, will discuss the relation of this pattern to the actual geomagnetic field insofar as this is known. The description will necessarily be very abbreviated since there is not enough time to dwell on any of the details; a rather general picture of the auroral particle influx will be presented, and some of the supporting gross observational evidence on the aurora and on the associated anomalous ionization will be reviewed. This general picture is intended to serve as a background for the subsequent discussions in the symposium.

Most people are familiar enough with the geographic location of the maximum of the auroral zone in the northern hemisphere, but they sometimes forget that this zone has some substantial width. In Figure 1 is shown the region in which nighttime aurora is seen overhead most frequently: also shown is the zone width between the half-maximum contours (1, 2). The geographic area over which auroras appear sometimes during the night for more than half of the nights of the year includes a large part of Finland, Sweden, and Norway, and almost all of Iceland, Greenland, Canada, and Alaska.

These data apply only to the nighttime visual aurora, or to the phenomenon that appears on all-sky camera records. At the moment, however, our interest is not so much in the visual phenomena as in the related anomalous ionization that can affect the propagation of radio waves. Does such ionization appear at the same place and time as the nighttime visual aurora, and if not, what is the temporal and spatial pattern? In this paper various data on aurora and the associated ionization phenomena are presented in the form of latitude versus time plots so that the systematic diurnal patterns can be seen. These data and the inferred precipitation pattern for the causative particles have already been described in some detail by Hartz and Brice (3) and the description here, although much condensed, will show the relationship of the various observations to the inferred particle pattern.

For convenience the general precipitation pattern for auroral particles in the northern hemisphere is shown here in Figure 2. This diagram represents a rather idealized average of the various gross data that appear later in the paper. For the sake of clari-

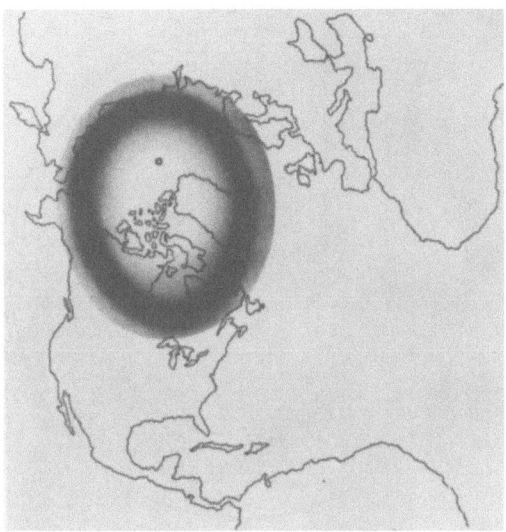

Figure 1. The location of the auroral zone in the northern hemi-
           sphere. The zone maximum, indicated by the most inten-
           se shading, corresponds to the regions over which auro-
           ras are observed on more then 90% of the nights. The
           edges of the shaded region in this diagram correspond
           to the half-maximum contours, and contours or isochasms
           of lesser magnitude are not portrayed. (After
           Feldstein (1)).

ty the two zones are portrayed somewhat too narrow; actually, the
data would suggest more overlap of the zones than is shown here.
The density of the symbols gives an indication of the average intensity of two types of particle precipitation. The zone having
the triangular symbols represents the discrete events associated
with a "splash" type of precipitation, whereas the diffuse events
that are associated with a "drizzle" type of precipitation are indicated by the dots. Later we shall discuss the characteristics
of these zones in more detail; before doing that we would like to
summarize some of the observations on the auroral phenomena in the
latitude-time domain.

    Data averaged over all conditions are probably more appli-
cable to a communications system than those for just one particular degree of disturbance and therefore we have included here only gross, statistical data on the occurrence of each of the phe-
nomean. Where practicable we use geomagnetic time in portraying
the data, since this gives the observer's angular position relative to the magnetospheric tail, and we make no significant dis-

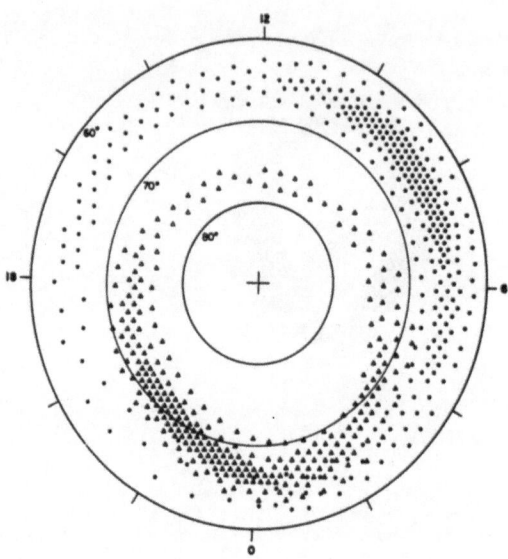

Figure 2.   An idealized representation of the two main zones of
            auroral particle precipitation in the northern hemi-
            sphere, where the average intensity of the influx is
            indicated very approximately by the density of symbols
            and the co-ordinates are geomagnetic latitude and geo-
            magnetic time. The discrete events and the associated
            splash type precipitation are represented by the tri-
            angles, and the diffuse events together with the drizz-
            le type of precipitation are indicated by the dots.
            (After Hartz and Brice (3)).

tinction between geomagnetic and invariant latitude: time differences of an hour or so are not considered as very significant,
not are latitude differences of only one or two degrees, especially in cases where different epochs are involved.

     It should be specifically noted that we prefer to compare ob-
servations at different locations in a quantitative sense in each
case, rather than to locate only the maximum feature on the latitude-time diagram.

     In general, a latitude-time diagram can be constructed if
homogeneous data are available from a north-south line of stations. Because of the elliptical shape of the northern auroral
zone the choice of the longitude for the stations influences the
final pattern, whereas this is not the case in the southern hemisphere; there the geographic auroral zone is approximately circular in shape. A majority of the observations included in this

paper are from northern hemisphere stations in the longitude sector that includes the major axis of the elliptical auroral zone. Also, the precipitation pattern portrayed in Figure 2 has been drawn so as to be compatible with data from these stations. However, these results may depart by up to 4 or 5 degrees from those obtained at other longitudes, or in the southern hemipshere.

2. OPTICAL AURORAL DATA

Feldstein's results of an all-sky camera study of the aurora during the years 1957 to 1959 are shown in Figure 3 (4). Taken as a whole these results show a fairly constant probability of occurrence throughout the day, with slightly higher values in the night hours. The geomagnetic latitude of the maximum is seen to shift systematically from about 67° at night to about 76° at noon. Some people have used the term "auroral oval" to describe this pattern, but an eccentric circle would seem to be a more appropriate description. In any event, these results fit the discrete auroral zone portion of Figure 2. Furthermore, because Figure 3 shows the percentage occurrence rather than brightness, it does not emphasize the fact that near midnight the aurora consists dominantly of the bright active forms that constitute the "breakup" phase (5).

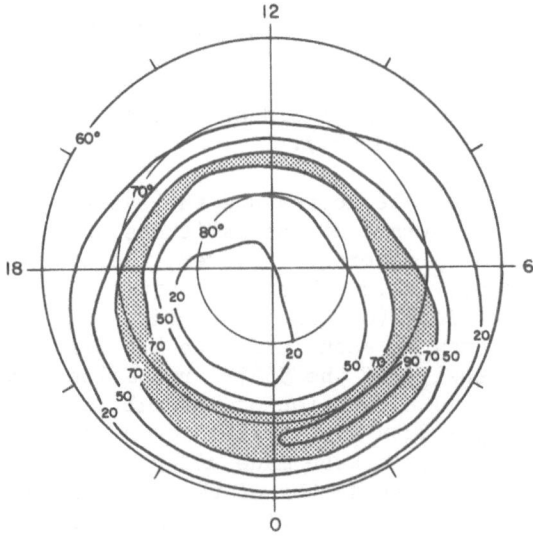

Figure 3. The percentage occurrence of zenithal aurora in the all sky camera records from the northern hemisphere for the period 1957-1959, plotted as a function of geomagnetic latitude and local time. (After Feldstein (4)).

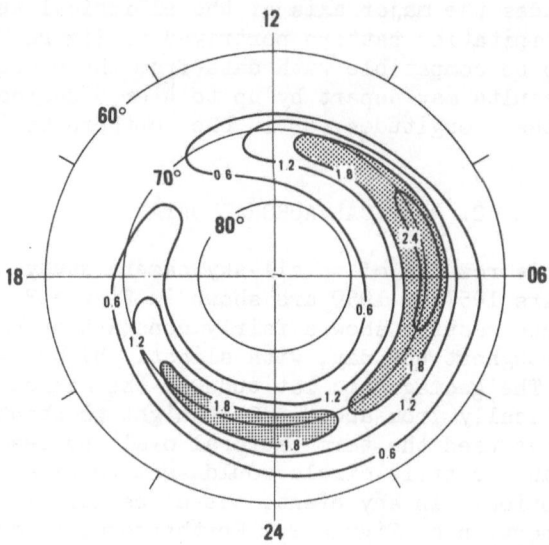

Figure 4. The mean intensity of the 5577 Å night sky emission in
           Kilorayleighs as a function of geomagnetic time and au-
           roral latitude for the southern hemisphere. Included
           are all data for clear nights when the solar zenith
           angle exceeded 102° for the years 1957-1959, and for
           which the all-sky camera recordings showed no aurora
           within 10° of the field of view of the spectrograph.
           (After Sandford (6)).

    This eccentric circle represents the average percentage oc-
currence of zenithal aurora at the latitudes and times indicated,
as recorded on all-sky cameras. However, there are other auroral
phenomena that do not follow this same pattern. The spectro-
graphic data of Sandford (6), in particular, trace out a rather
different zone, as may be seen in Figure 4. Here are shown contours of equal intensity of the 5577 Å radiation from the southern
hemisphere, plotted as a function of auroral latitude and geomagnetic time. These data are averaged over the period 1957 to
1959, and the observations were made under conditions when no
aurora was visible in that sector of the sky, or appeared in the
all-sky camera record. The "mantle aurora" -- as he termed this
phenomenon -- does not follow the eccentric circular pattern of
Figure 3, rather it seems to trace out an annular zone concentric
on the pole with a maximum at about 07 hours. This pattern is
closely similar to the diffuse auroral zone portion of Figure 2.

In the auroral data, then, there are apparently two diurnal zones for the two distinct types of aurora; these fit the two portions of the general particle precipitation pattern. In the next section we shall examine some of the available statistical data on the aurorally associated ionization as well as data on several other related geophysical phenomena that give direct, or indirect, evidence on the particle influx. We shall see that the two-zoned pattern is evident in most of the observations.

3. RADIO AURORA AND ASSOCIATED DATA

Statistical analyses of the type of radio wave absorption known as auroral absorption have resulted in the occurrence pattern depicted in Figure 5 for the northern hemisphere (7). Here the contours are of constant percentage occurrence of 30 MHz absorption, as measured with riometers, on a geomagnetic time-geo-

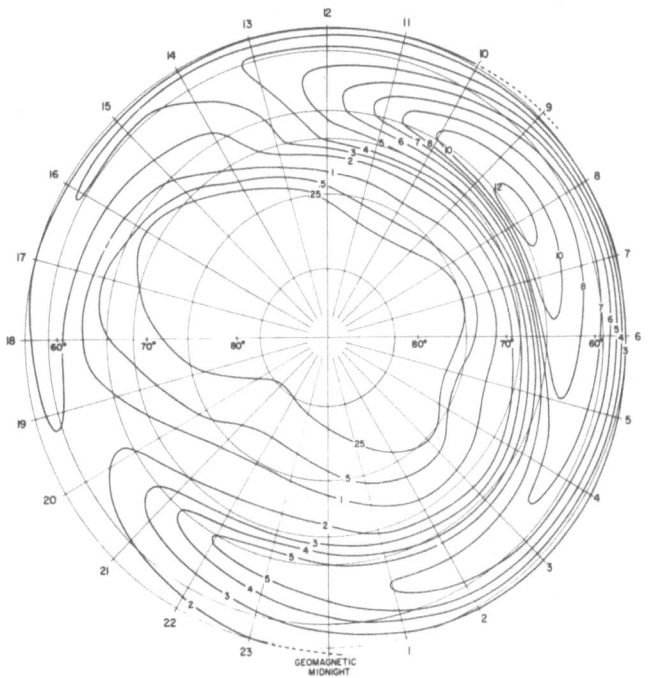

Figure 5. The percentage of the time that auroral radio wave absorption of 1.0 db or more occurred at 30 MHz. The data were obtained in the northern hemisphere during the period 1959-1961, and are plotted as a function of geomagnetic latitude and mean geomagnetic time. (After Hartz et al (7)).

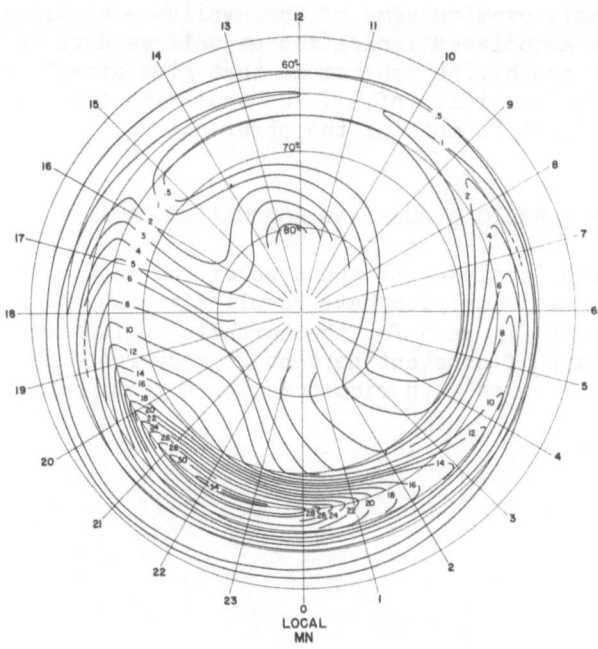

Figure 6. The percentage of the time that type A_2 radio aurora
           occurred in the period 1959-1961 on V.H.F. propagation
           circuits in Northern Canada. The data are plotted as
           a function of geomagnetic latitude and mean geomagnetic
           time appropriate for the mid point of the path.
           (Courtesy of E.L. Vogan (8)).

magnetic latitude diagram. A principal maximum is seen to occur
at about 08 hours and a general similarity to Figure 4 and to the
diffuse auroral zone is quite apparent.

    Radio reflection or scattering from auroral ionization has
been studied by many workers, but the lack of standardized observations and the dominant dependence on the geometry of the ionization irregularities make the intercomparison of data from different locations rather hazardous. At very-high-frequencies the
situation is somewhat better, especially for signals scattered in
the forward direction. In Figure 6 we show the results of a
statistical study of A_2 radio auroral signals, again in the form
of a geomagnetic latitude-geomagnetic time diagram (8). (This
characteristic type of scatter signal has been described by
Collins and Forsyth (24); it shows rapid fading, evidence of as-

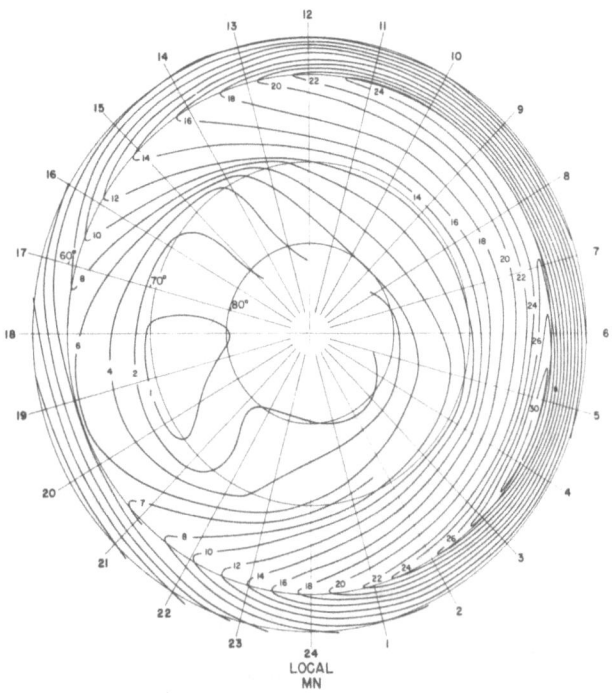

Figure 7. The percentage occurrence of type A$_3$ radio aurora, re-
          corded in the same way as in Figure 6. (Courtesy of
          E.L. Vogan (8)).

pect sensitivity implying scattering from irregularities aligned
along the geomagnetic field, and scatter heights of the order of
105 km). It may be seen that a dominant maximum appears just before midnight. On the other hand, the A$_3$ type of signal, whose
occurrence pattern is given in Figure 7, shows a maximum in the
morning hours, and a minimum in the late evening. (This type exhibits slower fading, no aspect sensitivity, and scatter heights
of the order of 85 km; by definition it appears during magnetically disturbed times, requiring a K index of 4 or greater (24).
Because of the uncertainties in the horizontal location of the
scattering ionization, particularly for the A$_2$ signals, there are
some reservations concerning the latitude information in these two
diagrams. However, the times of the maxima (and minima), along
with other characteristics of the signals, would identify the A$_2$
phenomena with the discrete auroral zone and the A$_3$ phenomena with
the diffuse auroral zone.

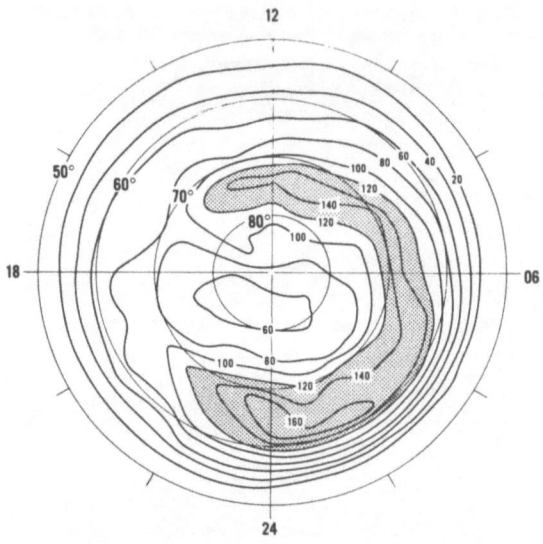

Figure 8. The mean hourly range, in gammas, for the X (or H)
           magnetometer component for all days in 1960 in the
           northern hemisphere: the data are plotted as a func-
           tion of local time and geomagnetic latitude.
           (After Loomer and Whitham (9)).

     Magnetometer data for the northern hemisphere are shown in
Figure 8, in the form of the mean hourly range in gammas as a
function of geomagnetic latitude and local solar time (9). The
maximum variation is seen to occur near midnight at about 65°,
although the contours extend around the morning side with only
slightly reduced values and with a gradual shift toward higher
latitudes. Not only does this suggest that the magnetometer data
belong to the discrete auroral zone, but also that the diffuse
auroral zone phenomena are not associated to any very significant
extent with local magnetic disturbances.

     Figure 9 shows statistical results on the occurrence of "au-
roral hiss" in the V.L.F. emissions data (10). This phenomenon
consists of short bursts of broad-band radio noise whose frequencies generally lie above about 4 kHz. The diagram shows a maximum
at night, just before midnight, along with indications of a latitude shift at later and earlier times: again the similarities to
the discrete auroral zone are obvious. Other studies of V.L.F.
emissions have shown that the "chorus" phenomenon is predominant
in the morning hours -- between about 08 hours and noon -- and
consists mainly of frequency components below about 2 kHz (11,12).

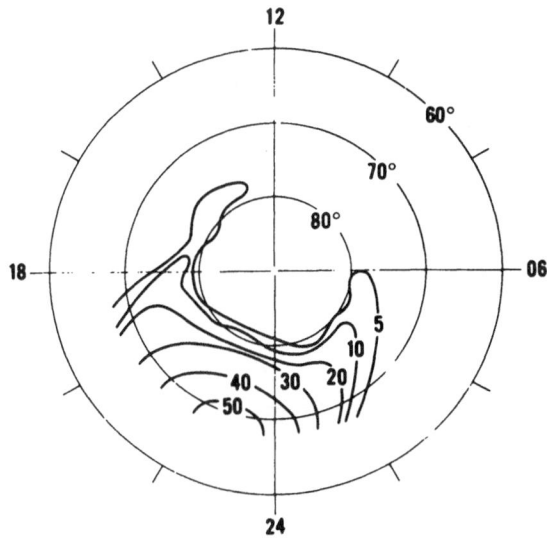

Figure 9. Contours of the percentage occurrence of auroral hiss
           bursts with spectral densities in excess of 10^{-15} w.m.$^{-2}$
           (Hz)$^{-1}$ at 8 kHz; plotted in a Mayaud (27) coordinate
           system. (After Jørgensen (10)).

     Two types of ionosonde data are included here because there
is reason to believe they are associated with the influx of auroral particles; these are sporadic-E and spread-F echoes. Figure
10 shows the top-side ionosonde data on the occurrence of spread F
at a height of 1000 km during quiet geomagnetic (Kp ≤ 2) conditions (13); comparable data are not yet available for all disturbance conditions. There is some indication of a diurnal maximum in the night hours (note that the gap between 20 and 22 hours
indicates a lack of observations not the absence of the phenomenon); furthermore, the pattern is asymmetric with respect to the
geomagnetic pole, appearing at a mean geomagnetic latitude of
about 80° near noon and about 70° at night. Clearly, this phenomenon follows the discrete auroral zone.

     The sporadic-E observations made with ground-based equipment
are not sufficiently homogeneous to permit a quantitative comparison between stations. We have averaged together five years of
data from three stations located near the maximum of the auroral
zone to produce the curve shown in Figure 11(a). This diagram
shows the diurnal percentage occurrence of intense sporadic-E
echoes -- usually designated "auroral sporadic E" (14). Again the
gross (diurnal) variation is that of the discrete auroral zone.

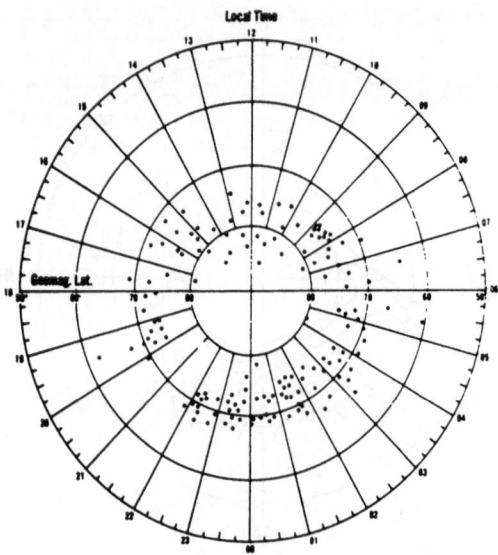

Figure 10. The distribution, in geomagnetic latitude and local
            time, of the maximum frequency range for the spread
            echoes observed by the Alouette I top-side ionosonde
            at a height of 1000 km. These data are for undis-
            turbed conditions ($K_p \leq 2$) during a four month period
            in 1963. (After Petrie (13)).

Another type of sporadic E that appears in the same latitude
range is the low altitude (75 to 95 km) type that has been identified by Olesen and Wright (15). Their results, shown here in
Figure 11(b), maximize in the morning and mid-day hours, and show
a diurnal variation quite similar to that of the diffuse auroral
zone.

There are several other types of data whose gross statistical
behviours show similarities to one or the other of the two characteristic diurnal auroral zones. Perhaps the most interesting
of these are the direct observations of the precipitating particles made with rocket- and satellite-borne instrumentation.

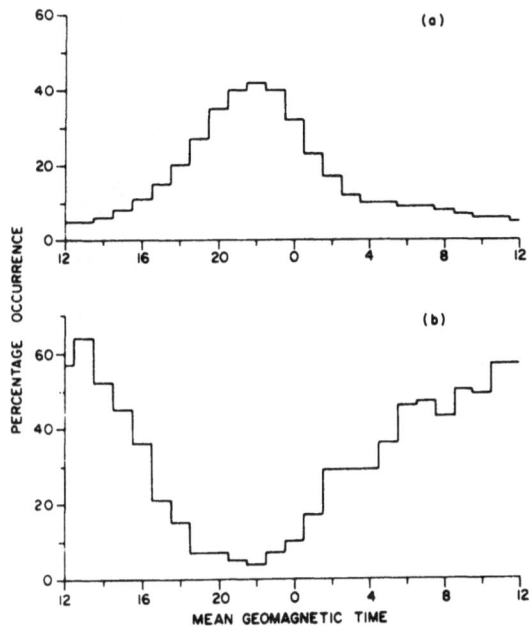

Figure 11. The diurnal variation in the percentage occurrence
of two types of sporadic E.
(a) Sporadic-E echoes that extend beyond 7 MHz on
ionograms from auroral-zone stations: these data
are the average of 5 years' observations at Pt.
Barrow (68.4°N), Churchill (68.7°N), and Ft.
Chimo (69.6°N). (Courtesy of E.L. Hagg and
D. Muldrew (28)).
(b) Low altitude, diffuse sporadic E at Narssarssuak
(71.2°N). (From Olesen and Wright (15)).

4.   DIRECT PARTICLE OBSERVATIONS

Statistical data on precipitated electrons with energies
greater than 40 keV have been obtained from the Alouette I satellite (16) at a height of 1000 km, and are portrayed here in Figure
12. Contours of equal percentage occurrence above a particular
flux threshold are plotted as a function of invariant latitude and
local time. The two most obvious features of this figure are the
pronounced morning maximum at a latitude of about 65° and the
broad minimum at about 22 hours. The general similarity to the
diffuse auroral zone is readily apparent.

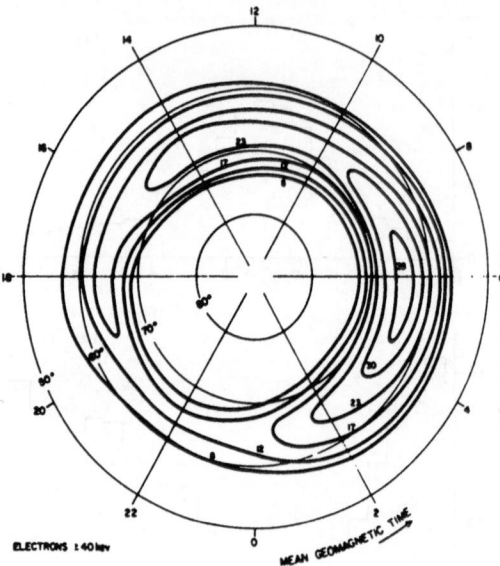

Figure 12. The distribution, in invariant latitude and local time,
            of the percentage of the Alouette I satellite recor-
            dings that showed a precipitation flux (i.e. pitch
            angles less than 45° at 1000 km heights) in excess of
            1.5×10^4 cm^{-2} sec^{-1} steradian^{-1} for electrons with
            energies greater than 40 keV. (Courtesy of I.B.
            McDiarmid and J.R. Burrows).

     While the interval just prior to midnight in general shows a
minimum flux of such energetic electrons, it is during the night
hours that intense bursts of electrons in this same energy range
have been observed with the Alouette I satellite on sporadic occasions (17). In addition, Anderson has obtained analogous observations in regions of the magnetosphere that correspond to the
same latitudes and times (18). Figure 13 shows the distribution
of the intense electron bursts found by McDiarmid and Burrows at
latitudes higher than the apparent boundary for trapped electrons;
despite the small data sample a concentration near midnight is
apparent, suggesting an association with the discrete auroral
zone.

     With reagard to the precipitated electrons of energy substan-
tially lower than 40 keV, no statistical or average pattern is yet
available even though numerous isolated reports, based mainly on
rocket data, have appeared of intense electron fluxes of energies
of a few keV in association with aurora. The satellite results of

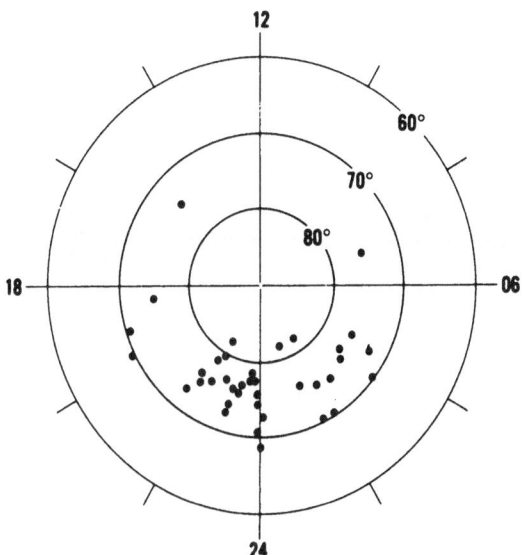

Figure 13. The distribution, in invariant latitude and local time, of the intense electron (energies greater than 40 keV) spikes observed at a height of 1000 km at latitudes greater than the boundary of the outer radiation zone. (From McDiarmid and Burrows (18)).

Fritz and Gurnett (19) are, perhaps, indicative of the situation during disturbances: they observed very intense bursts of precipitated electrons with energies in excess of 10 keV as indicated in Figure 14. Again a nighttime concentration is apparent; this and the association with discrete auroral events found in the rocket observations point to a relationship with the discrete auroral zone of Figure 2.

5. CHARACTERISTICS OF THE TWO PRECIPITATION ZONES

It is abundantly clear that all of the aurorally associated phenomena considered here appear in one or the other of the two zones that together make up the precipitation pattern. This gross pattern is essentially independent of the particular upper atmospheric phenomenon that one happens to be monitoring, although variations in intensity or probability are apparent from one phenomenon to another and from one time of day to another. It is this general unanimity among the varied sets of data that imparts a credibility not possible from one set alone, so that the preci-

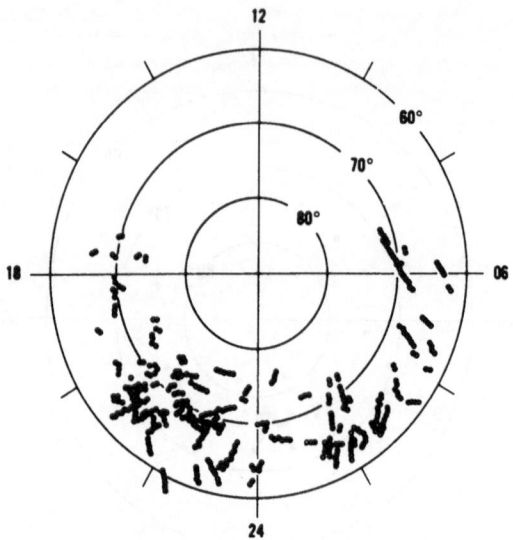

Figure 14. The distribution, in geomagnetic time and invariant
           latitude, of the intense fluxes (in excess of
           2.5×10^7 cm^{-2} sec^{-1} steradian^{-1}) of electrons with
           energies greater than 10 keV observed with the Injun
           III satellite. (After Fritz and Gurnett (19)).

pitation pattern can be considered to be largely free of equipment
and observing limitations. It is, then, instructive to compare
the general characteristics of the two zones, since it is clear
that more than one type of particle influx is involved (3). We
shall refer to these two classes as discrete and diffuse events
for convenience, and shall summarize the main properties of these
events here and in Figure 15.

     The discrete events include:
 (1) Discrete, localized, bright, often rapidly fluctuating auro-
     ral forms having characteristic heights greater than 100 km.
     These are the aurora with distinguishing form as seen by the
     eye or observable with the all-sky camera.

 (2) Abrupt or rapidly varying 'auroral' absorption events on rio-
     meter records. A typical example of such events is seen in
     the upper part of Figure 16. Such absorption events with
     abrupt onsets (changes in excess of 0.5 db per minute) are
     a nighttime feature as indicated by the histogram in Figure
     17, whereas the peak absorption (indicated by the curve)
     occurs in the late morning (20).

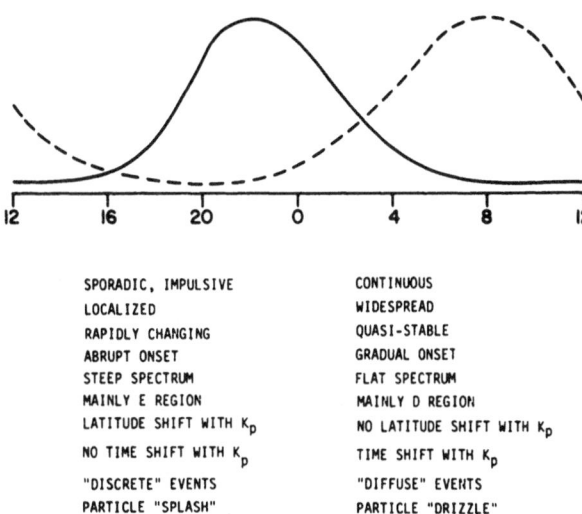

SPORADIC, IMPULSIVE	CONTINUOUS
LOCALIZED	WIDESPREAD
RAPIDLY CHANGING	QUASI-STABLE
ABRUPT ONSET	GRADUAL ONSET
STEEP SPECTRUM	FLAT SPECTRUM
MAINLY E REGION	MAINLY D REGION
LATITUDE SHIFT WITH K_p	NO LATITUDE SHIFT WITH K_p
NO TIME SHIFT WITH K_p	TIME SHIFT WITH K_p
"DISCRETE" EVENTS	"DIFFUSE" EVENTS
PARTICLE "SPLASH"	PARTICLE "DRIZZLE"

Figure 15.   Summary of the main properties of the two precipitation zones; an indication of the magnitude of the particle influx in each zone is shown as a function of
geomagnetic time, and the general characteristics of
the phenomena are listed for each zone.

(3) Rapidly fluctuating, almost impulsive, V.H.F. signals scattered by field-aligned ionization at heights of about 105 km.

(4) Intense, auroral sporadic-E echoes from heights of 100 km or
more.

(5) Spread-F echoes on ionograms.

(6) Bursts of relatively high frequency (> 4 kHz) V.L.F. emissions or 'auroral hiss'.

(7) A negative bay on a magnetometer, having a rapid or abrupt
onset (in excess of 40 gammas per minute) and a slower re-
covery (20).

(8) Impulsive (P_i) micropulsations of the earth's magnetic
field (21).

(9) Balloon observations of bursts of bremsstrahlung X-rays of
characteristically soft energies (22).

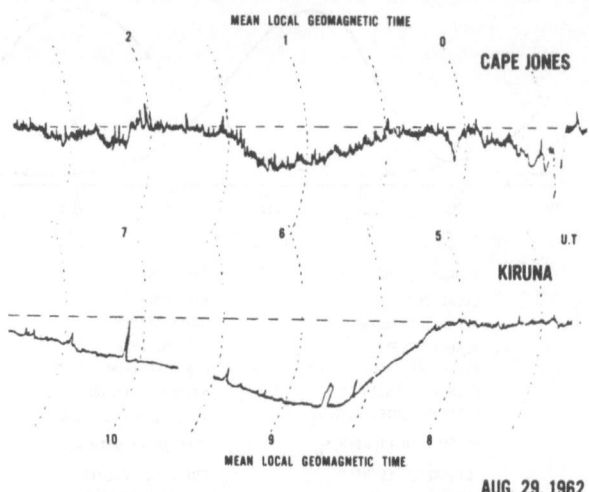

Figure 16. Typical examples of auroral absorption events on 30
            MHz riometer charts. The upper record shows a night-
            time record at Cape Jones (geomagnetic latitude
            65.4°N) with the abrupt and variable features. The
            lower record shows a morning event at Kiruna (geomag-
            netic latitude 65.3°N) recorded at the <u>same universal
            time</u>.

(10) Relatively short-duration bursts consisting of intense fluxes
     of soft electrons (energies of the order of a few keV), or,
     more likely, a flux with a steep energy spectrum such that
     the low energies predominate.

     By contrast, the diffuse events -- statistically at least --
are associated with:
 (1) Steady, diffuse, widespread, weak to subvisual mantle aurora.

 (2) Slowly varying riometer absorption such as is exemplified in
     the lower part of Figure 16. These events are also fairly
     extensive on a horizontal scale (23).

 (3) A steady or slowly varying mean signal level for V.H.F. for-
     ward scatter echoes from extensive regions of isotropic ioni-
     zation irregularities at heights of the order of 80 or 85 km
     (24).

 (4) Diffuse sporadic-E echoes from heights in the 80 or 90 km
     range.

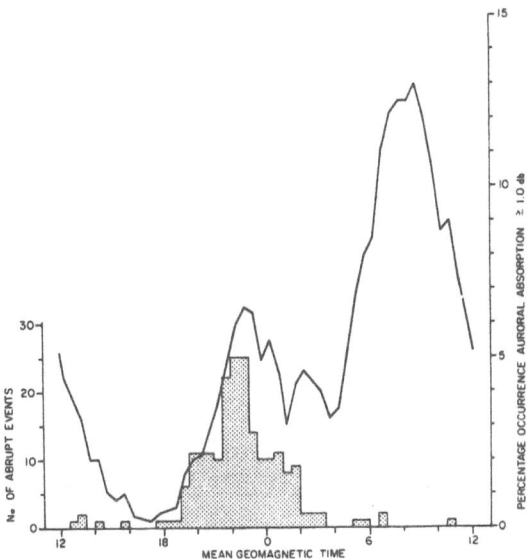

Figure 17. The diurnal variation in the percentage of the time that 30 MHz auroral absorption of 1.0 db or more occurred at Cape Jones (geomagnetic latitude 65.4°N). Also shown, for comparison purposes, is a histogram giving the diurnal variation in the number of abrupt absorption events in the Cape Jones riometer record; these are generally characterized by an onset in excess of 0.5 db per minute. (Courtesy of Montalbetti and Vogan (20)).

(5) Quasi-constant V.L.F. emissions (polar chorus) at frequencies below 2 kHz.

(6) No large, characteristic magnetometer deflections, but continuous type (P_c) geomagnetic micropulsations are prevalent and in many cases the long period pulsations also register on the standard magnetograms (21).

(7) Long duration, slowly varying, and characteristically hard balloon X-rays events (22).

(8) Consistent, moderately intense fluxes of electrons of energies at least 40 keV. The particle energy spectrum appears to be rather flatter (or less steep) than in the case of the discrete events.

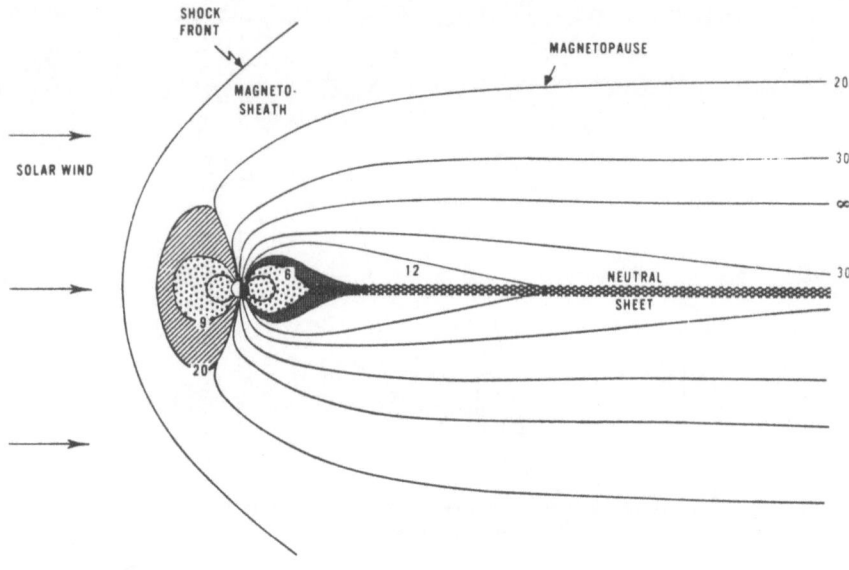

Figure 18.   A cross section of the magnetosphere to illustrate the
geometry of the precipitation zones. The numbers on
the field lines are L values which can be related to
invariant latitude by the relation $\cos^2 \lambda = 1/L$.

It is apparent that the particle precipitation associated
with the discrete events is intense, limited spatially, shows rapid changes, produces its maximum effects at heights greater than
100 km, and in general can be thought of as a "splash" (25). Such
splashes occur predominantly at night, with a peak occurrence near
22 hours geomagnetic time, and at a latitude that can be linked
directly to the neutral sheet in the magnetospheric tail by field
lines that go well out into the tail. The geometry of this situation is illustrated in Figure 18. The discrete events also occur,
albeit to a lesser degree, at times of the day well removed from
midnight, and the zone of occurrence can be associated everywhere
with the boundary between the so-called closed field lines and the
tail field lines. During highly disturbed conditions, or storms,
this discrete auroral zone shifts toward lower latitudes, but the
zone maximum still occurs just before midnight.

The precipitation associated with the diffuse events leads to
ionization effects that seem to concentrate at heights of the
order of 85 km. The particle influx is steady and widespread, and
can be described aptly as a "drizzle" (3). The geomagnetic latitude of the diffuse zone remains essentially constant throughout

the day, indicating that no direct connection with the magnetospheric tail is involved, except perhaps at night. The maximum
drizzle occurs near 08 hours at a geomagnetic latitude of about
65°: at this time of day the field lines connecting to this
drizzle region lie deep in the trapped particle zones or outer
Van Allen belt. The maximum drizzle must therefore constitute the
"sink" for the trapped radiation.

During highly disturbed conditions, or storms, the diffuse
auroral zone does not shift in latitude; rather the zone maximum
shifts earlier in time (5, 26). (For instance, the A_3 data, which
by definition associate with a K-index of 4 or more, show a peak
near 05 hours, rather than at the average time of the other diffuse events (24)).

6. IMPLICATIONS IN THIS PRECIPITATION PATTERN FOR COMMUNICATION SYSTEMS

In the main, any Arctic radio communication system that includes the anomalous aurorally associated ionization in its propagation path can be expected to suffer some adverse effects.
Whether or not these adverse effects constitute a serious limitation on the communications depends on such factors as the radio
frequency, the degree of disturbance, the sophistication of the
equipment, the time of day, etc. For purposes of this paper, we
shall disregard most of these factors and consider only some of
the general features of the physcial situation.

On the average, the auroral particles precipitate in a general pattern which is illustrated diagrammatically in Figure 2, and
the location of a particular radio propagation path relative to it
can be readily specified. It is immediately clear that certain
times of day are much more affected than others, and certain latitudes stand out as likely problem areas. The dominant nighttime
maximum of the discrete auroral zone appears at a geomagnetic latitude of about 68°: in geographic coordinates this corresponds to
the auroral zone maximum as shown in Figure 1. The dominant morning maximum of the diffuse auroral zone appears at a geomagnetic
latitude of about 65°: this would correspond to a zone in Figure 1
some three degrees lower in latitude than the auroral zone maximum.

If, now, an Arctic communication system happens to be so located that the propagation path always includes the anomalous auroral ionization, two main propagation effects can be expected. A
sufficient increase in ionizaiton in the lower ionosphere may introduce absorption along the path, leading to a reduction of signal or, in severe cases, to blackout. This limitation is mainly
associated with the diffuse auroral zone, but occasionally some

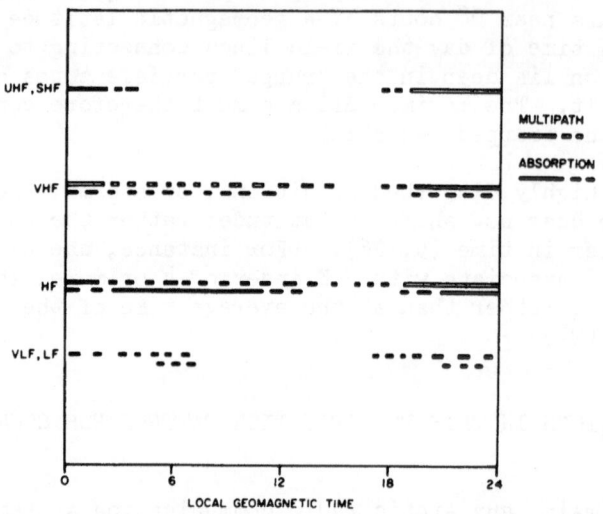

Figure 19. Summary of the effects due to absorption and ionozation irregularities (here identified only as multipath) on four general classes of communication systems, as a function of time of day. The assumption
has been made that the anomalous auroral ionization is
included in the radio propagation path.

short-lived absorption effects may occur in the discrete auroral
zone. The second effect is linked with ionization irregularities
or small scale structures; these may introduce frequency and time
dispersion on the signal (at the low and very low frequencies
phase fluctuations may be considered important). The discrete
auroral zone is principally associated with this propagation limitation.

The two effects are summarized in a qualitative way in Figure
19 for four general classes of communications and for the different times of day. In general the highest and the lowest frequencies can be expected to be most immune, and the afternoon
hours are seen to be least affected on all systems.

7. SUMMARY

On the average the auroral particles precipitate in a two-
zoned diurnal pattern which can be thought of as more-or-less
fixed in space with the earth rotating beneath it. The discrete
auroral zone shows a maximum just prior to geomagnetic midnight

and a geometrical association with the boundary region between the
"closed" and the "tail" geomagnetic field lines. The diffuse auroral zone has a maximum near 08 hours geomagnetic time and exhibits a geometrical association with the outer Van Allen radiation
belt. This pattern is illustrated in Figure 2, and the main
characteristics of the two types of events are summarized in
Figure 15. None of the uncertainties connected with the source
mechanisms are inherent in this pattern, nor is its validity dependent on the interpretation of only a single type of measurements.

The relevence of this precipitation pattern to a particular
communication system can only be described in the most general
terms because so many additional parameters may be involved in
such an application. The two general factors of the propagation
medium to consider are radio-wave absorption which is of greatest
importance in the diffuse zone and at the lower heights, and ionization irregularities which have their greatest effects in the
discrete zone and at heights above 100 km.

8. REFERENCES

1. Y.I. Feldstein, Investigations of the aurorae: Bulletin of
    the International Geophysical Year 4, 61, 1960.

2. Y.I. Feldstein and F.K. Solomatina, J. Phys. Soc. Japan 17,
    Supplement A - I, 223, 1962.

3. T.R. Hartz and N.M. Brice, Plan. Space Sci. 15, 301, 1967.

4. Y.I. Feldstein, Plan. Space Sci. 14, 121, 1966.

5. S.-I. Akasofu, Space Science Reviews 4, 498, 1965.

6. B.P. Sandford, J. Atmos. Terr. Phys. 26, 749, 1964.

7. T.R. Hartz, L.E. Montbriand and E.L. Vogan, Can. J. Phys. 41,
    581, 1963.

8. E.L. Vogan, private communication.

9. E.I. Loomer and K. Whitham, Publication of the Dominion Ob-
    servatory, Ottawa, XXVII, 75, 1963.

10. T.S. Jørgensen, J. Geophys. Res. 71, 1367, 1966.

11. H.M. Morozumi and R.A. Helliwell, A correlation study of the
    diurnal variation of upper atmospheric geophysical phenomena
    in the southern auroral zone. (To be published).

12. D.A. Gurnett, A satellite study of very-low frequency emis-
     sions, Paper presented at URSI Meeting, Washington, D.C.,
     April 18-21, 1966.

13. L.E. Petrie, Proceedings of AGARD/IRC Meeting on spread F and
     its effects upon radiowave propagation and communication,
     Copenhagen, Denmark, 26-29 August, 1964.

14. E.K. Smith and S. Matsushita, Ionospheric Sporadic E, The
     MacMillan Co., New York, 1962.

15. J.K. Olesen and J.W. Wright, J. Geophys. Res. 66, 1127, 1961.

16. I.B. McDiarmid and J.R. Burrows, Can. J. Phys. 42, 1135,
     1964.

17. I.B. McDiarmid and J.R. Burrows, J. Geophys. Res. 70, 3031,
     1965.

18. K.A. Anderson, J. Geophys. Res. 70, 4741, 1965.

19. T.A. Fritz and D.A. Gurnett, J. Geophys. Res. 70, 2485, 1965.

20. R. Montalbetti and E.L. Vogan, private communication.

21. J.A. Jacobs and C.S. Wright, Can. J. Phys. 43, 2099, 1965.

22. J.R. Barcus and T.J. Rosenberg, J. Geophys. Res. 71, 803,
     1966.

23. J. Ortner and W. Riedler, Nature 204, 1181, 1964.

24. C. Collins and P.A. Forsyth, J. Atmos. Terr. Phys. 13, 315,
     1959.

25. B.J. O'Brien, J. Geophys. Res. 69, 13, 1964.

26. T. Laaspere, M.G. Morgan and W.C. Johnson, Proc. I.E.E.E. 52,
     1331, 1964.

27. P.N. Mayaud, Ann. Geophys. 16, 278, 1960.

28. E.L. Hagg and D. Muldrew, private communication.

CHARACTERISTICS OF POLAR CAP ABSORPTION

Olav Holt

Norwegian Defence Research Establishment, Kjeller and

The Auroral Observatory, Tromsø, Norway

Abstract: The paper is a review of our knowledge of Polar Cap Absorption. The time and space variations of PCA, mainly as they appear on riometer records, are described. The relation of PCA to solar activity, the propagation of solar cosmic rays through interplanetary space, and how they interact with the geomagnetic field and produce ionization in the earth's atmosphere, are discussed. Finally, some comments on radio communication and PCA are made.

1. INTRODUCTION

During normal conditions the ionization in the D region is due to X-rays and ultraviolet radiation from the sun, and for the very lowest part to cosmic rays. The resulting electron density is rather low. The total absorption of ionospherically reflected radio waves sets the limit of lowest useable frequency, but presents no hindrance to the operation of ordinary communication circuits. There are, however, periods when the electron density increases to such high values that the reflected signals cannot be detected above the noise level. These are the conditions known to radio operators as short wave fade-outs or radio black-outs.

The higher electron density may be due to the increased X-ray intensity from solar flares which occur simultaneously over the sunlit side of the earth. Such events are called sudden ionospheric disturbances (SID), and are usually of short duration - from a few minutes to some tens of minutes.

At high latitudes, radio blackouts are much more frequent, and are very closely associated with the occurrence of visual aurora and geomagnetic storms. It is fairly well established that the increased ionization in these events is caused by energetic electrons, probably of solar origin. The events are rather localized, normally up to a few hundred km across the area covered. Sometimes they may be of a global scale in the east-west direction, but are still of limited extent in latitude. Such disturbances are termed Auroral Absorption (AA).

Much less frequent is the type of event known as Polar Cap Absorption (PCA). The absorption then occurs simultaneously over the entire polar caps, down to geomagnetic latitudes of ~ 65°. The duration of the events may be several days, and the absorption is sometimes very strong.

The occurrence of absorption with these characteristics is known to accompany increases in the intensity of cosmic radiation at ground level, following some strong solar flares. These events are extremely rare. The first one to be observed took place in 1942, and since then less than fifteen such increases have been recorded. The largest increase observed hitherto followed an intense solar flare on 23 February 1956. The associated ionospheric effects were described by Bailey (1957).

It was not until 1957, however, that it became clear that these events are just outstanding samples from a group of much less rare solar cosmic ray emissions. Reid and Collins (1959), from a study of riometer recordings at Fort Churchill (geomagnetic latitude 69°), concluded that several absorption periods had much the same characteristics as the one described by Bailey, except that no ground level events were observed. They concluded that the increased ionization in the D region was due to solar cosmic rays (protons and perhaps α-particles) emitted during some strong solar flares. Most of these events are not sufficiently intense, or the particles have not sufficient energy, to be observed at ground level. This was the beginning of systematic studies of Polar Cap Absorption, and in the following sections an attempt is made to review briefly our present knowledge of this subject.

2. OBSERVATIONS OF POLAR CAP ABSORPTION

There is little doubt, that up to now the most useful observations of polar cap absorption have been made by riometers. A riometer is a device which records continuously the cosmic radio noise level at a chosen frequency. The dynamic range of the instrument is about 15 db, and it turns out then, that for studies of ionospheric absorption, a frequency of ~ 30 MHz is suitable. For special purposes riometers in the range 5 - 60 MHz have been used.

   As an introduction to our subject, Figure 1 shows riometer results (Holt, 1963) from three different locations, Bjørnøya (geomagnetic latitude 70.4°), Skibotn (66.4°) and Trondheim (63.4°), during the PCA event of 16-18 August 1958. The riometers were operated at 27.6 MHz.

   A solar flare of importance 3+, occurred at the location S14W50 at 0433 on 16 August. The flare was followed by an SID which is clearly seen on the riometer records. About two hours later, the absorption started to increase at the two northern stations, and from 1500 UT also at Trondheim. A maximum is reached at about 1700, - a few hours later at Bjørnøya. The maximum absorption is about 13 db at Bjørnøya and Skibotn, and 5 db at Trondheim. Minimum absorption is reached around midnight. Part of this decrease is explained by photodetachment of electrons from negative ions. The photodetachment mechnism is active during day-time only, and will lead to a higher value for the equilibrium

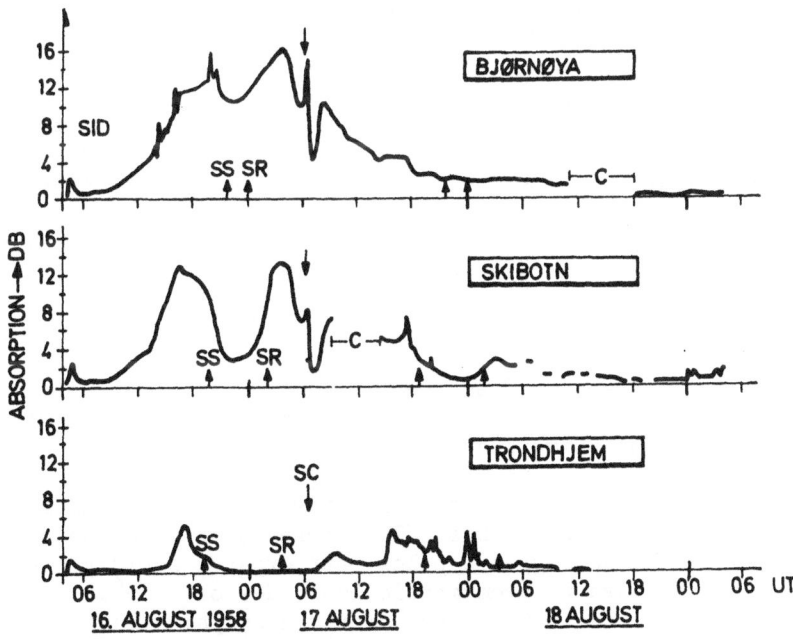

Figure 1. Absorption observed on 27.6 MHz riometers during PCA.

electron density. This is an interesting problem, but it is not specific of PCA, and we shall not dwell upon it here.

Ground sunset and sunrise are indicated on the figure. At Bjørnøya, however, there is still midnight sun in the D region on 16 August, and the variation in absorption may be taken to reflect the variation in the primary particle flux. It seems quite likely that part of the variation at the lower latitudes can also be attributed to this effect.

In the morning of 17 August the absorption again increases at Bjørnøya and Skibotn, reaching maximum between 0300 and 0400 of 16 db and 13 db respectively. In this period there is no increased absorption at Trondheim.

At 0622 on 17 August there was a sudden commencement magnetic storm (SC). Immediately following this, there was a short, but strong increase in the riometer absorption. This is the SC type of absorption, here superimposed on a PCA event. After this short increase, the absorption decreased considerably for a period of more than one hour. This decrease in absorption is often being observed during PCA events, particularly at stations close to the lowest latitude of the absorption region. When the absorption again increases, the increase is also seen at the low latitude of Trondheim. A maximum of 10 db at Bjørnøya, 7 db at Skibotn and 2 db at Trondheim is reached 2 - 4 hours after the SC magnetic storm. The absorption then decreases at all stations until about 1500 UT, when a sudden strong increase is observed at Trondheim. No such effect is seen at the two northern stations, and in the period following this increase, the absorption is almost as strong at Trondheim as at the higher latitudes. This may be auroral absorption, superimposed on the PCA, with the auroral zone shifted southwards in the disturbed geomagnetic field.

The conditions gradually return to normal during 18 August.

This event has been discussed in some detail, because it shows many of the features that may be said to be typical of a PCA. These may be summarized as follows:

Some time after the solar flare, the absorption begins to increase on the riometer records. The increase is first seen at locations close to the magnetic poles, but expands rapidly over the whole polar cap down to geomagnetic latitudes of ~ 65°. The time delay from the flare to the onset of the absorption may be as short as $\frac{1}{2}$ hour, and in some rare events more than 12 hours. An average delay is between 2 and 4 hours. In some events a further expansion to lower latitudes is observed a few hours after the onset.

During this phase, there is usually no magnetic activity. This behaviour of the ionosphere and the geomagnetic field is consistent with the view that the ionization in the D region is caused by protons (and possibly heavier nuclei) of energies in the 5 - 100 MeV range. A moderate flux of such particles (far less than that required to give any magnetic effects) is sufficient to provide the electron density necessary to explain the observed absorption. About 24 hours after the flare, a SC magnetic storm occurs. This is sometimes followed by a temporary decrease in the absorption, of duration ~ 1 hour. After the SC, the absorption expands to still lower latitudes, < 60°.

Most events show a maximum of absorption before the SC, but some have their maximum at some time after the start of the magnetic storm.

Of course, information about the ionosphere during PCA, has been obtained by other methods, as well. Some of them will be mentioned here:

The first instrument developed for systematic studies of the ionosphere, was the <u>vertical-incidence ionosonde</u>.

The echo height for pulsed signals in the HF band is recorded as a function of frequency. Information about the absorption in the D-region is obtained from the minimum frequency on which an echo can be observed (f min). This depends on the transmitted power as well as on the sensitivity of the receiver, and will therefore be different for different ionosondes, even for otherwise equal conditions. Moreover, the echo strength, and thus f min, depends on the reflection coefficient of the higher layers from which echoes are returned.

For most ionosondes, no echo can be detected if the absorption is greater than about 50 db. The highest frequency on which an echo can be observed is the critical frequency of the F-layer. If absorption at this frequency exceeds 50 db, no echo is observed on the record. This is tabulated as a <u>black-out</u> in the sounding data, and will certainly occur in PCA's like the one shown in Figure 1. The information obtained is then only the qualitative one that a blackout has taken place. However, because of the fairly dense and large net of ionosondes built up during the IGY, useful information on the morphology of PCA has been obtained from such data.

Another rather sensitive, but again only qualitative indication of absorption is provided by the VHF scatter communications. The scattering is due to small-scale irregularities in the lower ionosphere. If the absorption takes place below the scattering region a decreased signal strength is observed. However, the increase in ionization may result in an increase in the scattering coefficient, and a corresponding enhancement of the signal. This will often be the case during PCA. Because of this combined dependence on the reflection mechanism and the absorption, the information is qualitative only. Its usefulness is due to the sensitivity and the rather large dynamic range. It was from VHF scatter circuits that Bailey (1957) obtained the largest amount of data in his pioneer work on PCA.

Another interesting feature of PCA is the effect on LF and VLF radio waves. Phase measurements at 70 kHz reported by Belrose and Ross (1961), show that the reflection takes place at a lower level than during normal conditions. Amplitude measurements sometimes show a value larger than normal, sometimes less. It seems that the distance of propagation rather than the frequency determines this. These effects are far from well understood, but they will be dealt with in other papers at this meeting and we shall leave it here.

In a few cases, direct measurements of electron density in the D region during PCA have been made (Belrose (1964), Kane (1962), Parthasarathey et al (1963)). These show that the increase in electron density takes place at a lower height than during auroral absorption. This is also in good agreement with the assumption that protons are responsible for the ionization.

3. THE SOLAR COSMIC RAYS

3.1 The Solar Activity during PCA

As has already been stated, there is a good empirical correlation between the occurrence of solar flares and PCA. The obvious physical explanation is, that high energy particles that are emitted from the flare area, propagate through interplanetary space and in their interaction with the earth's amosphere cause the PCA. There are very few strong PCA events for which it has not been possible to identify a flare that may be held responsible. In the case of some rather weak PCA's observed near sunspot minimum, longlived particle emissions from solar M regions rather than flares seem to be the source of ionization.

Another solar phenomenon which is strongly correlated to PCA is the broadband centimeter wavelength emissions classified as type IV solar radio bursts. Nearly every PCA event is preceded by a type IV burst. It is now commonly accepted that these emissions can be explained as synchrotron radiation from high energy (relativistic) electrons in a solar magnetic field. This requires an acceleration mechanism for the electrons near the sun, and it seems that this should also accelerate protons and other charged particles to similar energies. Little is known about this mechanism. The relation between Type IV bursts and PCA has recently been studied by Švestkova and Švestka (1966).

The visual flare and the radio bursts have a duration of the order of 1 hour. As we have mentioned, however, the PCA effect may last for several days. Unless protons and α-particles continue to be emitted from the sun long after the other signs of activity have stopped (which seems rather unlikely), the particles must be stored somewhere in the solar system. It is therefore necessary to look into what happens to the particles on their way from the sun to the earth.

3.2 Propagation of Solar Cosmic Rays in Interplanetary Space

It is now well known that the solar corona extends to well beyond the earth's orbit. The gas is streaming away from the sun, and is therefore often called the solar wind. The particle density at the orbit of the earth is of the order of 10 cm^{-5}, and the gas velocity is there ~ 1000 km/sec during normal conditions. The kinetic energy density in this stream is high enough that it may carry with it a weak magnetic field, which will then be stretched out radially from the sun. Because the sun is rotating, the field lines will become spirals, like water from a garden hose. The earth thus becomes magnetically connected to a point on the western half of the sun.

The direct path travel time from sun to earth for protons of energies between 10 MeV and 100 MeV is from more than one hour to less than half an hour. Since the particles will be guided along the spiralling field lines, the transit will be a little longer, but not nearly enough to account for intervals of 2-4 hours or more, that are sometimes observed, between the visual flare and the onset of PCA.

It is quite likely that in interplanetary space there will be comparatively small regions of stronger magnetic fields. These will serve as scatterers for the energetic particles, and the problem becomes one of diffusion. An effective mean free path of about .05 AU (Hofmann and Winckler (1963), Hakura (1966)) would make the average time for 100 MeV protons to travel from sun to earth about 4 hours. For α-particles of the same energy, the time would be 8 hours. There are indeed many details in the observations from the early phase of PCA that confirm a relatively simple diffusion model. We shall not go into these details here, but it is interesting to notice that the delay in the onset of PCA at Trondheim in the event shown in Figure 1, would readily be ex-

plained as the delayed arrival of α-particles. These will, because of their larger momentum, be able to penetrate the geomagnetic field to lower latitudes. The diffusion theory is also con-
current with the observation that a PCA originating in a flare on
the eastern side of the sun usually has a longer delay from flare
to onset than those from western flares. Particles from eastern
flares would be further away from the tube of field lines connecting sun and earth, and must somehow diffuse into this tube. It
should be mentioned here that Hakura (1967) has suggested that the
very first phase of PCA, when only the area near the magnetic
poles is covered, may be due to solar electrons. In one event,
instruments in the IMP-3 satellite detected an increase in high
energy electron flux, shortly before the arrival of energetic
protons lasting for about 2 hours.

The above mechanism does not explain how the proton and α-
particle bombardment can last for several days, though. Before we
go into that, let us look at what happens at the time of the SC
magnetic storm, which almost invariably takes place about 24 hours
after the onset of PCA. At the time of the flare a plasma cloud
is ejected from the sun. The particles in the cloud are not PCA
particles, but have much lower energies, and consist of both positive particles and electrons. Their density is high enough that
the stream will carry with it the relatively strong magnetic field
in the flare area, and stretch it into space and/or wipe up the
existing field in front of the stream. When this plasma reaches
the earth, it will also interact with the geomagnetic field to
give the SC storm. The deformation of the field will be such as
to allow the protons and α-particles to penetrate to lower latitudes, thus explaining the expansion of the PCA at this time of
its development.

Several theories have been put forward in the last years to
explain how an event can continue for days after the sudden commencement storm. It is impossible in a brief lecture like this to
do justice to them all. Most of them have two things in common, -
they are rather speculative, and they do not account for all of
the observational facts.

One model, which has become associated with the name of
T. Gold (1959), is also known as the magnetic tongue. The magnetic field dragged out from the bipolar flare area takes the shape
of a long tongue, in which charged particles may become trapped,
in the same fashion as in the Van Allen belts of the geomagnetic
field. If the earth is situated in the tongue, the event will
last as long as the trapping exists, or until the earth leaves the
tongue. It should be noted here that although the plasma stream
does not consist of PCA particles, there is no reason why such
particles should not be trapped in the tongue created by the
stream.

A theory presented by Haurwitz, Yoshida and Akasofu (1965)
is, as far as I can see, an elaboration (or refinement) of the
magnetic tongue model. It accounts for many of the asymmetries in
the developments of PCA events originating in western compared to
eastern flares. Figure 2 shows the main features of the theory.
A few hours after the flare the plasma cloud has just begun to
stretch the tongue radially from the sun, but the more energetic
of the PCA particles have already reached the orbit of the earth
by the diffusion mechanism described above. (Haurwitz et al did
not really include a magnetic tongue, only a plasma cloud moving
out from the sun). If the earth is in a position to see the flare
on the western part of the sun, the PCA will have started. The
figure also shows the situation just before the SC storm. The
tongue has swept up the normal spiralling field lines, and pushed
PCA particles ahead of the magnetic edge thus created. Some of
the particles remain trapped in the tongue. If the flare is on
the east side of the sun, the particle flux before the SC is
rather weak, whereas the bulk of the particles will reach the
earth after the SC. This is in good agreement with observations.
Only PCA from eastern flares show a post SC maximum. For western
flares, however, very few particles will hit the earth after the
SC. Haurwitz et al suggested that the decrease in absorption
shortly after the SC is caused by the lack of energetic particles
behind the magnetic front. This decrease was very clearly seen in
the event shown in Figure 1, and indeed all observational evidence
is that the decrease only occurs during western flare PCA. It

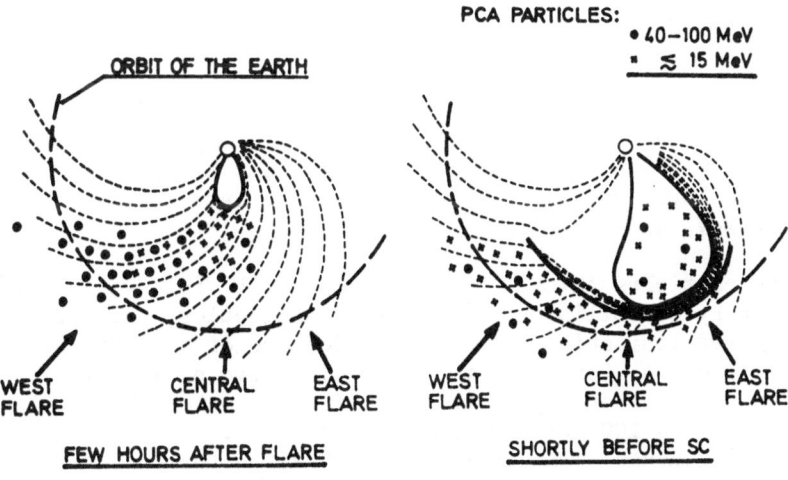

Figure 2. Idealized model of solar cosmic ray propagation.

seems not quite clear, though, why this decrease should be only
temporary.

Here we will leave the interplanetary propagation of the particles, and look at their behaviour near the earth.

3.3 Solar Cosmic Rays in the Earth's Magnetic Field and Atmosphere

Charged particles meeting with the geomagnetic field will of
course experience a force, depending on their charge and velocity,
with a resulting acceleration depending on their masses. The
rigidity of a charged particle is defined as

$$R = \frac{pc}{Ze} \quad ; \quad (p = mv)$$

Here m is the particle's mass, v its velocity, and Z the
number of electronic charges e. c is the velocity of light in
free space. Thus, rigidity means momentum (times the velocity
of light) per charge unit. It can be shown that particles with
the same rigidity will behave in the same way in a magnetic field.
The calculation of the resulting path of a particle is very complex. Solutions have been given by Størmer for the case of a dipole field. One of the analytic results of the theory is the
existence of a cut-off rigidity. This means that particles of
rigidity less than a critical value R_c cannot reach the earth at
magnetic latitudes lower than λ, the relation between the two
being given by

$$R_c = 14.9 \cos^4 \lambda$$

where R_c is measured in BeV. As an example, a proton of kinetic
energy 100 MeV has a rigidity of 400 MeV, and a corresponding
cut-off latitude 67°. An α-particle of the same energy has rigidity 870 MeV, and cut-off at $\sim 62^\circ$.

The particles allowed by the geomagnetic field will lose
their energy by collisions with the air molecules, thus producing
the observed ionization. A striking feature of this process is
that the protons (and α-particles) will give up a very large
fraction of their energy in the last few kilometers before they
are stopped. A 100 MeV proton may penetrate to about 30 km at
vertical incidence. The depth of penetration for a 10 MeV proton
is ~ 70 km. For off-vertical incidence, the energy will of course
be deposited at higher altitudes.

The energy spectra of solar cosmic rays have been measured by
instruments in balloons, rockets, satellites and space probes.

Perhaps with the exception of the very first phase of PCA, the flux is isotropic over the upper hemisphere. The energy spectrum varies in a manner that is in general agreement with the mechanism for sun-earth travel of the particles as outlined in the last section. Given the energy spectrum and the angular distribution, the electron production rate may be calculated. The processes of electron ion recombination is, however, not well known, and it is therefore difficult to make a detailed comparison of particle flux and the measured electron density profiles mentioned before. Below 40 km the electrons will disappear very rapidly, by formation of negative ions. Also the absorption per electron per cc is quite small for the normal riometer frequency at these low heights. Therefore, the main part of the absorption is due to ionization by particles of less than ~100 MeV, and the cut-off latitude for these particles determines the low latitude border for the PCA.

We are now back where we started - at the observed ionospheric effects during PCA events. It must be realized that this review is far from complete. Many interesting features of the PCA complex, some of them definitely beyond the scope of this meeting, have not even been mentioned. For more extensive reviews, reference is made to papers by Reid (1965) and by Bailey (1964). Hartz (1965) has recently given a review of particle propagation in the sun-earth environment.

Before finishing, some comments on radio propagation during PCA may be appropriate.

4. BRIEF COMMENT ON RADIO COMMUNICATION AND POLAR CAP ABSORPTION

The most serious effect of PCA on communication circuits occurs in the HF-band. As mentioned in section 2, the signal level on both VLF, LF and VHF scatter propagation may even increase during PCA. The importance of HF-communication via the ionosphere may be considerably reduced in the future through the advent of communication satellites, but it seems likely to be in use for many years still.

Nothing can probably be done to prevent PCA events, but a forecast service is not impossible. Some sort of long term forecast is already provided by the 11-year sunspot cycle. Being dependent on the flare activity, the occurrence frequency of PCA shows a good general correlation with the sunspot number. This will probably not be of much practical help, though. Much effort is spent in the US (and very likely in the USSR) on the possibility of more accurate flare predictions than obtained by the general trend of the sunspot number. This is very important in connection with the radiation hazard in manned space flights, but would also be useful for communication purposes.

Short term forecasts, based on flare observations, are quite feasible at present. If a major flare is observed, accompanied by a type IV radio burst, there is at least a 50 per cent chance that it will be followed by a PCA. As we have seen, the time delay to the onset of the PCA will depend to a certain degree on the position of the flare. It may be less than one hour for western flares, and more than 12 hours for some eastern flares.

Also variations in the absorption during the development of the PCA might be taken advantage of. For instance, we have seen that for western flares, there is very often a marked decrease in absorption shortly after the SC storm. The day to night variation briefly mentioned in section 2, could possibly give an opening during the night, - at least for the weaker events.

Perhaps these aspects deserve more attention than they are presently receiving.

5. REFERENCES

1. D.K. Bailey, J. Geophys. Res. 62, 431, 1957.

2. D.K. Bailey, Plan. Space Sci. 12, 495, 1964.

3. J.S. Belrose, In: Propagation of Radio Waves at Frequencies below 300 kc/s, Ed. W.T. Blackband, Pergamon Press, London, 1964.

4. J.S. Belrose and D.B. Ross, Can. J. Phys. 39, 609, 1961.

5. T. Gold, J. Geophys. Res. 64, 1665, 1959.

6. Y. Hakura, NASA Goddard Space Flight Center, Report X-641-66-521, 1966.

7. Y. Hakura, NASA Goddard Space Flight Center, Report X-641-67-116, 1967.

8. T.R. Hartz, In: Physics of the Earth's Upper Atmosphere, Prentice-Hall, Inc. Englewood Cliffs, N.J. 1965.

9. M.W. Haurwitz, S. Yoshida and S.-I. Akasofu, J. Geophys. Res. 70, 2977, 1965.

10. D.J. Hofmann and J.R. Winckler, J. Geophys. Res. 68, 2067, 1963.

11. O. Holt, NDRE Report No 46, Norwegian Defence Research Establishment, Kjeller, Norway, 1963.

12. J.A. Kane, In: Radiowave Absorption in the Ionosphere, Ed. N.C. Gerson, Pergamon Press, London, 1962.

13. R. Parthasarathy, C.M. Lerfald and C.G. Little, J. Geophys. Res. 68, 3581, 1963.

14. G.C. Reid, In: Physics of the Earth's Upper Atmosphere, Prentice-Hall, Inc, Englewood Cliffs, N.J. 1965.

15. G.C. Reid and C. Collins, J. Atmos. Terr. Phys. 14, 63, 1959.

16. L. Fritzova Švestkova and Z. Švestka, Bull. Astron. Inst. Czechoslovakia 17, 249, 1966.

THE IONIZATION MAXIMUM IN POLAR LATITUDES

Charles F. Power[*] and Charles M. Rush

University of California, Los Angeles, California

U.S.A.

Abstract: Swept frequency sounders in satellites constitute excellent means for synoptic studies of the electron density distribution above the F2 peak. The essential part of the material in this presentation is based on data from Alouette I. It is concerned with the morphology and variability of a characteristic enhancement in the electron content observed at high latitudes. An attempt to relate the location and the structure of the enhancement region to solar-geomagnetic effects is described. Derived seasonal variations of the southern boundary and diurnal changes in the average position are given.

1. INTRODUCTION

The launch of the Alouette I satellite into a circular, polar orbit on September 29, 1962, marked the successful beginning of attempts to monitor, synoptically, the ionospheric electron density from the F_2 peak out to about 1000 km. To date, over two million sweep-frequency ionograms have been produced from the Alouette I soundings. Most of these measurements were made in rapid succession, one for each degree of latitude, from pole to pole. In this paper some of the results obtained from topside soundings taken at high geomagnetic latitudes during the first year of the experiment will be presented.

---

[*] Presently at Headquarters, Ait Weather Service, Scott Air Force Base, Illinois, U.S.A.

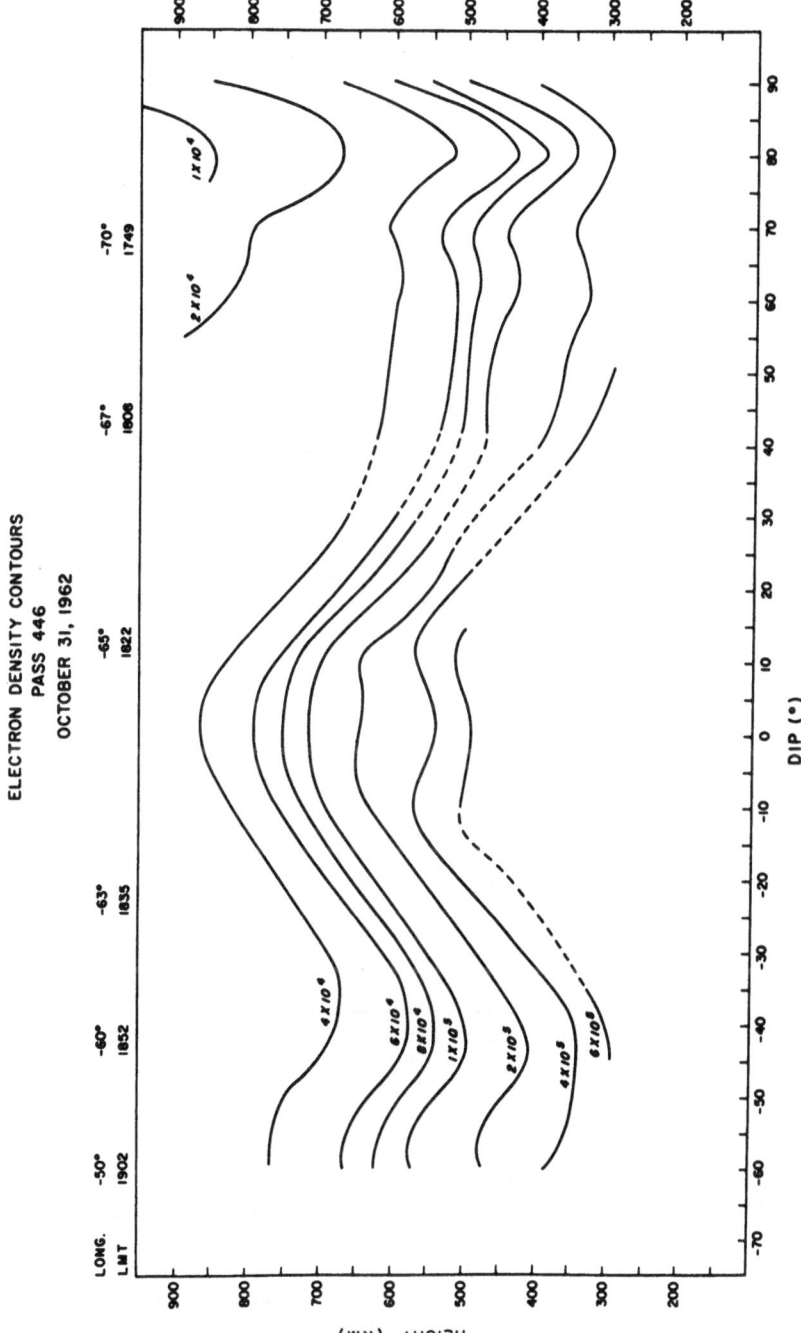

Figure 1. Latitudinal and vertical distributions of electron density on October 31, 1962

The results are based on the analyses of over 15,000 ionograms obtained during 302 transits of the satellite across the
western hemisphere. Meridional cross-sections of the topside ionosphere were constructed from the height-versus-electron density
profiles deduced from the ionograms. Cross-sections of a similar
nature, but displaying constant plasma frequency as a function of
height and geopgraphic latitude have been constructed by Nelms (1)
for about forty Alouette I transits during geomagnetically quiet
periods covering several months of 1962-1963. Although it is considerably more difficult to construct cross-sections from the
electron density profiles deduced from each ionogram, they are
easily interpreted and may be more physically significant. Figure
1 is an example of one such cross-section, showing two characteristic features of such charts: the equatorial anomaly, centered
over the magnetic equator, and the high-latitude enhancement in
ionization, near the magnetic poles. We concentrate here on the
latter.

The absence of read-out stations near the south pole pre-
cludes an investigation of this nature over that region. A poleward decrease in ionization is evident on the southern half of
most cross-sections, to about 70°S dip latitude. As data are un-
available at higher latitudes, it is not known if enhancements in
electron density, such as are indicated in Figure 1, occur regularly near the south pole.

2. MORPHOLOGY

We have found a north polar enhancement to be a regular feature of the ionosphere throughout the year during all local times.
It was identified on all but eight of the 302 cross-sections analyzed. The general features of electron density distribution
near the polar maximum are apparent in Figure 1. There is usually
a monotonic decrease in electron density with increasing dip latitude north of about 40° dip. This extends through all levels between the satellite and the F_2 peak. The decrease continues until
the axis of the 'trough' is reached at an average dip latitude of
71°N. The density then increases from this point poleward, and
usually attains a maximum value at about 85°N. This maximum often
exceeds the equatorial maximum in electron density at the same
altitudes.

Although we are mainly concerned here with the nature of this
polar maximum, it is convenient to refer to it in terms of the adjacent 'trough'. The 'trough' in ionization at high dip latitudes
was first identified by Muldrew (2) from a study of the critical
frequencies, fxF_2, observed on Alouette I ionograms. Thomas et al
(3) have also presented a preliminary study of this feature,
emphasizing the adjacent polar maximum. The maximum is often

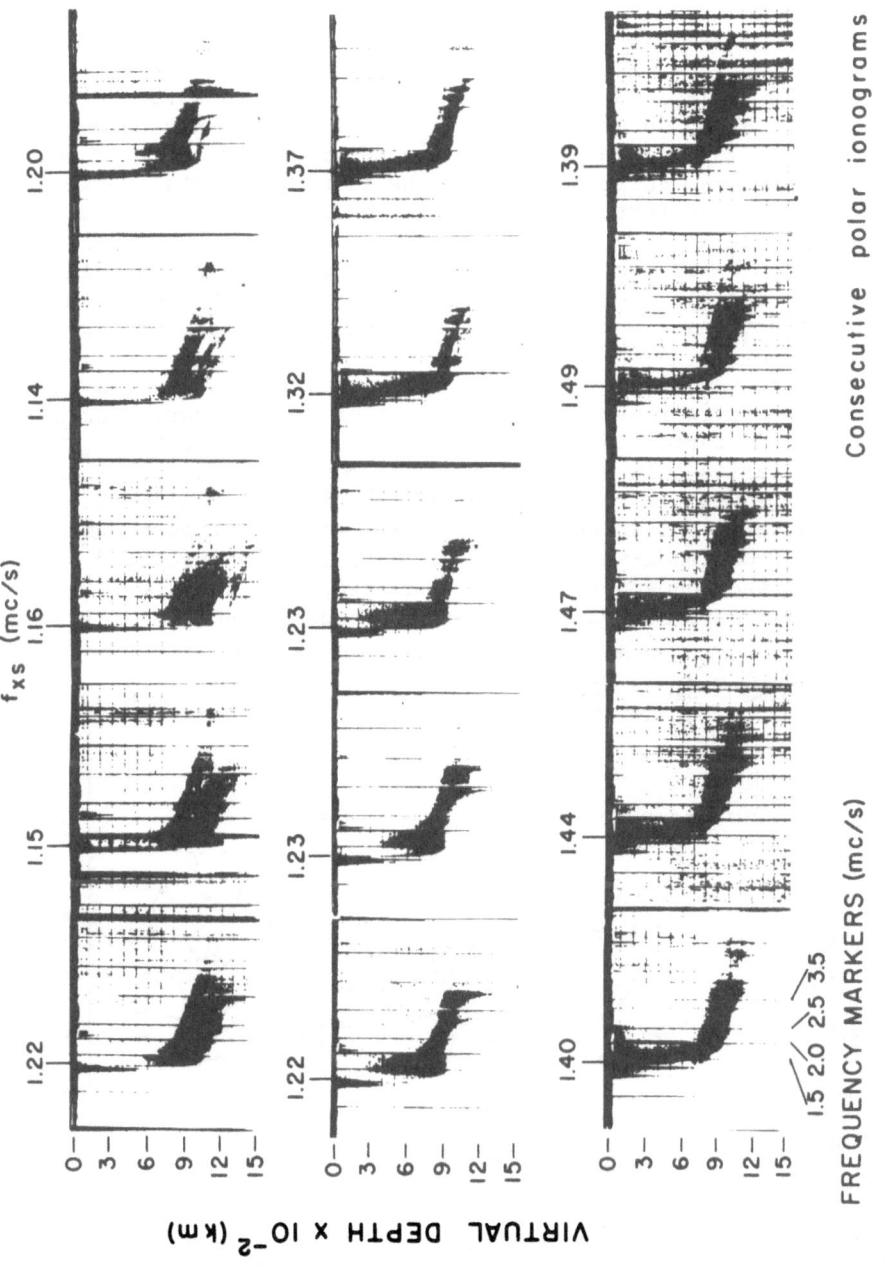

Figure 2. Consecutive polar ionograms on October 27, 1962

poorly defined due to naturally occurring irregularities, difficulties in scaling the associated 'spread' ionograms, and the absence of data beyond about 85°N dip latitude (the 'turn-around'
latitude of the Alouette I orbit). However, the axis of the
trough, which is actually the leading edge of the enhanced region,
is usually quite well defined. An example of one such case is
illustrated in Figures 2 and 3.

Figure 2 displays fifteen consecutive ionograms recorded at
high latitudes on October 27, 1962. The virtual depth below the
satellite is given in hundreds of kilometers to the left of each
row of ionograms, corresponding to the faint horizontal markers on
each record. The vertical lines are the frequency markers as indicated on the ionogram in the lower left hand corner. Scaling is
made difficult because of the presence of considerable spread of
the echoes and also because of the occasional occurrence of multi-
tiple reflections. Such features are typical of polar ionograms.

The frequencies at which the extraordinary wave traces have
zero range are indicated along the tops of each ionogram, as determined from three independent readings. As the electron gyrofrequency varies only slightly over the range of interest, changes
in these frequencies reflect real changes in electron density at
the Alouette I orbit. An examination of the variation of this
frequency from ionogram to ionogram indicates the irregular behaviour of the ionospheric electron density near the dip-pole.
The ionograms are separated in time by about 18 seconds or 100 km.

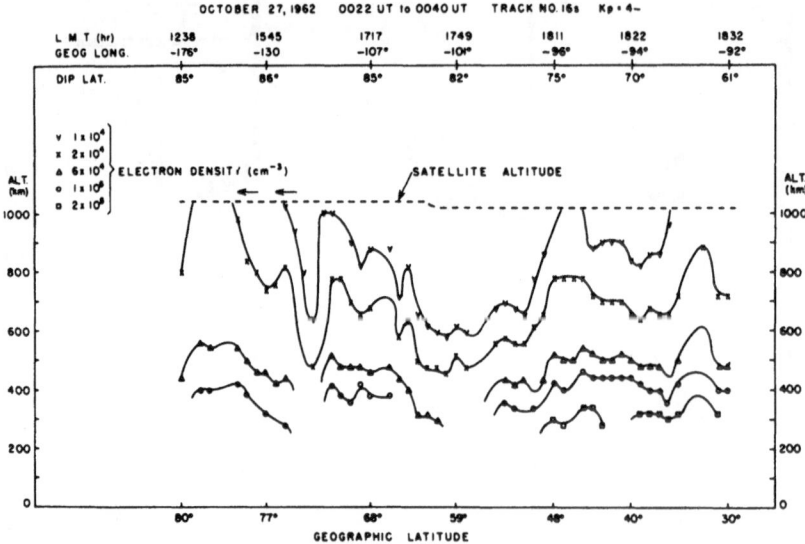

Figure 3. Latitudinal and vertical variations of electron density
           as determined from the ionograms of Figure 2.

The first ionograms correspond to the primary minimum in electron density over the entire pass. A sharp increase in electron density on the poleward side of the 'trough' is indicated by the jump in the zero range frequency from the 8th to the 9th frame and a sharp decrease is indicated from frame 14 to frame 15.

A vertical cross-section through the polar ionosphere, constructed from the electron density profiles corresponding to 60 consecutive ionograms, including those shown in Figure 2, is shown in Figure 3. The electron density in the region from about 82° dip to about 73° dip latitude were deduced from the 15 ionograms of Figure 2. The first few frames correspond to the primary minimum. The abrupt changes mentioned occur between 77° and 76° dip and between 74° and 73° dip latitude.

The maxima and minima at orbital altitudes, expected on the basis of the changes in the zero range echoes seen on the ionograms are seen to extend through nearly all levels between the satellite and the F_2 peak.

Cross-sections similar to this one were constructed for eight consecutive passes over polar regions on October 26-27,1962. Taken together, these charts revealed the presence of some global-scale features of the topside ionosphere. In order to investigate

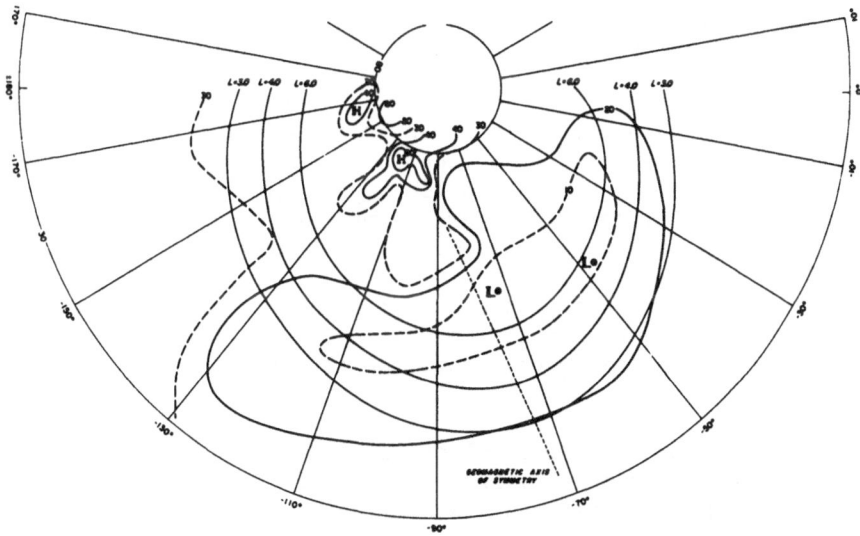

Figure 4. Horizontal cross section, at 700 km, of the polar ionospheric electron density ($\times 10^{-3}$ cm^{-3}) distribution on October 27, 1962

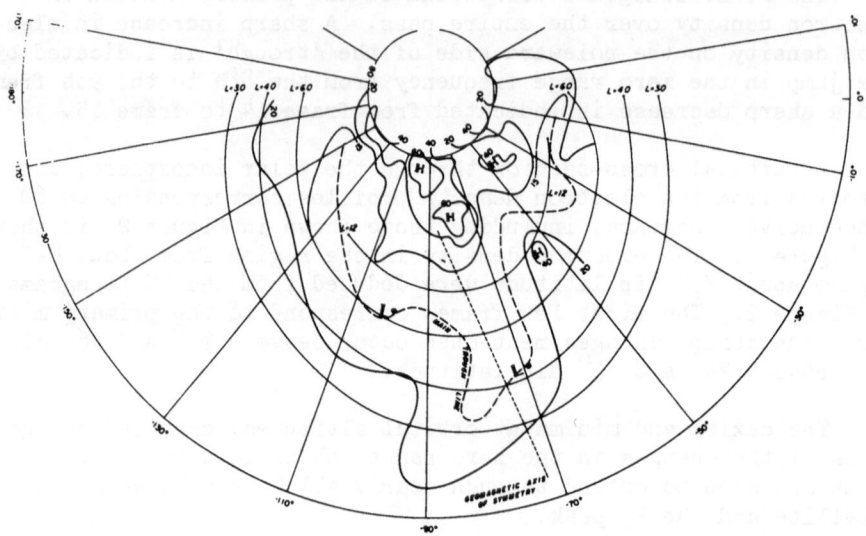

Figure 5. Horizontal cross section, at 700 km, of the polar iono-
          spheric electron density ($\times 10^{-3}$ cm^{-3}) distribution
          on April 4, 1963

these features at various levels, horizontal cross-sections at
constant altitudes were prepared from all data profiles corresponding to the eight passes.

     One such cross-section is shown in Figure 4 for the altitude
700 km. The data supports the location of a main trough at the
position indicated by the dashed line. A region of electron density enhancement is centered near the magnetic pole and is bounded
by the L=12 isoline, where L is McIlwain's parameter. An elongation of the trough occurs in the vicinity of the axis of symmetry
of the ovals represented by the isolines of L. It is interesting
to note that the central position of the main trough coincides
approximately with the central position of the region of maximum
auroral occurrence at about L=8.

     Figure 4 shows a horizontal cross-section at 700 km, con-
structed in the same manner, from data obtained on April 4-5,
1963. The principle features are similar to those found six
months earlier.

3. VARIABILITY

The following sections are devoted to the results of a study of solar-geomagnetic, seasonal, and diurnal effects on the position and structure of the enhanced region.

3.1 Solar-Geomagnetic Effects

The main interest in the detailed study of the polar maximum lies in the possibility that this maximum is in some way the result of coupling between the magnetosphere and the polar ionosphere. Neither the preliminary study we have made nor similar studies so far reported, have enabled the precise nature of this coupling to be isolated.

However, on the assumption that the maximum may be governed by outer magnetospheric and interplanetary phenomena, it would appear reasonable to correlate its strength and location with parameters, such as K_p, which are grossly related to the strength of the solar wind and hence to the size of the magnetospheric cavity, etc. In the course of our investigation, relations were sought between the following parameters:

(a) trough axis position

(b) amplitude of the polar enhancement

(c) the solar-geophysical parameters K_p and 10.7 cm solar flux

Any possible relations between these quantities were expected to depend on time and hence, longitude, and were therefore tested for different longitude sectors.

The results were disappointing in the sense that no clearcut relations between any two of these could be substantiated on the basis of the data sample available for this study. It should be pointed out, however, that although no obvious relation between trough axis position and K_p was found in the gross sense, it was discovered that during sudden commencement storms, at least, the trough is displaced to the south of its average position.

3.2 Seasonal Effects

The average positions of the low latitude boundary of the polar enhancement during the various seasons are summarized in Table I. Some care must be taken in considering these results for two reasons:

Fall	Winter	Spring	Summer
OND	JFM	AMJ	JAS
69°N	74°N	71°N	73°N

Winter	Summer	Equinoctial
NDJF	MJJA	SOMA
69°N	71°N	74°N

Winter Half	Summer Half
Oct. 1962 - Mar. 1963	Apr. 1963 - Sept. 1963
71.0°N	71.5°N

Table I. Seasonal Variations of the Average Positions of the
Low-latitude Boundary of the Polar Enhancement

(1) The orbital plane of the Alouette I satellite precesses some
two degrees per day with respect to the earth–sun line. Thus
data are available for all local times once each season
(90 days). Hence, each month of a particular season is represented by data collected during a different eight-hour
period from the other two months. Some months have more data
than others, which may cause any diurnal variation in the
position of the trough to be reflected in the seasonal averages.

(2) The possible relation between trough axis position and K_p
during sudden commencement storms may account partly for the
fact that the average latitude of the trough is found significantly farther to the south during the seasons containing
the months of November and December, 1962. This is because
the sum of the K_p for this period was generally higher than
that during any other two months of the year.

For these reasons it is felt that perhaps the most meaningful
averages are those for the two six-month periods centered on the
two solstices. The abundance of data for all local times should
vitiate the effects listed above, while the symmetry with respect
to the sun's declination is retained. These results suggest
little dependence of the southern boundary of the enhancement of
the solar declination.

3.3 Diurnal Effects

The average position of the trough during various times of
the day for the two six-month periods are summarized in Table II.
Both sets of data show the trough occurring at higher latitudes on
the day side of the earth. It is interesting to note that the

October, 1962 - March, 1963

Day	Night	Dawn	Dusk
0800-1600L	2000-0400L	0400-0800L	1600-2000L
71°N	69°N	74°N	69°N

April, 1963 - September, 1963

Day	Night	Dawn	Dusk
0800-1600L	2000-0400L	0400-0800L	1600-2000L
74°N	72°N	65°N	64°N

Table II. Diurnal Variations of the Average Positions of the Low-latitude Boundary of the Polar Enhancement

variation of the trough position in each case is in the same direction and is of the same magnitude as the diurnal distortion of the field lines at about 1000 km due to the solar wind (4,5).

Data from the energetic particle detectors aboard the Alouette I, gathered during the same period, support a similar day-night variation of the high latitude boundary of trapped and precipitated electrons having energy above 40 keV (6). The high latitude boundaries of the trapped 40 keV electrons generally fall within a few degrees of the day and night positions of the trough given in Table II. However, the high latitude boundary of precipitated 40 keV electrons lies well within the boundaries given in Table II. Further, McDiarmid et al (7) and Frank (8) show an appreciable softening of the spectrum at these higher latitudes. If such softening does occur it may well be that the peak intensity of much lower energy trapped and precipitated electrons occurs at latitudes corresponding to peak polar electron density enhancements.

The average positions in the dawn-dusk sectors defy such easy comparisons, however. The elongation of the boundary in these directions is marked during the summer half of the year but is absent during the winter half. The average positions of the trough at dawn from October through March are higher than those at dusk from April through September. This is especially interesting in view of the findings of Frank (8), that the average intensities of 40 keV electrons detected by the Explorer 14 in the magnetospheric tail were about ten times higher on the morning side during Winter, 1962, than on the evening side during Summer, 1963.

These same instruments detected a 'catastrophic' two-order-of-magnitude decrease in the intensity of 40 keV electrons at about L=5 on December 20, 1962. The electron density cross-sections constructed during this period show the positions of trough axes to be much lower than average (35°N to 45°N dip).

4. CONCLUSIONS

Satellite sounders such as the Alouette I are seen to provide an excellent means of obtaining detailed synoptic information concerning vertical and meridional distributions of electron densities of F_2 altitudes and above. A typical pass across one hemisphere yields as much data from the F_2 region as a network of 150 or more bottomside sounders, spaced about one degree of latitude apart along a meridian, operating nearly simultaneously. Information concerning maximum usable frequencies, ionospheric tilts, and irregularities in electron density distribution can be readily determined for radio communication purposes, from cross-sectional "snapshots" of the ionosphere such as that shown in Figure 1.

The use of small samples of topside sounder data to specify the nature of "mean" seasonal or diurnal ionospheric configurations must be undertaken only with greatest care. If a large fraction of the data available is included in investigations of this type, the results should be correspondingly more significant.

Several comparisons have been made between ionospheric variations at high latitudes and variations in energetic particle distribution within the outer magnetosphere. These comparisons suggest the existence of significant ionospheric-magnetospheric coupling at high latitudes. The precise nature of the required coupling mechanism has not been isolated at this point. However, the impression was gained that the high-latitude ionosphere will have to be studied as a synoptic feature, at different elevations, ranging from the F_2 peak to very high levels in the magnetosphere, such as are accessible to the Alouette II satellite, launched during IQSY.

The present study is to be taken as a prelude to a more detailed investigation of this nature.

Acknowledgement: We are grateful to Professor S.V. Venkateswaran for many helpful suggestions and for uncounted hours of rewarding discussions on subjects covered in this paper. The data analyses for this investigation were carried out at the Department of Meteorology, University of California, Los Angeles, California, with financial support from the Naval Ordnance Test Station, Chinalake, California, and through a grant from NASA NsG 237-62.

5. REFERENCES

1. G.L. Nelms, Electron Density Profiles in Ionosphere and Exosphere, In: J. Frihagen, John Wiley & Sons, Inc, 1966, p 358.

2. D.B. Muldrew, J. Geophys. Res. 70, 2635, 1965.

3. J.O. Thomas et al, Electron Density Profiles in Ionosphere and Exosphere, John Wiley & Sons, Inc. 1966, p 322.

4. J.M. Malville, J. Geophys. Res. 65, 3008, 1960.

5. E.W. Hones, J. Geophys. Res. 68, 1209, 1963.

6. D.C. Rose, Radioation Trapped in the Earth's Magnetic Field, Gordon and Breach, 1966, p 192.

7. F.B. McDiarmid et al, Can. J. Phys. 41, 1332, 1963.

8. L.A. Frank, J. Geophys. Res. 70, 1593, 1965.

NORMAL AND ABNORMAL ABSORPTION AT HIGH LATITUDES

W. R. Piggott

Radio and Space Research Station, Slough, England

Summary: The object of this paper is to summarise the main features of absorption and absorption theory which apply to problems
of radiocommunications at high latitudes. The criteria for simplification is that any errors should be small compared with the
variability of the total absorption.

The simple theory of absorption shows that it is due to the
loss of energy when electrons oscillating under the influence of a
radio wave make collisions with other particles. The absorption
coefficient, K, per unit distance in an absorbing region depends
on:

(a) the mean energy picked up by the electron from the wave

(b) the number of collisions per unit volume per unit time

(c) the speed with which the wave tranverses the medium

When the refractive index, μ, is near unity, K can be written
in the non-deviative form:

$$K \;=\; \frac{2\pi e^2}{mc} \cdot \frac{1}{\mu} \cdot \frac{N\nu}{\nu^2 + (\omega \pm \omega_L)^2} \;=\; \frac{1}{\mu} \cdot K_o \left\{ \frac{\nu^2}{\nu^2 + (\omega \pm \omega_L)^2} \right\} \qquad (1)$$

where e, m are the electron charge and mass, N is the number of
electrons per cc, ν the collision frequency, ω the angular frequency ($\omega = 2\pi f$) and ω_L the gyrofrequency about the component of
the magnetic field directed along the wave direction. When the
refractive index is small or there is much bending in the wave

path, the frequency variation of K is mainly determined by that of μ and is rapid compared with the frequency term in equation (1). A more useful form is then:

$$K = \frac{\nu}{2c} M(\frac{1}{\mu} - \mu) \qquad (2)$$

where M is a factor between, usually, 1 and $\omega/(\omega \pm \omega_H)$. At vertical incidence the frequency variations of μ and $1/(\omega \pm \omega_L)^2$ are roughly comparable when μ is about 0.6.

Refraction in the ionosphere is usually accompanied by focus and scatter phenomena whose effects are frequently indistinguishable from those of deviative absorption and may greatly exceed it. Thus computed deviative losses seldom agree with observations to better than a factor of two, or sec i, whichever is the bigger. The frequencies at which they are maximum obey Martyn's theorem:

$$f = f_0 \sec i \qquad (3)$$

where i is the angle of incidence and f_0 a critical frequency.

Similar absorption bands occur in the D region, usually at frequencies where $\nu/\frac{dN}{dh}$ is a maximum at vertical incidence, and also obey equation 3. The rate of change of absorption with frequency can be very great and only calculable if both parameters, f_0, cos i are known to be better than 1%. Practical field strength calculations are therefore based on:

(a) identifying the absorption bands

(b) calculating the absorption away from these bands using non-deviative theory.

The accuracy depends on the skill with which the equivalent non-deviative absorption has been evaluated.

In practice the electron density N and collision frequency ν change with height by a factor of about 10^4 in the main absorbing region but K only changes by a factor near 10. For most normal conditions $\nu^2 \ll (\omega \pm \omega_L)^2$ and the total absorption is proportional to $\int N\nu \{\frac{(\omega \pm \omega_L)^2}{\nu^2 + (\omega \pm \omega_L)^2}\}$ ds. Thus collisions are less effective when $\nu^2 \to (\omega \pm \omega_L)^2$ or are more frequent.

Equation 1 gives an absorption coefficient at any height which is constant for $\nu^2 \gg (\omega \pm \omega_L)^2$ and decreases at higher frequencies. The overall absorption shows a maximum by both day and night when the equivalent frequency, cos i, is in the medium wave

band. The decrease in absorption at low frequencies is due to the
longer waves being reflected low in the absorbing region. The
frequency of the maximum is determined by the structure of the
lower ionosphere and behaves like an absorption band associated
with a critical frequency.

It is convenient to divide the absorption in three arbitrary
classes:

(a) normal absorption

(b) regular abnormal absorption

(c) absorption events

Normal absorption at HF obeys equation (1) with
$\nu^2 \ll (\omega \pm \omega_L)^2$. The absorption layer is solar controlled and is
first formed when the solar zenith angle $\chi = 92°$. It varies with
time of day, season and position roughly proportional to
$(\cos \chi)^{0.75}$ with a time delay of about 15 minutes in the daytime.
While most of the absorption disappears quickly in the evening
when $\chi = 90°$, the remainder roughly obeys a recombination law with
an initial time constant of 1 hour, corresponding to a slowly de-
caying E layer with a critical frequency mainly near 0.5 MHz in
the late night. The distribution is skew with large deviations
more common than given by a normal distribution. The quartile
range divided by median absorption is near 15% at temperate latitudes, decile ranges being usually twice the quartiles. This absorption varies with solar cycle so that:

$$L = L_o (1 + 0.004 R) \qquad (4)$$

where R is the sunspot number and the total non-deviative absorption L at oblique incidence varies in daytime so that:

$$L = nL_o (\cos \psi)^{0.75} \sec i /(f+f_L)^2 \qquad (5)$$

where n is the number of hops, f the working frequency and f_L is
numerically equal to half the gyrofrequency, $f = f_H/2$, for almost
all trajectories at oblique incidence.

There are two types abnormal absorption to be considered,
both of which add to the normal absorption:

(a) winter anomaly at non auroral latitudes

(b) auroral and polar cap absorption.

The winter anomaly at HF is really two anomalies:

(i) the absorption is much larger than expected from the value of cos χ usually approaching the absorption in midsummer or even greater in years of solar minimum.

(ii) the absorption varies much more from day to day than does normal absorption. Thus the quartile range in the winter anomaly months is three times that in summer months. For F region reflections the diurnal variations become more rapid as the excess absorption increases giving (cos χ)^n with n usually between 0.9 and 1.3 but occasionally up to 1.5. The excess absorption still obeys equation (1) but it is likely that this is due to a balance between additional losses where $\nu^2 \rightarrow (\omega \pm \omega_L)^2$ and a decreased value of μ at greater heights in the D region. There is evidence that at least some of the anomalies are associated with stratospheric phenomena.

Nearer the auroral zone the temperate latitude winter anomaly dies out, even in fairly quiet periods, and is replaced by an even more variable absorption which is indirectly linked with magnetic activity. The diurnal variation, in contrast with winter anomaly, is slow corresponding to $L = L_o(\cos \psi)^{0.2}$, at least on the lower frequencies. Taking the polar regions as a whole the noon absorption falls by a factor near three when the latitude is increased so that χ = 90°.

The morphology of the auroral zone regular absorption is still not clear, there is some evidence for a peak in absorption just outside the zone of maximum auroral activity but it is likely that it changes with longitude. In the Halley Bay area there is good evidence for a photodetachment type of mechanism giving a sudden increase in absorption as χ decreases through 100° in winter. The absorption varies considerably from day to day, is associated with magnetic activity particularly in years of high activity but is usually delayed by one or more days.

Events will be dealt with by other speakers. Here we want only to stress:

(a)   the spiral type incidence of blackout at constant time with a concentration over the North American sector

(b)   the effect of shift in local time of particle activity with latitude and longitude which can cause the absorption to be a maximum before or sometimes after noon near auroral latitudes

(c)   the importance of ionization where $\nu^2 \ll (\omega \pm \omega_L)^2$ at high frequencies and large angles of incidence during PCA's.

There is much need for a systematic analysis of high and temperate latitude absorption data and of changes in the prediction
methods to accomodate the winter and auroral anomalies.

COLLISION FREQUENCY IN THE HIGH LATITUDE D-REGION

E. V. Thrane

Norwegian Defence Research Establishment

Abstract: This paper deals with collision frequencies in the ionospheric D-region. A description is given of the methods which exist for studying the collision frequencies: laboratory investigations and radio wave propagation experiments. It is shown that the results obtained by the two techniques are in satisfactory agreement.

1. INTRODUCTION

The air in the D-region is only weakly ionized, the ratio of the number density of free electrons to the number density of neutral particles being in general less than 10^{-7}. Collisions of free electrons with neutral molecules will therefore be much more numerous than collisions with ions, and it is the average frequency $\bar{\nu}$ of such electron-neutral collisions which will determine the absorption a radio wave suffers when it passes through the D-region. It is therefore of major importance for radio wave propagation via the ionosphere to measure $\bar{\nu}$ and to study its temporal and spatial variations.

In this lecture a brief review will be given of the present knowledge of the D-region collision frequencies. This knowledge comes mainly from two sources: a) Laboratory measurements of collision cross sections of free electrons in air combined with measurements of temperature, pressure and composition of the air in the mesosphere b) Radio wave propagation experiments in the ionosphere using rocket or ground based techniques. The two approaches will be discussed separately, and it will finally be shown that they do seem to give consistent answers.

2. THE NEUTRAL ATMOSPHERE

A knowledge of the state of the neutral gas is necessary to understand the electron collision processes in the D-region, and to put things in perspective, it is useful for a moment to look at the height variation of the neutral atmosphere as given in the US Standard Atmosphere (Sissenwine et al 1962) (Figure 1). The D-region coincides with the mesosphere, and is the region between the temperature maximum (300°K) at 50 km and the temperature minimum (about 180°K) at 85-90 km.

When discussing D-region collision frequencies there are, as I will show later, two parameters of particular importance, namely the height variation of atmospheric composition and the height variation of the total pressure.

It seems firmly established that up to about 100 km winds and turbulence provide an effective mixing of the atmospheric gases, so that even if the air in the D-region is very tenuous, it has essentially the same composition as air at ground levels.

The atmospheric pressure cannot be measured directly at these heights, but it can be derived from the hydrostatic equation, if the pressure at some reference level and the height variation of

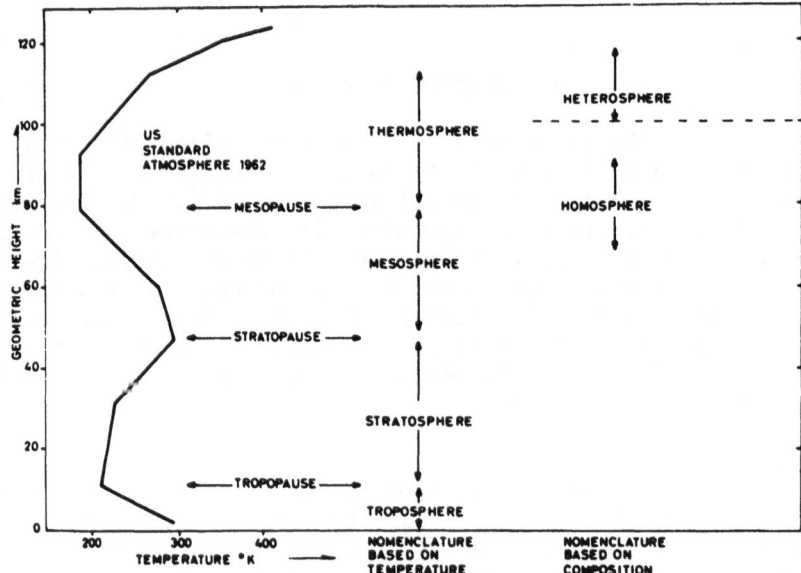

Figure 1. Height variation of atmospheric temperature from the US Standard Atmosphere (Sissenwine et al 1962)

density or temperature above this level are known. The standard
atmosphere published by Cole and Kantor (1963) gives atmospheric
pressures derived in this manner which shows (Figure 2) that
there is a seasonal change of D-region pressures and that this
change increases with latitude. At 60°N the summer pressure exceeds the winter pressure by a factor of about 1.6 in the region
50-80 km. The curves in Figure 2 represent average values, and it
should be remembered that the upper atmosphere is not static, but
in an active dynamical state. At low latitudes the mesosphere
seems to be fairly stable, thus the US Standard Atmosphere states
that there is less than 5% chance of the density at any level de-
viating by more than 20% from the mean. In winter at high and
middle latitudes, however, great day to day changes in temperature
have been observed.

3. THE DEFINITION OF THE MEAN COLLISION FREQUENCY

The free electrons in the D-region are produced by the primary ionizing agents or by detachment from negative ions, with
initial energies considerably larger than the mean thermal energy
of the neutral gas particles. Several authors (Rumi, 1962;
Sears, 1963; Belrose and Hewitt, 1964) have therefore suggested
that the D-region electrons may not be in thermal equilibrium with
the neutral gas. Dalgarno and Henry (1965) have considered this
problem on a theoretical basis and they conclude that the fast
electrons must cool very rapidly through inelastic collisions with
heavy molecules. We can therefore assume that the D-region elec-

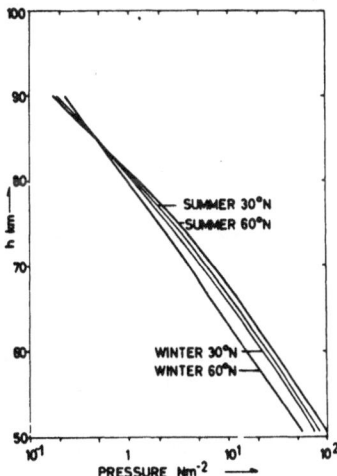

Figure 2. Height variation of atmospheric pressure
           (Cole and Kantor 1963)

trons have a Maxwellian velocity distribution characterized by the neutral gas temperature.

The rigorous definition of the mean collision frequency of D-region electrons must therefore be

$$\overline{\nu} = \int n\sigma(v) \ v \ f(v) \ d\tau \qquad (3.1)$$

when n is the number density of neutral particles, $\sigma(v)$ the collision cross section of electrons with speed v, and $f(v)$ the normalized Maxwell-Boltzmann distribution function. Integration is over all velocity space.

To use this expression the function $\sigma(v)$ must be determined. In the next section I will show that laboratory results lead to a mean collision frequency which is proportional to pressure.

4. LABORATORY DATA ON COLLISION CROSS SECTIONS OF THERMAL
    ELECTRONS IN AIR

Most of the data on electron collision cross sections in air have been obtained using the diffusion method due to Townsend and Tizard (1913) in which electrons are made to drift slowly through the gas under the influence of a uniform electric field.

Such measurements have been made in air, and also in oxygen and nitrogen, by a number of workers.

In N_2 the electron collision cross section has been measured down to room temperatures by Phelps and Pack (1961) and by Crompton and Elford (1963). They find

$$\sigma(v) = a v \qquad a = constant \qquad (4.1)$$

Combining (3.1) and (4.1) we obtain

$$\overline{\nu} = n \ a \ \overline{v^2} = n \ a \ \frac{3kT}{m} \qquad (4.2)$$

since $\frac{1}{2} m \overline{v^2} = \frac{3}{2} kT$ (m is the electron mass).

We note that $\overline{\nu}$ is proportional to the product of the temperature T and the number density n of neutral particles, that is

$$\overline{\nu} \propto p \qquad (4.3)$$

where p is the pressure.

Putting the value of a from the laboratory measurements and the values of k and m into Equation (4.2) we obtain for nitrogen

$$\bar{\nu} = 1.05 \cdot 10^6 \, p \, (N_2) \quad (s^{-1}) \qquad (4.4)$$

where p is the pressure in $N \, m^{-2}$.

Now it has become customary to work in terms, not of the mean collision frequency $\bar{\nu}$, but of the collision frequency ν_M of mono-energetic electrons with energy kT. For these electrons we have of course $\frac{1}{2} m \, v^2 = kT$ and thus from (4.2) we have simply that

$$\nu_M = \frac{2}{3} \bar{\nu} \qquad (4.5)$$

Hence for nitrogen the relation between ν_M and the pressure becomes

$$\nu_M = 7 \cdot 10^5 \, p(N_2) \quad (s^{-1}) \qquad (4.6)$$

This relation, then, is based on reliable measurements of collision cross sections at temperatures close to those existing in the D-region.

For air, unfortunately, it is extremely difficult to measure collision cross sections at the temperatures of interest here, because of the strong tendency for the electrons to attach to neutral oxygen molecules and form negative ions. Several attempts have been made to extrapolate measurements made in air at high electron energies down to thermal energies, but as has been pointed out by Smith et al (1965), such extrapolations are very uncertain. Hence very little is known about electron collision cross sections in O_2 at D-region temperatures.

However, nitrogen is the most abundant atmospheric constituent and it may be reasonable to assume that the contribution to the collision frequency from collisions with oxygen molecules is small.

Phelps and Pack (1961) have made an estimate, which is often quoted, of the contribution to the collision frequency in air from collisions with O_2 molecules. According to these authors the velocity dependence of the electron collision cross section in oxygen is of the same form as in nitrogen, and for air the contribution to ν_M from oxygen should be about 15%. For air their results lead to

$$\nu_M = 6.34 \cdot 10^5 \, p \quad (s^{-1}) \qquad (4.7)$$

where p is the total atmospheric pressure in $N \, m^{-2}$. Since the composition of the air in the D-region is essentially the same as for the air at ground levels, estimates of the height variations of ν_M in the D-region may be obtained from Equation (4.7) using the standard pressure curves given in Figure 2. A seasonal and latitudinal change in ν_M (h) should therefore be expexted from these arguments.

5. COLLISION FREQUENCIES DEDUCED BY RADIO METHODS

The other approach used to study the electron collision frequencies in the lower ionosphere is to deduce the parameter ν_M from radio wave propagation experiments, for example measurements of absorption or Faraday rotation. A magnetoionic theory expressing the complex refractive index of the radio wave in terms of ν_M and the electron density N must then be adopted.

For example, for a radio wave travelling along the earth's magnetic field the absorption per meter in db may be written (Barrington and Thrane 1962)

$$\kappa_{o,x} = \frac{8\sqrt{\pi}}{3mc} N e^2 \int_0^\infty \frac{\nu(u)}{\nu(u)^2 + (\omega \pm \omega_H)^2} \left(\frac{u}{kT}\right)^{\frac{3}{2}} e^{\frac{u}{kT}} d\left(\frac{u}{kT}\right) \quad (5.1)$$

where $\nu(u)$ is the collision frequency of electrons with energy u, ω and ω_H are the angular frequency and gyro frequency respectively and subscripts o and x refer to the magneto-ionic modes. To use this expression something must be known or assumed about $\nu(u)$. The laboratory measurements quoted in the previous section suggest that $\nu \propto u$ which is equivalent to $\sigma \propto v$. In this case (5.1) gives the Sen and Wyller (1960) formula

$$\kappa_{o,x} = const \frac{N}{\nu_M} C_{5/2} \left(\frac{\omega \pm \omega_H}{\nu_M}\right) \quad (5.2)$$

where $C_{5/2}$ is the semiconductor integral tabulated by Dingle et al (1957).

Thus if κ_o and κ_x is measured, ν_M and N can be deduced[*]

It is interesting to compare the values of ν_M derived from propagation experiments with the values deduced from laboratory measurements and pressure data.

---

[*] (If $\nu(u)$ = constant is assumed in (5.1), the formula reduces to the well known Appleton-Hartree expression for κ. In this case an effective collision frequency, ν_{AH}, can be deduced from the measurements. According to Molmund (1959), $\nu_{AH} = \frac{5}{2} \nu_M$ when $\frac{\nu}{\omega} \ll 1$ and $\nu_{AH} = \frac{3}{2} \nu_M$ when $\frac{\nu}{\omega} \gg 1$)

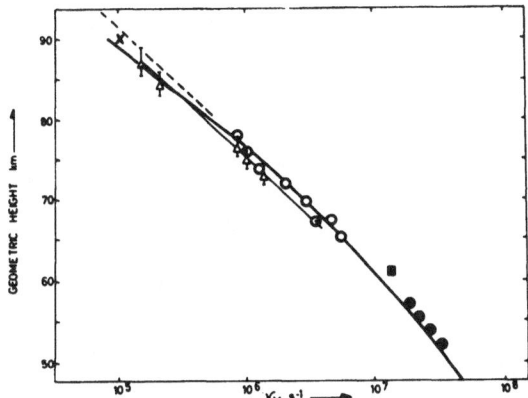

Figure 3. Height variations of ν_M in winter and spring
           (Thrane and Piggott 1966)
           The thick line shows the height variation deduced
           by fitting the US standard pressure curve to the
           points by the RMS method.

The available data on D-region collision frequencies deduced
by radio methods have been reviewed by Thrane and Piggott (1966).
Most of the data were obtained at subarctic latitudes $(60^\circ N)$. At
this latitude the values of ν_M fall into two distinct groups, summer and autumn, and winter and spring.

In Figure 3 the data for winter and spring have been collected and compared with a pressure curve from the US Standard Atmosphere (Sissenwine et al 1962). The pressure curve has been
moved horizontally on the log scale until the best fit was obtained. The scale heights of collision frequency and pressure
agree very well. Figure 4 shows a comparison between the standard
pressure curve and the summer values of collision frequency. The
scale heights are again in excellent agreement. The dotted line
in Figure 4 is the pressure curve from Figure 3. The summer values of ν_M exceed the winter values by a factor of 1.6, and the
seasonal change in the measured values of ν_M is highly significant.

The seasonal change of collision frequency at higher latitudes is illustrated in Figure 5. Here collision frequencies measured by rocket techniques at Andøya in Northern Norway (Jespersen et al 1964, and Jespersen et al 1966) are compared with pressures measured at Ft Churchill by Norberg and Stroud (1961). In
this case the best fit between collision frequency and pressure
curves is obtained when

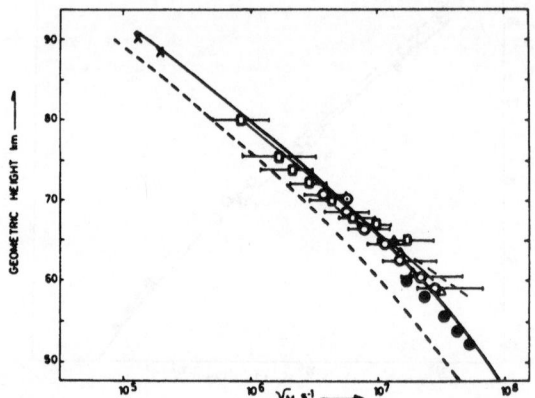

Figure 4. Height variations of ν_M in summer and autumn
(Thrane and Piggott 1966)
The thick line in Figure 3 is shown as a broken
line, and the thick line is the US standard pressure
curve fitted to the points by the RMS method

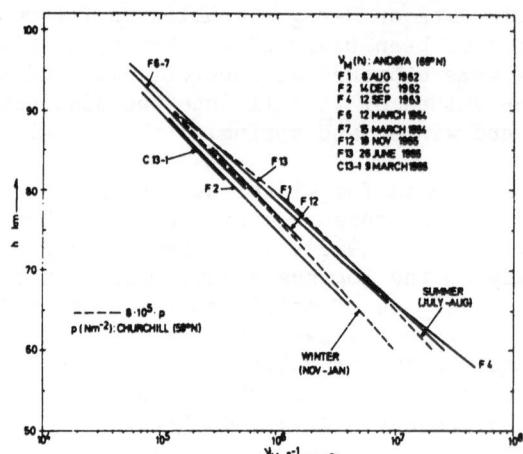

Figure 5. Collision frequencies measured by rocket techniques in Northern Norway (Jespersen et al 1964,
Jespersen et al 1966) compared to pressure measurements (broken lines) published by Norberg
and Stroud (1961)

$$\nu_M \; = \; 8 \cdot 10^5 \; p \quad (s^{-1}) \qquad\qquad (5.7)$$

6. CONCLUSION

We conclude that the D-region collision frequency has the same height variation and the same seasonal variation as the atmospheric pressure.

The laboratory measurements of collision cross sections of electrons in air and in nitrogen indicate that

$$\nu_M \; = \; 6.34 \cdot 10^5 \; p \quad (s^{-1}) \qquad\qquad (4.7)$$

while a direct comparison between observed collision frequency and pressure gives

$$\nu_M \; = \; 8 \cdot 10^5 \; p \quad (s^{-1}) \qquad\qquad (5.7)$$

The agreement must be regarded as satisfactory, although the relation (4.7) gives collision frequencies which are 25% smaller than those observed. This may be due to an underestimation of the collision cross section of electrons in oxygen. The fact that the observed collision frequencies are proportional to pressure over a wide height range, suggests that the adopted form of the velocity dependence of the electron cross section in air is correct.

The experimental ν_M values in Figures 3 and 4 show fairly little scatter. However, Belrose et al (1966) has shown that collision frequencies deduced from partial reflection measurements during winter in Ottawa show a day to day variation which can be greater than the seasonal change.

7. REFERENCES

1. R.E. Barrington and E.V. Thrane, J. Atmos. Terr. Phys. 24, 31, 1962.

2. J.S. Belrose, I.A. Bourne and L.W. Hewitt, Presented at conference on ground based radio wave propagation studies of the lower ionosphere, Ottawa, April 1966.

3. J.S. Belrose and L.W. Hewitt, Nature 202, No 4929, 267, April 1964.

4. A.E. Cole and A.J. Kantor, Air Force Survey in Geophysics No 153, 1963, L.G. Hanscom Field Bedford Mass.

5. R.W. Crompton and M.T. Elford, Proceeding of the Sixth International Conference on Gaseous Ionization Phenomena, Paris 337, 1963.

6. A. Dalgarno and R.J.W. Henry, Proc. Roy. Soc. 288, A 521, 1965.

7. R.E. Dingle, D. Arndt and S.K. Roy, Appl. Sci. Res. 6B, 155, 1957.

8. M. Jespersen, A. Haug and B.Landmark, In: Ed. J. Frihagen, Electron density profiles in ionosphere and exosphere, North Holland Publ. Co. 1966, p 27.

9. M. Jespersen, O. Petersen, R. Rybner, B. Bjelland, O. Holt and B. Landmark, In: Ed. E.V. Thrane, Electron density distributions in ionosphere and exosphere, North Holland Publ. Co. Amsterdam, 1964, p 22.

10. P. Molmud, Phys. Rev. 114, 29, 1959.

11. W. Norberg and W.G. Stroud, NASA Technical Note D - 703, April 1961.

12. A.V. Phelps and J.L. Pack, Phys. Rev. Letters 6, 111, 1961.

13. G.C. Rumi, IRE Trans. Antennas Propag. 10, 594, 1962.

14. R.D. Sears, J. Geophys. Res. 68, 5135, 1963.

15. H.K. Sen and A.A. Wyller, J. Geophys. Res. 65, 3931, 1960.

16. N. Sissenwine, M. Dubbin and H. Wexler, US Standard Atmosphere US Government Printing Office, Washington 25, D.C. 1962.

17. R.A. Smith, I.A. Bourne, R.G. Loch, C.S.G.K. Setty, T.N.R. Coyne, P.H. Barratt and B.S.N. Prasad, Final Report Contract AF19 (604-6177 AFCRL-65-460 Air Force Cambridge Research Laboratories Bedford Mass.

18. E.V. Thrane and W.R. Piggott, J. Atmos. Terr. Phys. 28, 721, 1966.

19. J.S. Townsend and H.T. Tizard, Proc. Roy. Soc. A 88, 336, 1913.

ELECTRON DENSITY OBSERVATIONS DURING AURORAL ABSORPTION RELATED

TO RADIO WAVE COMMUNICATION PROBLEMS

Mogens Jespersen
Danish Space Research Institute

Bjørn Landmark
Norwegian Defence Research Establishment

Abstract: This report deals with results from propagation experiments for measurements of the density and collision frequency
of the free electrons during conditions of auroral absorption.
The results are discussed with emphasis on their importance for
problems of radio wave communication at high latitudes.

1. INTRODUCTION AND OUTLINE

The importance of research rockets (and satellites) in
modern geophysical research has become increasingly clear during
recent years. There are so many of the fundamental problems in
the understanding of, for example, the physical processes in the
ionosphere which can only find their solution through rocket or
satellite experiments, that it seems difficult for a research
group today to formulate a research programme without being able
to make use of these new research tools.

It was against this background that our groups desired to
formulate a research programme in which we could also make use of
rockets. It is stressed here that we by no means consider that
the rocket experiments will make the ground-based experiments unnecessary; on the contrary, it is our opinion that rocket experiments should be an integrating part in an experiment in which full
use is made of conventional methods.

During the last few years the main efforts of our groups have
been directed towards studies of the polar D region during disturbed conditions. By means of a number of riometer stations, the

morphology of the polar blackouts has been studied, and by more
sophisticated methods, such as the cross modulation and D region
backscatter techniques, information has been deduced on distributions of electron density and collision frequency in the D region.
It was therefore natural that in the first rocket experiments we
also wished to measure electron density and collision frequency
profiles in the D region. The results which can be obtained by
means of rockets are, of course, more reliable and accurate than
those determined by ground-based techniques. An important consideration is therefore that by comparing the results from rocket
observations with those obtained simultaneously by means of
ground-based techniques, the value of the latter will be increased.

In section 2 of this report, the results of electron density
measurements in rockets that have been obtained and analyzed so
far, will be presented. The bearing of these results upon some
problems of radio wave communication will then be discussed in
section 3.

2.   ELECTRON DENSITY AND COLLISION FREQUENCY OBSERVATIONS

The first research rocket in a cooperative research programme
between NASA and Scandinavian research groups was launched from
the Andøya rocket range on 18 August 1962. This rocket was in-
strumented with a multifrequency absorption and phase experiment.
Signals were transmitted from the ground and received in the rockets. This method was chosen because the propagation experiment
has been proved to be reliable, and because it was felt to be the
easiest to start with.

The experimental technique has been modified in later rockets,
and will be discussed briefly in section 2.1.

In 2.2 results will be presented of electron density observations from rocket flights during conditions of auroral absorption.

2.1 Experimental Method

In the first five firings a multifrequency inverse Jackson-
Seddon type experiment was utilized. A detailed description of
this technique has been given elsewhere (1).

In our later rocket payloads, however, we have made use of a
simple Faraday rotation experiment. The principle of this method
is simply to transmit a linearly polarized wave from the ground,

and receive this on a linearly polarized antenna in the rocket. By comparing the position of the minima in the amplitude record with spin recordings, the relative phase change between the magneto-ionic components can be deduced. The difference in absorption of the two magneto-ionic components can be deduced by comparing the amplitude maxima and minima in the pattern.

This method, which is extremely simple, has been found to give quite reliable results.

2.2 Results

In Figure 2.1 is shown as an example the results of electron density observations in the rocket Ferdinand VII. The riometer absorption at this time was 1.2 db at 27.6 MHz, and the rocket was launched on 15 March 1964 at 0345 MET (GMT + 1 hr). In the illustration the observed ν-curve is plotted with increasing values towards the left.

It can be concluded from Figure 2.1 that in this case most of the radio wave absorption (at HF and VHF) would occur in the height range where it is proportional to $N\nu$. If the absorption was equally important at all heights, then $N \propto 1/\nu$. As the curves in the illustration are plotted on logarithmic scales, they should then be parallel. The illustration therefore shows that most of

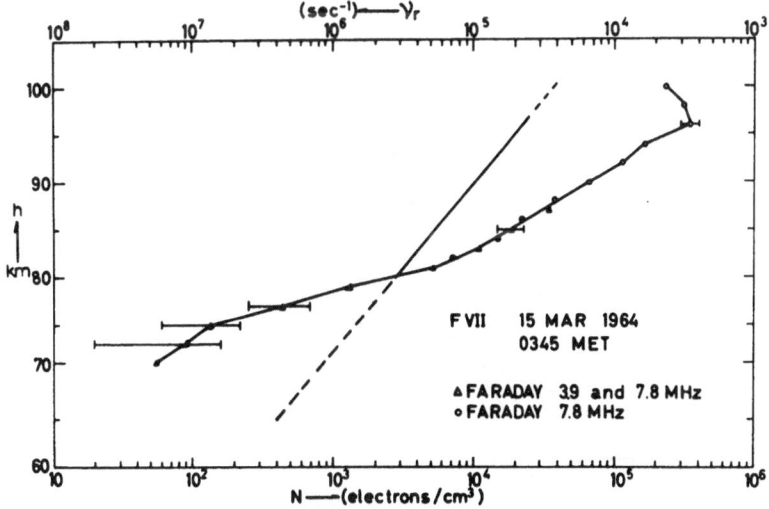

Figure 2.1   Results of electron density and collision frequency observations in Ferdinand VII

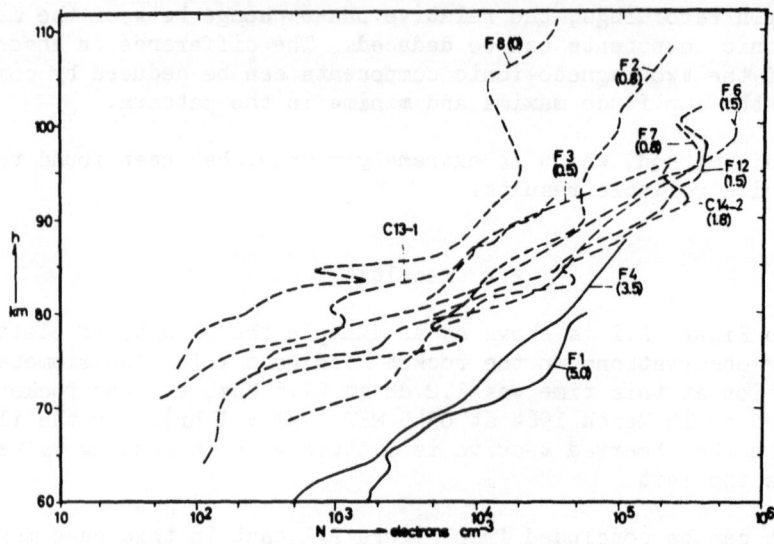

Figure 2.2   Electron density profiles observed during auroral
             absorption events

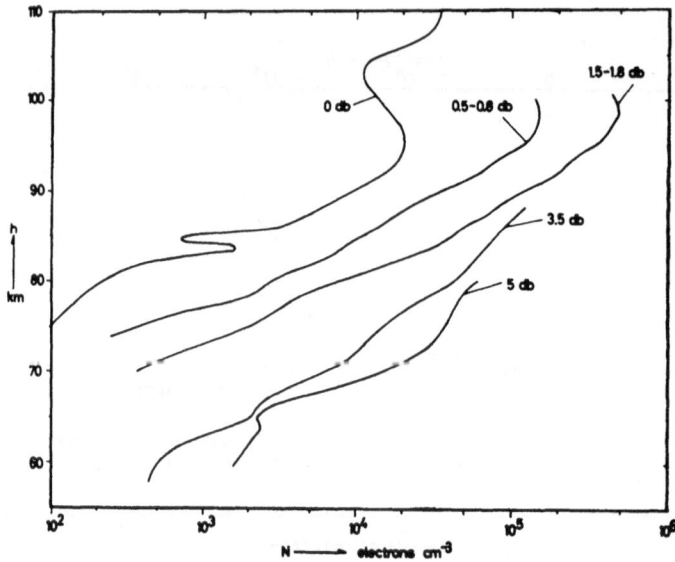

Figure 2.3   Average electron density profiles for different
             amounts of auroral absorption

the absorption occurs well above 80 km, and that it peaks around 100 km where a peak in the electron density is observed.

In Figure 2.2 the profiles observed so far during auroral absorption events are shown. The amount of absorption observed at a riometer at 27.6 MHz in each case is also given.

In Figure 2.3 the results of Figure 2.2 have been used to determine the electron density profile for different amounts of riometer absorption. Average curves are shown for the profiles when the absorption was between 0.5 and 0.8 db, and between 1.5 and 1.8 db.

The electron densities for absorption between 1.5 and 1.8 db are about 3 times those for absorption between 0.5 and 0.8 db, indicating that the average form of the particle spectrum was the same. The marked difference for day and nighttime results at the lower levels must be due to a different negative ion to electron ratio for day and nighttime conditions.

3. POSSIBILITY OF COMMUNICATION DURING AURORAL ABSORPTION CONDITIONS

The results of the preceding section show that in most cases of auroral absorption, the major part of the radio wave absorption occurs at fairly high levels in the ionosphere, say above 85 km. The results also show a steady increase of the electron density up to heights of at least 100 km. In fact, the observed layer looks more like a low sporadic E-layer than a D-layer. It may therefore be of importance to evaluate the possible use of this layer for communication purposes.

3.1 The MUF

From Figure 2.3 we estimate that for nighttime conditions a riometer absorption of about 1 db corresponds to a peak electron density of about 3×10^5 electrons per cm^3. During day-time about twice the amount of riometer absorption would be observed.

A riometer absorption of about 1 db for nighttime conditions will according to experience be sufficient to make normal HF-communication very diffcult, and can perhaps be taken as a threshold value.

An electron density of 3×10^5 electrons per cm^3 corresponds to a critical frequency of

$$f = \sqrt{81\,N} \cdot 10^{-3} \quad (\text{MHz})$$

$$= \sqrt{81 \cdot 3 \cdot 10^{5}} \cdot 10^{-3} \quad (\text{MHz})$$

$$\approx 5 \text{ MHz}$$

If a circuit of 1000 km is considered, the factor for this low layer will be fairly high, say between 4 and 5. It will therefore be possible to have communication up to a frequency of 20 to 25 MHz.

3.2 The LUF

It is well known that the lowest useful frequency for the circuit is that at which the received field strength is equal to that necessary for the service. The received field strength is determined by a number of factors. We shall, however, limit our discussion to the absorption of the waves because this is the factor that is different from normal radio circuits.

In order to make an estimate of the absorption, a full wave computation must be carried out. Such an analysis has been performed and the results are presented in Figure 3.1. The amount of absorption as a function of frequency at vertical incidence for waves reflected from two different layers has been calculated. These layers have been chosen to have the same form as the day and

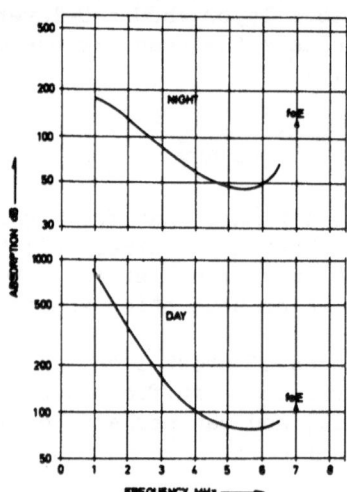

Figure 3.1 Absorption as a function of frequency for waves
reflected from auroral absorption type layers

nighttime profile of Figure 2.3, and in both cases a peak electron density of 6×10^5 electrons per cm^3, corresponding to a critical frequency of 7 MHz. The expected riometer absorption would be about 1.8 db.

From Figure 3.1 the total absorption at 5 MHz is estimated to be 45 db and 85 db for night and daytime conditions.

Martyn's absorption theorem can be used in order to make an estimate of the absorption to be expected for an oblique circuit. This theorem states that the absorption of a frequency f at an angle of incidence i equals cos i times the absorption of the frequency f cos i at vertical incidence, or

$$Abs_{f,i} = \cos i \cdot Abs_{f \cos i,o}$$

In our case f cos i = 5 MHz , where $M = \dfrac{1}{\cos i} \approx 4.5.$ The frequency for the oblique circuit, f, will then be 4.5×5 MHz = 22.5 MHz, and the absorption for the oblique circuit will be:

Nighttime conditions:

Absorption at 22.5 MHz:

$$A = \frac{1}{4.5} \cdot 45 \text{ db} = 10 \text{ db}$$

Daytime conditions:

Absorption at 22.5 MHz:

$$A = \frac{1}{4.5} \cdot 90 \text{ db} = 20 \text{ db}$$

Although the absorption to be expected for our circuit is high, it is felt that communication may still be possible. On a number of radio circuits in use, which seem to give reasonably good reliability during undisturbed conditions, an improvement of 20 db is signal strength would seem possible by improving antenna systems etc.

4. CONCLUDING REMARKS

The rocket profiles presented in this paper may be somewhat biased towards conditions when the absorption is steady and it´may well be that at other times the spectrum of the particles is much harder. Our results show, however, that in many cases of auroral absorption it should be possible to have communication over paths of the order of 1000 km at a frequency around 20 MHz. The absorption will be high, and a well engineered circuit will therefore be essential

The possibility discussed in this paper might have practical application for communication between Denmark and Greenland, and between Norway and Spitzbergen.

We would like to suggest that a test circuit be established in order to evaluate the usefulness of this proposal.

Acknowledgements: We would like to thank Mr W. R. Piggott and Dr K. G. Budden for help in carrying out the numerical calculations of the absorption. These were done on EDSAC II in Cambridge and Dr Budden has compiled the programme.

5. REFERENCES

1. M. Jespersen et al, Electron Density and Collision Frequency Observations in the Lower D-Region during Auroral Absorption, In: E. Thrane, Electron Density Distribution in Ionosphere and Exosphere, North Holland Publishing Co, Amsterdam, 1964, p 22.

PHYSICAL PROBLEMS OF PARTICULAR RELEVANCE TO ARCTIC COMMUNICATIONS

J. H. Meek

Defence Research Telecommunications Establishment

Ottawa, Canada

Abstract: The complicated physical processes governing the polar ionosphere often interrupt high latitude radio communications. For polar circuits presently used prediction techniques are not adequate and need revision. Specific requirements and means for improving communications in the region concerned are discussed. It is stressed that the possibility for utilizing anomalous ionization structures, such as E_s and the high latitude trough, for sustaining propagation should be further exploited. There is need for a program to establish empirical prediction patterns by synoptic and morphological studies.

1. INTRODUCTION

Radio communications to and from high latitude stations are subject to interruption due to the effects of charged particle precipitation into the ionosphere producing the various well-known geomagnetic and ionospheric disturbance effects. These effects have been observed for some years and the statistics of their occurrence, temporally and spatially, have been studied. However, one is not able yet to trace the path of a particular disturbance from the magnetosphere down to the ionospheric reflecting and absorbing layers and present a morphological picture of the state of these layers for a particular time. On the other hand, present routine data give information which could provide better predictions in polar regions then are now offered to communicators.

There is a requirement for:

1.   Reliable statistical maps of the polar ionosphere for commu-
     nication planning.

2.   A method of testing the ionosphere to help to choose the
     proper circuit frequency and bandwidth characteristics.

3.   A short term forecasting method to assist in planning traffic
     handling.

2. SHORT RANGE COMMUNICATIONS

In spite of the vagaries of the polar ionosphere, the principal method of communicating within the auroral zone is with medium and high frequency transmissions reflected by way of the
ionosphere. If a delay time of up to several hours is permissible,
failure to communicate arises only infrequently.

For short range (100-1000 miles) communication within the
auroral zone, one can get along with one or, at the most, two frequencies. Day-to-night variation of F region ionization over the
polar cap is small during most of the year although, admittedly,
the fine structure of the ionospheric layers is more vaiable than
at temperate latitudes. In addition, communication can be
achieved during many disturbances by reflection from the bottom
side of the D region absorbing layer, with proper choice of low or
medium frequencies. Another possible system using VHF has just
been proposed by Landmark.

If the required antenna installation is warranted, one can
largely avoid the ionosphere disturbances problem with VLF communications by ground wave or by taking advantage of the earth-ionosphere cavity for longer distance circuits.

Problems in predicting and choosing frequencies for polar region communications arise from the following:

(a) The F region critical frequencies (foF2) are determined from
     ionograms using temperate latitude rules of scaling.

(b) The tabulations of foF2 from polar ionosondes are not easily
     related to equivalent MUFs for communications, due principal-
     ly to the frequent occurrence of spread F. The proper method
     for determining foF2 from a spread F ionogram is not known
     nor is it possible to determine if an ionogram obtained at an
     isolated Arctic station is typical of the ionosphere environ-
     ment.

(c) The F1 region is known to be dominant in Arctic regions all day during the summer and yet most existing prediction systems ignore this fact.

(d)   The spatial and temporal characteristics of PCA and auroral absorption, now quite well known, have not yet been related to radio communications prediction methods, or to actual radio communication circuit fluctuations.

(e)   Enhancement of MF radio signals during disturbances, while observed for years, has not been related clearly to the state of the lower ionosphere.

3.   RADIO NOISE

Natural radio noise is weak in polar regions making possible the use of very low power transmissions. This does mean, however, that interference from distant transmitters is more important. Most significant and perhaps most neglected is the local man-made noise - from motor generators, electrical appliances, motorized vehicles, etc. The problem can be solved but is seldom considered when outfitting an Arctic outpost.

Low noise background, low signal level requirements and the lack of interfering transmissions mean that simple transmitting and receiving equipment may be used, and antennas may be simple. A dipole antenna strung on short poles, or between vehicles or merely stretched out on the snow surface, suffices. The latter is particularly valid since dry snow and ice are non-conducting and the effective ground plane is some distance below the surface.

4.   ANOMALOUS PROPAGATION

Rather than avoiding them, one should consider, for communications purposes, making use of the irregularities which occur in the ionosphere.

For example:

(a)   Meteor trails provide a reliable and predictable means of reflecting HF and VHF signals.

(b)   The characteristics of polar Es are now fairly well known statistically. With the information now available, progress could be made in determining the growth in extent and intensity and movement of this type of ionization - for application to radio communications.

(c) The conscious use of F region anomalies such as the high latitude trough. The latter has been responsible for several
propagation effects previously not understood. Such possibilities involve non-great-circle propagation.

The use of these irregularities does, of course, require propagation along paths which do not coincide with the great circle
path between transmitter and receiver. This is an interesting
problem for the antenna engineer since the direction of arrival of
the wanted signal (except for meteor propagation) is not predictable. The frequent availability of such modes of communications in the Arctic warrants consideration from a systems point of
view.

One must mention the ionospheric scatter system as a brute
force means of ensuring communication across the auroral disturbance zone. However, it is costly and is subject to major ionospheric disturbances.

5. COMMUNICATIONS WITH POLAR STATIONS

Internal Arctic communication often involves limited facilities such as man-portable less-than-optimum equipment. Communication to the Arctic is principally directed to major communities
where permanent installations are possible. The latter allow better exploitation of the ionosphere. This is balanced by the fact
that the most difficult region of the earth's ionosphere (auroral
disturbance zone) must be crossed.

The auroral zone has characteristics and limits statistically
known, but these have been of little assistance in planning or
predicting radio communications. The history of an auroral zone
disturbance is known only qualitatively. Relative to communication over a particular circuit, neither the occurrence can be predicted nor can its effect on the circuit performance be estimated.
Much observation and analysis is still required.

Problems arise from (a) the presence of the F region trough
and the ionospheric turbulence to the north of it, (b) the presence of Es which may cut off the radio wave from the F region
and make long distance communication impossible at times, (c) the
presence of the intensely absorbing D region is responsible for
extended periods of radio blackout. At such times, increasing the
transmitter power does not make a significant improvement but,
during minor or moderate disturbances, a change of frequency may
be helpful.

The effects of the aurorally disturbed ionosphere can be minimized by using propagation paths for transauroral communications

which are near normal to the zone. Use of a relay station well to
the south of the zone will extend communication time into disturbance periods significantly, especially on circuits where the direct line between stations is lateral to the auroral zone.

6. IONOSPHERIC SOUNDERS

During disturbed periods, it is not obvious to the communicator whether he should increase or decrease frequency or if a change will help. No prediction system exists to help him. Each disturbance is different and the best action depends as well upon the
particular circuit path. An effective solution is by use of a
circuit pulse sounding system. This allows one to determine all
frequencies on which communications may be possible at the time.
Of course, it does not assist in forecasting the future state of
the ionosphere.

7. PROPOSED ANALYSIS PROGRAM

The causes of the disturbances and variations in the polar
and auroral ionosphere are now geophysically understood. Very
little is known of their detailed temporal and spatial characteristics. A great deal of synoptic analysis has been done by
Japanese workers using existing ionosonde data. This work has not
yet reached the point where a geophysical or even an empirical
prediction pattern can be proposed. Even when this is done, the
information must be related to radio communications circuit performance.

Enough information does exist to enable one to plan observational programs involving groups of ionospheric observing stations
(not necessarily ionosondes of the existing type). These should
be set up in configurations to determine the synoptic features of
each ionospheric layer as it may affect radio communications
(rather than from the geophysical point of view) and to follow
the changes in the intensity of the ionospheric layers. The obvious analogy is the meteorological forecasting system.

8. HIGH SPEED COMMUNICATIONS

Up to this point, the ionosphere has been considered from the
point of view of voice or morse code or standard RTT types of circuits. The future lies, of course, in high speed digital types of
messages with minimum redundancy. The ever present fine structure
turbulence and rapid ionization fluctuations in the polar ionosphere will limit a polar communication system perhaps more than
the sporadic auroral ionosphere disturbances.

9. SUMMARY OF PHYSICAL PROBLEMS OF THE IONOSPHERE

F-F2 region   (a) Small day-to-night variation is an advantage.

               (b) Present foF2 tabulations do not lead to correct MUFs.

               (c) The effect of the F region trough must be considered.

F1 region   Is dominant for Arctic ionospheric communications in the summer daytime.

E region   (a) The frequent presence of Es should be exploited for use in radio communications.

               (b) Its effect as a possible cutoff for F region propagation should be investigated.

D region   LF and VLF observations of the lower ionosphere need to be related to long distance communication performance. The possibility of using VHF reflected off the lower edge of the D region should be investigated.

Auroral and   A synoptic and morphological observing program
Polar   should be conducted to determine in detail the
Disturbances   state of the polar and auroral ionosphere during the progress of individual disturbances. Empirical prediction patterns should be determined to be applied to specific trans-auroral radio communications.

DISCUSSION

(Discussion of papers by Piggott, Thrane and Landmark)

The main controversy in this session revolved around the definition of "winter anomaly". Piggott maintained that the winter
anomaly is independent of magnetic disturbance and, therefore, is
distinct from "auroral absorption". This view was challenged by
Belrose and Hartz who maintained that there is a magnetic association and that the winter anomaly seen in North America is almost
certainly an effect of particle precipitation. Piggott's replay
was that auroral absorption tends to peak in the morning and contains fine time structure whereas the winter anomaly essentially
followed the quiet day pattern the latter being enhanced by a factor of about 4. Piggott also pointed out that the winter anomaly
is non-existent when the solar zenith angle is greater than about
90°.

In reply to a question by Lied concerning the cause of the
winter anomaly, Thrane maintained that the effect could not be
explained on the basis of an increase in collision frequency. According to Thrane any increase in ν is fractional and not a factor
of 2 to 5 required to explain the anomaly. There was some incon-
clusive discussion as to the role of electron temperature versus
neutral temperature. It was pointed out that the question of
whether the winter anomaly was due to an increase in electron density or collision frequency (Holt) had been solved by Bowhill's
rocket measurements. These measurements showed that sufficient
extra electrons were produced between about 80 and 90 km to explain the observed absorption. Belrose argued in favour of synoptic rather than odd rocket measurements. He presented evidence to
show that the day to day variability (factor of 2) during the winter in collision frequency completely overshadowed any seasonal
variation, and showed that this was consistent with meteorological
data. It was pointed out that a great deal of information on this
subject is contained in the proceedings of the Ottawa Meeting on
the D region which can be obtained by writing to J. S. Belrose,
or to any one of the other DRTE people.

(Discussion on Meek's Paper)

The discussion of physical problems in Arctic Communications
was greatly influenced by the possible economic returns for any
effort expended. The general view seemed to be that regularities
in behaviour existed which are not as yet exploited and although a
large amount of data have been collected and studied since the end
of the war little has been done to simplify the results and present them in a form suitable for engineers and operators. The

needs of simple communications systems in which delay was not of
critical importance were very different to those of complex systems aiming at providing near 100% service with small time delay.
Similarly the great advances in technique since the war have scar-
cely ever been applied to scientific studies of high latitude phenomena. These offer the possibility of relatively easy and cheap
solution of at least some long standing problems.

In general questions which appear simple to the engineer and
users involve scientific problems of great complexity and complete
answers are most improbable. Doubt was expressed about whether
partial solutions would be used sufficiently to justify the work
involved in producing them.

The possibilitiy of providing useful predictions, (planning
time scale), warning (short time; hours to day or so), and immediate (hour to hour) information for improving high latitude propagation was discussed at some length. The work involved is very
considerable. While in principle there is already enough data to
sustain extensive post mortem **researches**, there is difficulty in
interesting the people with enough background to study these problems effectively. Some engineers in particular, felt that progress would only be made by real time experiments, preferably
based on a meteorological type of approach and close liaison between the forecast groups and the users. The economics of such an
effort appeared very different to those involved in high reliability, high capacity circuits and those operating on minimum cost.
Actually there has been very little improvement in the use of HF
in the Arctic for the last 20 years and it was clear that some effort to improve this was greatly overdue. In particular the use
of abnormal modes of propagation to fill gaps in normal modes and
more knowledge of the most probable way of reopening circuits as
quickly as possible could be valuable.

There was a considerable discussion on the relations between
meteorological and ionospheric forecasting. The main differences
were due to difference in time scale in the two fields. Rapid
means of communication, similar to the WMO network, existed and
were underused so that an attempt to make ionospheric forecasts
was technically feasible.

Aspects of LF and VLF Communication

SOME ASPECTS OF THE VLF OMEGA NAVIGATION SYSTEM AS APPROPRIATE TO

THE ARCTIC ENVIRONMENT

B. Burgess

Royal Aircraft Establishment, Farnborough, Hants, U.K.

Abstract: Long range navigation using the VLF OMEGA radio navigation system gives promise of reliable and accurate (of order 2 km
errors) navigation on a world wide basis. The areas of the world
most likely to give the most variable position lines are those of
high latitudes where ionospheric disturbances are most frequent.
While ionospheric disturbances, such as those associated with solar proton events, can seriously affect navigation where highly
accurate fixes are required, many users will probably be able to
use Omega with a slightly reduced accuracy and reliability even
during periods of very severe ionospheric disturbances. A limited
amount of experimental evidence is presented to substantiate these
indications.

1. INTRODUCTION

When dealing with communication or navigational radio systems
in the VLF to HF bands, which operate over distances of 1000 km or
more, the influence of the ionosphere on the propagation of the
radio waves, which carry the information necessary for communication or navigation, becomes very marked. While it is true that
this type of long range circuit can operate only because the ionosphere exists, the variability of the ionosphere is the major
cause of the unreliability of such systems.

In the case of long range radio navigational aids, which this
paper will deal with, it is usually the relative phase of transmitted signals that is of prime importance and it is the prevail-
ing ionospheric conditions, which govern the variability of the
phase delay of the transmitted signals, that ultimately define the

accuracy and reliability of such systems. Radio aids to navigation
such as Decca (70-130 kHz) and Loran C (100 kHz) give remarkable
accuracy while the received signals are within ground wave propagation range. When ionospheric propagation starts to make a contribution to the received signal the accuracy of the system begins
to deteriorate.

Very low frequency radio waves (VLF, 3-30 kHz) have the property that they can propagate on a world wide basis with small attenuation rates of order 2 to 3 db/Mm. Further, they have the remarkable property that the phase delay suffered by signals propagating over a given large distance follows a predictable diurnal
pattern. When used in a navigation technique this property allows
an inherent accuracy of approximately 1 to 2 km. However, it is
still the variability of the ionosphere which governs the ultimate
accuracy of navigational aids using VLF radio waves. In this context, because the ionosphere is known to be more liable to disturbing influences at high latitudes than at lower latitudes, the use
of such aids will probably be accompanied by a degradation in performance in these regions, and this degradation will have to be
ascertained.

At present there seems likely to be only one proposed VLF navigational aid that will become operational; that is the U.S.
Navy's OMEGA system. Already four transmitters, out of eight
planned, are radiating part of the frequency - time multiplex
transmissions on a Research and Development basis. This is allowing an evaluation of the aid to be performed under varying ionospheric conditions. In order to discuss some of the differences
that will be encountered in using Omega in high latitudes it will
be necessary to consider briefly (1) the Omega system itself, (2)
some navigational requirements of various operators and also (3) a
resumé of the basic characteristics of VLF propagation. This sets
the scene for discussing various measurements of signal characteristics, made using the Omega transmissions, that would be appropriate to high latitudes. Their interpretation in terms of navigational performance is brought out in the discussion. Indications
are that although lines of position, obtained when using Omega
under high latitude conditions have a greater variability compared with a low latitude environment, a performance acceptable to
many users can be obtained even during periods of severe ionospheric disturbances.

2.   OMEGA NAVIGATIONAL AID

The OMEGA radio navigational aid (1. 2) is based on the principle of phase comparison of signals transmitted from a complex of
eight VLF transmitters situated over the earth's surface. The
phase comparison of signals from two such transmitters allows a

family of hyperbola of constant phase difference to be set up
which in turn represent lines of position to a navigator.

The frequency band of operation is 10-14 kHz and with baseline distance between transmitters of order 5000 to 6000 n.
miles, the 8 station geometry provides line of position (l.o.p.)
divergence not greater than 15%, and angle crossings of the l.o.ps
of not less than 60 degrees. The basic Omega frequency of 10.2
kHz allows l.o.ps to be spaced by approximately 8 n.m. along the
baseline.

The basic limitation of any phase comparison system is its
inability to differentiate between multiples of 360 degrees.
This is overcome in Omega by radiating additional frequencies to
10.2 kHz from all stations on a time-multiplex basis. The fre-
quency-time format of transmissions from an eight station complex
is given in Figure 1. In turn, each station transmits three basic
frequencies viz. 10.2, 13.6, 11 1/3 kHz in sequence over a period
of approximately 3 seconds, the transmissions from each station
being delayed by approximately one second on each other as indicated in Figure 1.

Every 10 seconds the sequence is repeated, the starting time
of these 10 second intervals being referred to 00.00.00 hrs. UTC.
The 13.6 kHz transmission is phase modulated at a frequency of
226 2/3 Hz and thus it is possible to synthesize the following
frequencies from any Omega transmission, 226 2/3 Hz, 1133 1/3 Hz,
and 3.4 kHz which are respectively 45, 9 and 3 times the basic

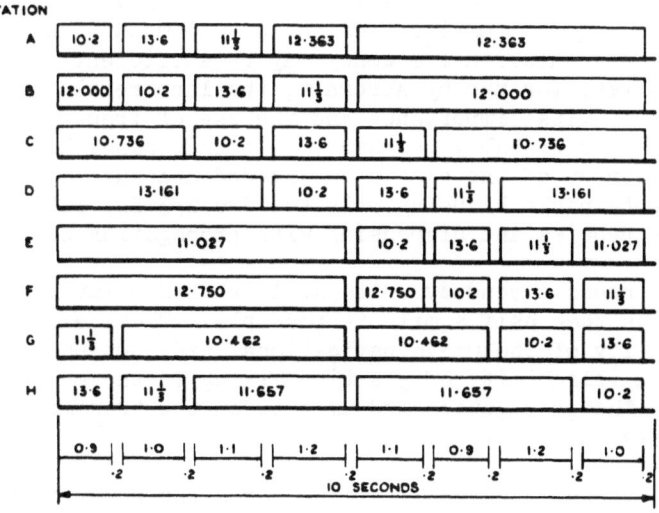

Figure 1. Omega transmission format.

frequency of 10.2 kHz. Thus new families of hyperbolic l.o.ps can
be generated with spacings on the baseline of 360, 72 and 24 n.m.
in addition to the existing 8 n.m. l.o.ps (10.2 kHz), and allows
ambiguities to be resolved in the finer mesh of l.o.ps.

The intended radiated power of each station is some 10 kW,
and the frequency and phase of all transmitters are synchronised
and locked to each other by a complex of monitoring procedures.

Omega is being developed by the U.S. Navy to meet an operational requirement of navigation anywhere in the world. These
requirements also meet the needs of a number of non-military bo-
dies and it is upon this concept of use by both military and non-
military organisations that Omega is being considered for implementation.

3. NAVIGATIONAL REQUIREMENTS

Omega is a navigational aid that can be used by a wide variety of navigators. It has been proved capable of fixing position
with an r.m.s. accuracy of one n. mile or better in locations
where the geometry of the system is good. To resolve positions
unambiguously to this accuracy requires sophisticated equipment,
full use of the Omega format and diurnal propagation corrections.
However, if the requirement is less stringent than that quoted
above, then much simpler and hence less costly equipment can be
used to obtain position fixes.

For marine use, where the vessel is slow moving, the ability
to resolve the 8 n.m. ambiguities is not paramount and reception
of only one transmission frequency may be sufficient for most purposes. However, for use by aircraft, the lane resolution ability
of Omega is of vital importance and the use of frequencies as low
as 226 2/3 Hz is probably necessary. In this case, however, the
ultimate accuracy of the aid is not needed and sufficient accuracy may be obtainable by using only the 226 2/3 Hz, 1133 1/3 Hz and
3.4 kHz frequencies and dispensing with propagation corrections.

If one needs a rendezvous capability then relative accuracy
of order less than half a mile is possible. These examples show
the flexibility of the Omega navigational aid, but underlining the
accuracy and reliability of position fixing is the problem of how
variable is the propagation medium between the transmitting and
receiving stations.

This paper will direct itself towards the problem of aircraft
navigation and hence will be deeply involved in the problem of
ambiguity resolution. It is assumed that an aircraft can identify
the l.o.p.s produced by the 226 2/3 Hz modulation transmissions.

The next problem is to resolve the 1133 1/3 Hz l.o.p. ambiguities
by using the 226 2/3 Hz l.o.p. This can only be achieved if the
variability of the 226 2/3 l.o.p. due to propagation conditions is
smaller than a lane width at 1133 1/3 Hz (i.e. 72 n. miles). The
exact measure one uses to determine this ability to resolve lanes
will depend on the ultimate accuracy of the final position fix.
Similar considerations lead to the resolution of the 3.4 kHz l.o.
p.s by the 1133 1/3 Hz l.o.p. and finally the variability of the
3.4 kHz l.o.p. determines the accuracy of the final position fix.

In use of the aid for aircraft navigation the correction for
diurnal propagation effects is likely to be embarrassing and it is
shown in the next section that the use of the difference frequencies 226 2/3, 1133 1/3 Hz and 3.4 kHz allows the diurnal correction to be ignored for a small penalty in accuracy and reliability. Similarly the effects of sudden ionospheric disturbances, polar cap disturbances etc, will tend to compensate themselves.

In order to appreciate the contribution of various sources to
the variability of the measured phase over a given propagation
path, it is worth outlining the modes of propagation of VLF radio
waves and observing why the 10-14 kHz end of the VLF band is optimum for world wide navigational purposes.

Before discussing this aspect it is important to note that
the l.o.p.s can only be given accurately if the phase velocity of
the carrier signals, or the effective group velocity of the centre
frequency from which the difference frequencies are derived, are
accurately known. Limited experimental determinations of these
quantities have been undertaken (3, 4) but their variations with
sunspot cycle, seasons, conductivity of land masses is not yet
well understood.

4. PROPAGATION CONDITIONS

At the wave frequencies of interest, the wavelengths measure
some 20 to 30 km., and it is convenient thus to treat the earth
and the ionosphere as the two boundaries of a spherical waveguide
and to consider propagation to great distances in terms of waveguide modes (5, 6).

A vertical electric antenna situated on the surface of the
earth will excite transverse magnetic waveguide modes (T.M. modes).
For each mode there are three parameters which govern the characteristics of the mode, i.e. the attenuation rate, α db/1,000 km,
the phase velocity of the mode, usually quoted relative to the
velocity of light v/c, and the excitation factor Λ which is approximately the ratio of the power launched into the earth ionosphere waveguide to that launched into a flat waveguide with per-

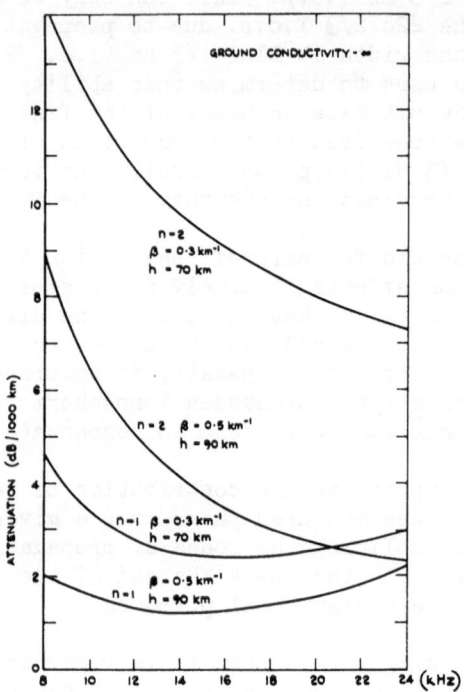

Figure 2. Plot of attenuation rates of TM_1 and TM_2 modes for
day and night conditions (Wait 1964).

fectly conducting boundaries. To indicate how these parameters
vary with frequency, figures 2, 3 and 4 show theoretical values
derived by Wait and Spies (7) of α, v/c and Λ for modes 1 and 2,
for a representative model of the ionosphere boundary for both
daytime and nighttime conditions with the lower boundary of the
waveguide seawater (effectively a perfect electrical conductor).
In these curves the ionospheric boundary is represented by a conductivity parameter ω_r which varies in an expronental manner with
height

$$\omega_r = (\omega_r)_o \exp \beta(Z-Z_o)$$

Z being a measure of distance above a reference level Z_o
where the conductivity parameter has a value $(\omega_r)_o$ and β is a constant. The value of β chosen by Wait to give a reasonable fit to
the best experimental profiles of the conductivity of the lower
ionosphere is 0.3/km for day and 0.5/km for night.

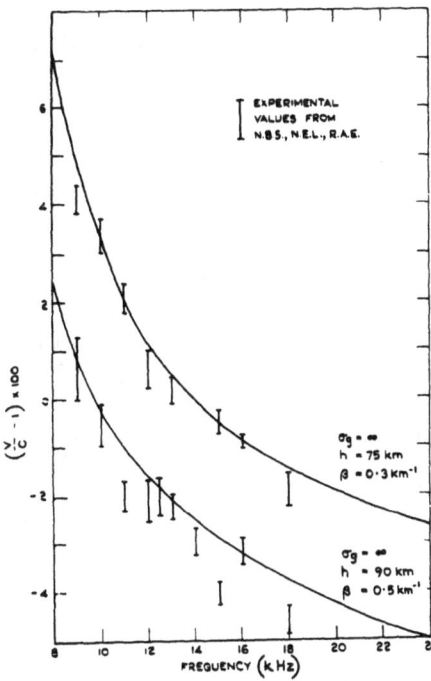

Figure 3. Plot of phase velocity against frequency (Wait 1964).

From these curves a number of general features are immediately discernible which experimental results can test. Mode 2 suffers a larger attenuation than mode 1. The difference is more
marked for the lower frequencies. Also the difference in the
phase velocity between night and day conditions of mode 1 is larger at the lower than at the higher frequencies. This indicates
that at day and night the diurnal phase delay pattern should (1)
increase in magnitude linearly with distance and (2) have a larger
magnitude the lower the frequency. Both these indications are
borne out by the experimental results. For distances closer to a
transmitter the effects of the second mode should become more marked. This is especially so at the higher frequencies since at
these frequencies, as may be seen from figure 4, the second mode
is more easily excited in the waveguide; at 20 kHz for example the
excitation factor for mode 2 is some 10 - 15 db greater than for
mode 1 while at 10 kHz they are more or less equally excited.

A striking feature of the diurnal phase delay pattern is the
phase steps which occur when sunrise passes across the propagation

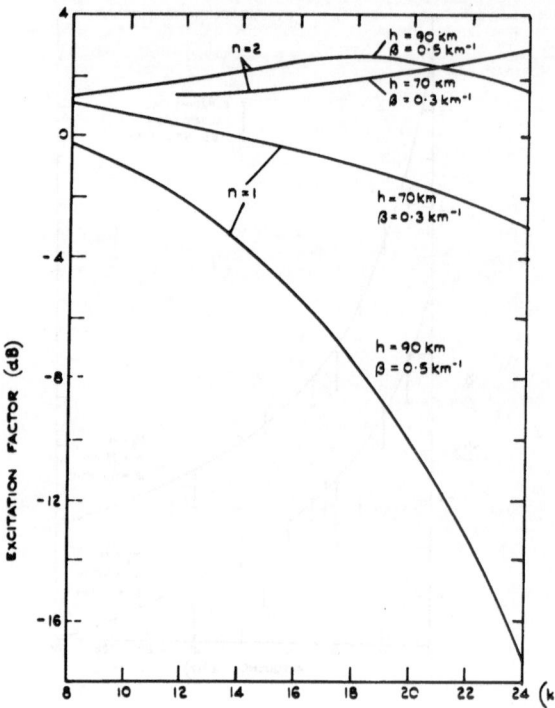

Figure 4. Plot of excitation factor of TM_1 and TM_2 modes for
          day and night conditions (Wait 1964)

path. This is more marked the higher the frequency. At some of
the higher frequencies, above 10 kHz, it often occurs that during
this period the signal almost completely disappears and the phase
of the resultant signal passes to a value which may be an integral
number of cycles of phase different from the expected values, thus
causing the recorded phase of the signal to be this number of cycles different from the expected value for the rest of the period
of observation. That this phenomena is explicable in terms of a
well excited second mode under night conditions was first shown by
Crombie (8) and amply verified by Walker (9). At the higher frequencies under night conditions, the two modes (TM_1 and TM_2) propagate to great distances. When sunrise occurs on the path, there
is effectively a discontinuity in the dimensions of the waveguide.
At this discontinuity mode 2 becomes converted to mode 1, the amplitude and phase of this converted mode 1 being dependant on the
distance of the sunrise discontinuity from the transmitter. The
resultant signal in the daytime part of the waveguide will thus
suffer regular amplitude fading and corresponding phase variations

as the sunrise waveguide discontinuity traverses the path.

From theoretical considerations it is seen that the amount of converted mode 1 should decrease with decrease in frequency and this sunrise phenomena should be of small proportion at the lower frequencies near 10 kHz. This is borne out by experimental data. Though this effect at the 10 kHz frequency is small it might well prove to be one of the limiting factors regarding accuracy of a navigation aid. Attention to this point will be drawn later.

From the above experimental and theoretical considerations of VLF propagation the lower frequency band of 10-14 kHz has definite advantages over the higher frequencies for navigational purposes. The main advantage is that due to the frequency dependance of the attenuation and excitation factors of the modes the second mode perturbing effects are much smaller in this lower frequency band.

At frequencies below 10 kHz the attenuation rates increase rapidly to high values. This coupled with the economics of designing efficient ground transmitting aerials at this end of the band effectively puts a lower limit of about 8 kHz to the optimum band for navigation use.

These considerations together with the maxim of using as low a frequency as possible for lane resolution has led to the Omega World Wide Navigational System choosing its main frequency of transmission as 10.2 kHz.

Great advantage would accrue if some means could be found which would allow navigation to be performed without the necessity of compensating for diurnal variations in the waveguide wavelength. This would apply particularly in aircraft navigation where an accuracy of navigation approaching 5 n.m. would be of great value.

Consider figure 3 where the variation of the phase velocity of mode 1 for day and night conditions is plotted against frequency. From this curve it is possible to derive the diurnal phase delay (measured in degrees) suffered per 1000 km of path assuming only a single mode of propagation. This is shown in figure 5. It is evident that in the 10-14 kHz band this diurnal phase delay parameter as measured in degrees is, to a first order, independent of frequency, while for frequencies of order 20 kHz is rapidly increasing with frequency. Thus if a transmitter radiates two frequencies in this band and the phase delay of the difference frequency monitored, this phase delay should differ little between daylight and night conditions. Figure 6 shows the results of monitoring at Farnborough two such frequencies of 10.2 and 13.6 kHz radiated from Omega Trinidad. The diurnal variation of the phase delay of the difference frequency shows an rms deviation of order 6 μsec (7 degrees of phase) which used in a hyperbolic type of

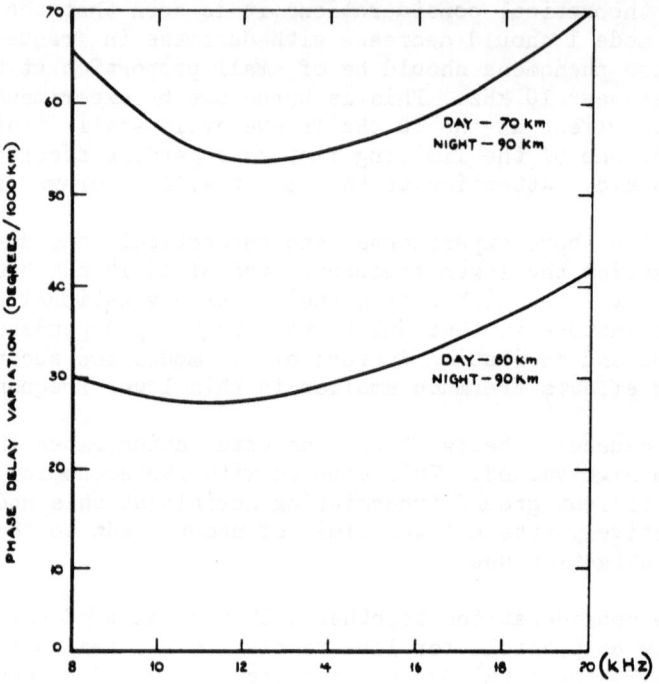

Figure 5.   Diurnal phase delay variation against frequency for
            two day-night models (TM$_1$ mode only).

navigation aid would correspond to fixing accuracy of 1 to 2 n.m.
It will be noted from a careful study of the figure that the largest variations occur during the period when sunrise or sunset occurs along the path. This would indicate that second mode effects
are responsible for these large variations and a reduction in the
frequency difference could lead to smaller variability. Note in
this case that the lane resolution is dependent on the magnitude
of the beat frequency.

5. HIGH LATITUDE INFLUENCES

     When considering the propagation of VLF radio waves over
great distances it is the properties of the waveguide boundaries
which determine the properties of the propagation parameters and
how they differ from one section of the waveguide to another. In
high latitudes the properties of the waveguide wall differ from
non high latitudes in the following way. On the earth's surface
there are large areas where the conductivity of earth is consider-

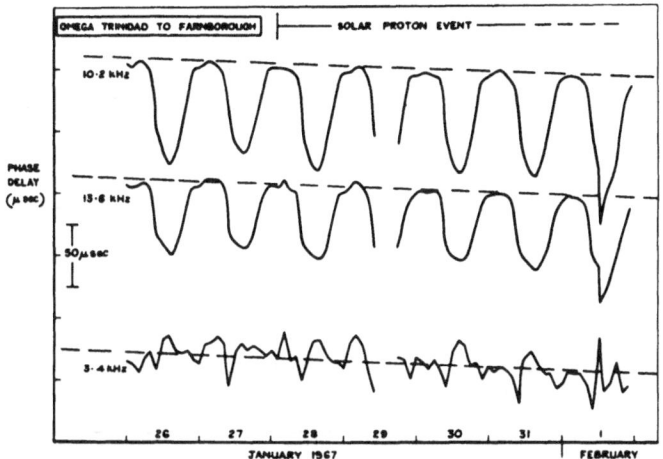

Figure 6. Plot of phase delay of 10.2 - 13.6 and derived 3.4 kHz
           signals radiated from Omega Trinidad, received at
           Farnborough, U.K.

ably lower than in other areas of the world. Over ice cap regions
and areas of permafrost the conductivity can be 2 or 3 orders of
magnitude lower than the land masses and 5 orders of magnitude
lower than the sea surfaces. This introduces a discontinuity into
the waveguide with the possibility of waveguide mode conversion.
Also attenuation of the radio waves propagating over these regions
can be substantially increased.

    The ionospheric regions at high latitudes are known to be
much more variable than at other latitudes. In particular the
ionospheric D and lower boundary of the E region, which govern the
propagation of VLF radio waves, are very susceptible to disturbances by solar particles, and VLF radio measurements form a sensitive means of studying such phenomena. The variability of the
waveguide surface with time and location will reflect in an increased phase variability of the received signal and if a transmitter is sited in these regions the effect of high latitudes will be
noticed on signals received well outside the disturbed regions.

    In order to appreciate the amount of extra phase variability
inserted by ionospheric variations in high latitudes one must
establish some norm against which comparisons can be made, both
for the carrier frequencies and difference frequencies relevant to
the Omega system. Information on these points are not easy to
come by, especially that corresponding to the difference frequencies.

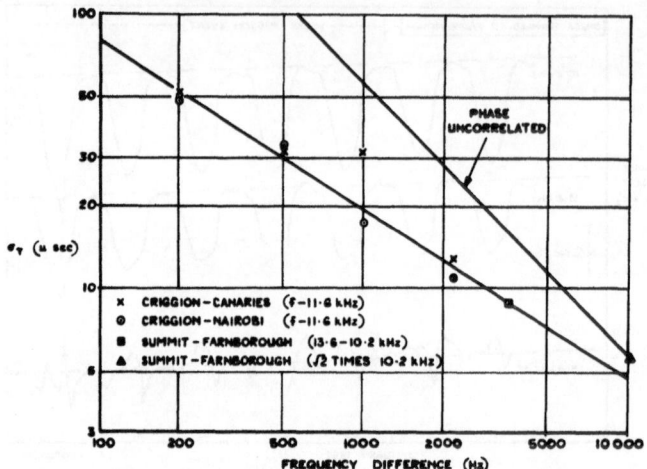

Figure 7. Plot of r.m.s. variation of phase delay, σ_T, as a function of frequency above 11.6 kHz.

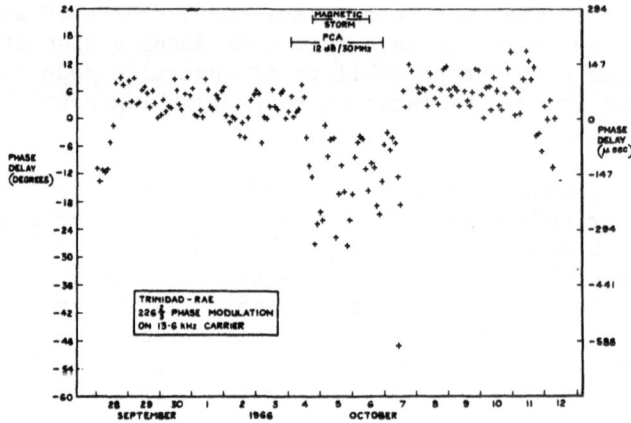

Figure 8. Plot of phase delay variability about daily mean of 226 2/3 phase modulation on 13.6 kHz carrier. (Trinidad - Farnborough U.K.).

Figure 7 shows the results of measurements undertaken in the
latter quarter of 1965 on some special time duplex frequency transmissions from station GBZ (Criggion, Wales) received at Nairobi
and the Canary Island. For frequency differences of the order 200
and 1000 hz the r.m.s. variations in phase delay were some 50 and
20 μsec respectively. Measurements on a frequency difference of
3.4 kHz on transmission of 10.2/13.6 kHz from Panama had indicated
a r.m.s. value of under 10 μsec. These values corresponded to an
overall variability of the carrier frequency of some 4-5 μsec after removal of the mean diurnal phase delay variation. (Note the
difference frequency values included diurnal variations).

The above measurements on the Criggion transmissions referred
to sample measurements covering 3-4 days at a time. Recent experimental values from the 226 2/3 Hz phase modulation on 13.6 kHz
from Omega - Trinidad have shown a phase variability of the same
order as the Criggion measurements when taken over a few days but
a much larger variability if results over prolonged periods
(15 days say) are used (see figure 8). What then are the effects
using Omega in northern latitudes?

First let us consider the problem of propagation across the
very low conductivity land masses of the ice cap and permafrost.
In the summer of 1965 the transmissions from NPG (18.6 kHz) and
GBR (16 kHz) were monitored in an R.A.F. Comet aircraft during a
flight over the Greenland icecap. Figure 9 shows the plot of sig-

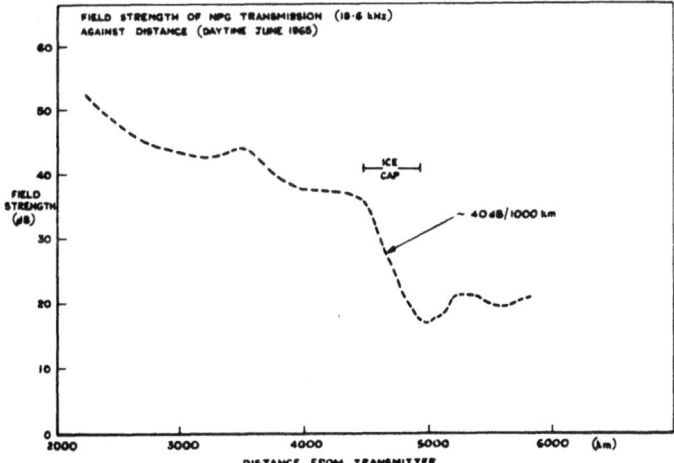

Figure 9. Plot of signal strength variation with distance of
           NPG transmission (18.6 kHz), showing influence of
           Greenland ice cap.

nal level for the 18.6 kHz transmission as a function of distance
from the transmitter under daylight conditions. The attenuation
rate of the 18.6 kHz signal is increased abruptly from its normal
day value of some 2 db/Mm to some 40 db/Mm for propagation over
the ice cap. On 16 kHz the attenuation rate increased to a value
of some 30 db/Mm. Taking the theoretical work of Wait (6) as a
means of estimating the corresponding attenuation rate at 19 kHz,
values in the region 30 to 40 db/Mm are found. Consequently any
reception of VLF signals which have to propagate over ice caps are
liable to suffer appreciable attenuation compared with propagation
path-lengths of similar length which are free from low conductivity land masses. For example areas to the east of Greenland and
also in western Europe are liable to receive very weak signals
from Omega Haiku. Similar conditions prevail for the use of Omega
Norway in parts of the United States and Canada.

Turning our attention now to the effect of the lower ionosphere at high latitudes on the phase variability of the propagated signals, one has to consider two aspects. Firstly there is
the generally greater variability of radio propagation in northern
latitudes and secondly the greater disturbing effect on propagation both in terms of magnitude and time scale of solar particle
effects at high latitudes. To illustrate these points some results of the monitoring of signals from Omega Norway at Ottawa,
Farnborough and Barbados will be discussed.

Figure 10 shows the variability of the phase delay of 3.4 kHz
for the path Omega Norway to Ottawa. It is difficult to discern a

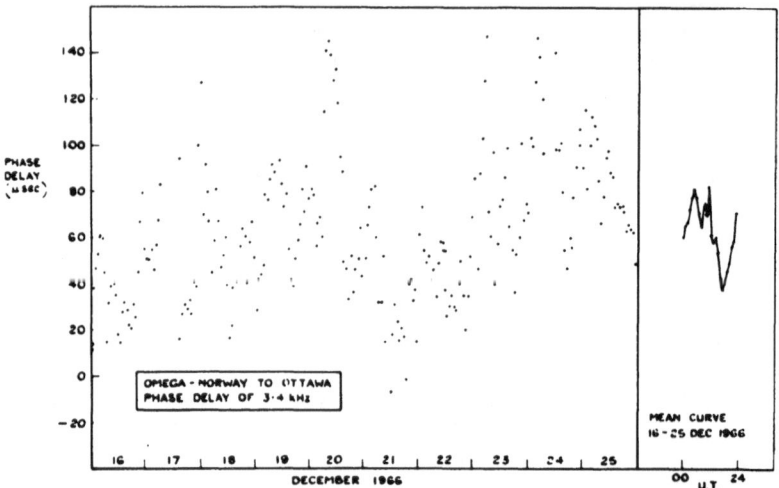

Figure 10. Plot of phase delay variation of derived 3.4 kHz sig-
            nal (13.6 10.2 kHz); Omega Norway to Ottawa.

consistent diurnal phase delay pattern on the carrier frequencies
of 10.2 and 13.6 kHz, from which the 3.4 kHz phase delay was derived, by just glancing at the records. A mean curve for a period
of 10 days, however, reveals a diurnal pattern which is consistent
with the proportion of the propagation path in daylight. The
r.m.s. variation of the phase delay of the 3.4 kHz difference frequency is some 30 μsec or five to six times the magnitude for a
similar length path from Omega Trinidad to Ottawa. While being
considerably worse than non high latitude paths this phase delay
variability still allows a l.o.p. to be determined to an accuracy
of 5 n.m. with a probability of 0.68. At present no information
exists, for this type of path, for the phase delay variability for
226 2/3 and 1133 1/3 Hz difference frequencies.

If one considers the phase delay suffered by a carrier frequency over a long transmission path, major phase delay perturbations occur due to solar X-rays at the time of solar flares and
also to solar proton events, when solar protons enter the polar
regions of the earth's atmosphere. At high latitudes the sun's
zenith angle remains appreciable large and hence sudden phase anomalies do not reach the magnitude observed at low latitudes. However, solar proton events cause extremely large phase delay perturbations on propagation paths that pass through polar regions
which are not observed on low latitude paths.

Consider the solar proton event which started in the early
hours of January 28th, 1967 and continued for at least 4 or 5
days. Omega signals from Norway monitored at Barbados (figure 11)

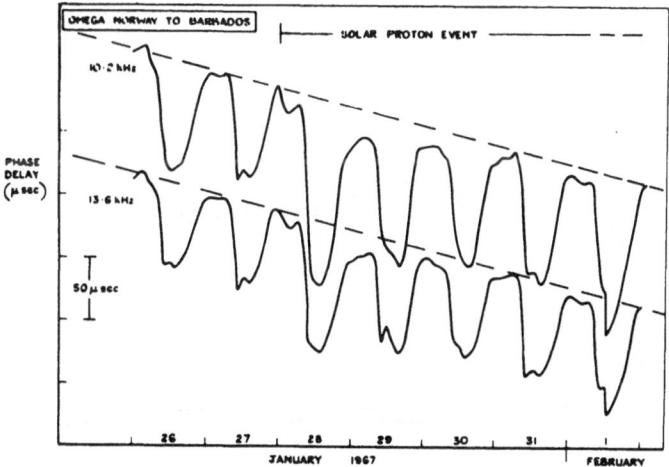

Figure 11. Plot of phase delay of 10.2 and 13.6 kHz transmission
            from Omega Norway recorded at Barbados.

showed that the carrier frequencies suffered decreases of phase
delays of up to 60 µsec during day conditions compared to normal
conditions. Use of these frequencies from Norway to determine
l.o.p.s in the West Indies area would have resulted in large ex-
cursions in the l.o.p.s over half a lane width after diurnal corrections had been applied, hence wrong resolution of lanes. The
3.4 kHz phase delay variation for the Omega Norway/Barbados path,
as shown in figure 12 does not readily indicate any marked increase in variability over the period 28th January to 1st February
due to the solar proton event. The r.m.s. variation at 3.4 kHz
for this period being some 7 µsec compared with 5.5 µsec for the
Omega Trinidad/Farnborough path for the same period. On this latter path the carrier frequency phase delay measurements gave no
indication of the solar proton event (figure 6). While no measurements have been given of reception of Omega signals in high
latitudes from transmitters situated outside these latitudes,
qualitatively one would expect similar results as have been obtained on reception of Omega Norway signals outside high latitudes.

To date measurements indicate that navigation by means of VLF
using the Omega system in high latitudes while giving l.o.p.s
which have a greater variability than obtained outside high latitudes nevertheless give a performance which is acceptable to many
users. In particular when navigating by means of the difference
frequencies the large phase delay perturbations introduced by
solar disturbances are appreciable reduced when compared to perturbations on the carrier frequencies. However, the amount of in-

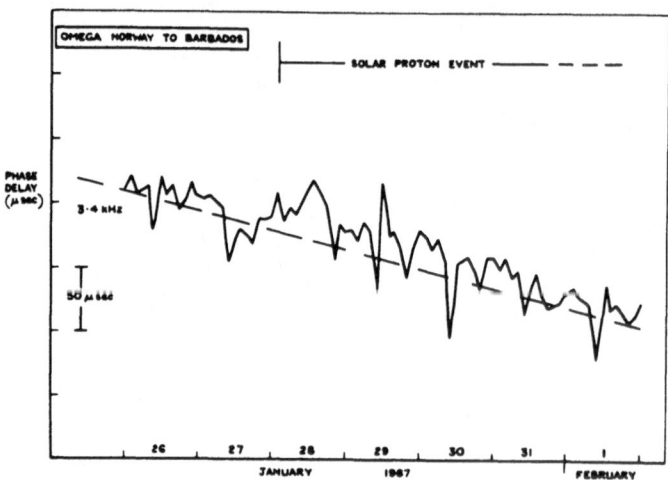

Figure 12. Plot of phase delay of 3.4 kHz derived signal
            (13.6 10.2 kHz) of Omega Norway recieved at
            Barbados.

formation obtained to date on which these tentative conclusions are based are rather limited and more thorough evaluation of Omega is needed under high latitude conditions in order to assess its reliability under varying conditions especially those pertaining to solar activity at the maximum of the sunspot cycle.

6. REFERENCES

1. J.A. Pierce, OMEGA, IEEE Trans. Aerospace and Electronic Systems, Vol. AES-1, No. 3, 206-15, Dec. 1965.

2. E.R. Swanson and M.L. Tibbals, The Omega Navigation System, J. Inst. Navigation 12 No. 1.

3. E.R. Swanson, Private communication, 1966.

4. B. Burgess, The influence of propagation conditions on the design of VLF radio wave long range navigation aids. Unpublished Ministry of Aviation Technical Report, October 1965.

5. K. Budden, The waveguide mode theory of wave propagation. Logos Press, Londen, 1961.

6. J.R. Wait, Electromagnetic waves in a stratified media, Pergamon Press, Oxford, 1962.

7. J.R. Wait and K. Spies, Characteristics of the earth-ionosphere waveguide for VLF radio waves. NBS Tech. Note No. 300, 1964.

8. D.D. Crombie, Periodic fading of VLF signals received over long paths during sunrise and sunset. Radio Science 68 D, 1964, pp 27 - 34.

9. D. Walker, Phase steps and amplitude fading of VLF signals at dawn and dusk. Radio Science 69 D, 1965, pp 1435 - 43.

PERIODIC FADING OF VLF SIGNALS RECEIVED OVER A POLAR PATH DURING SUNSET AND SUNRISE

S. Westerlund

Kiruna Geophysical Observatory, Kiruna C, Sweden

Abstract: VLF radio signals from the transmitter NPM/26.1 kHz
have been recorded in Kiruna, Sweden. The signals show very
strong mode interference with frequent cycle slippage. From the
data it is concluded that by night the first and second order
modes have almost exactly the same attenuation and a phase velocity difference of 0.51%, which is in agreement with Wait and
Spies' calculations for a nighttime ionosphere of height 90 km and
a lower waveguide boundary of infinite conductivity (sea water).
When the same model is used for the excitation factors the mode
conversion factor is determined and is found to depend strongly on
the angle between the shadow line and the great circle path.

1. INTRODUCTION

Quite often the regular phase transition of VLF radio signals
from night to day and reverse from day to night is stepwise, where
each step corresponds to a phase shift of approximately half a
wavelength or 180°. Each phase jump, which can occur in a time
interval of only a few minutes, is accompanied by a signal amplitude dip. The explanation to the phenomenon was given by Crombie,
1964. During night the propagation of the second order mode is
relatively strong, and neglecting higher order modes, the field-
strength on some distance from the transmitter is determined by
the first and second order modes. During daytime only the first
order mode can propagate to any distances. When a signal from a
transmitter in darkness, arrives at the shadow line the second order mode is converted into a first order mode. The signal on the
sunlit side of the shadow line thus consists of the first order
mode, which is transmitted directly from the nighttime region of

the waveguide, plus a contribution, which results from conversion
of the second order mode into the first at the shadow line. Depending on the relative phase of the two modes when they reach the
shadow line, the signal transmitted into the sunlit region of the
waveguide is either enhanced or decreased due to interference.

In an exact theory all modes must be taken into account.
Also, conversion takes place from each mode into all other modes.
However, Crombie's theory has proved to give good explanations
of experimental results (1, 2, 3, 4). This report presents
further experimental evidence supporting Crombie's mode conversion
theory.

2. EXPERIMENTAL RESULTS

The transmitter NPM-26.1 kHz (21.4°N, 158.2°W) has been recorded at Kiruna Geophysical Observatory, Kiruna, Sweden (67.8°N,
20.4°E) continuously from 1 January to 19 October, 1966 (1-20 May
the transmitter was off). The receiver used was a TEXTRAN phase
tracking receiver model 599G, which was connected to a tuned loop.

The great circle path NPM-Kiruna lies almost exactly in
north-south and passes very near the geographic north pole. The
lower boundary of the wave guide consists to more than 80% of sea
water.

Figure 1 shows phase and amplitude of NPM, recorded in
Kiruna, for some selected days during 1966. As seen the sunset
and sunrise during summer produce the typical phase steps and amplitude minima. In the evening (in Kiruna, which is the same as
evening in GMT) the amplitude dips are very pronounced, often ex-
ceeding 20 db and the whole phase step transition often takes a
time of only some minutes and in almost every case less than half
an hour. In the morning (GMT) the effect is much weaker. The
evening dips occur when the sunrise line approaches NPM along the
path and the morning dips when the sunset line moves away from
NPM. The transmitter is thus in darkness during both morning and
evening step transitions, and this is the case from beginning of
March to beginning of October when the sunset and sunrise lines
are parallel to the path and sunrise occurs over the whole part of
the path on one side of the north pole simultaneously with sunset
on the other part.

For the period March-September the times in each day when
phase steps occurred have been plotted as functions of the time of
the year in Figure 2. The time of a phase step was determined by
the time of maximum rate of phase change, which practically always
agreed with time of amplitude minimum. The morning phase steps
were only strong enough to be distinguished during the summer, and

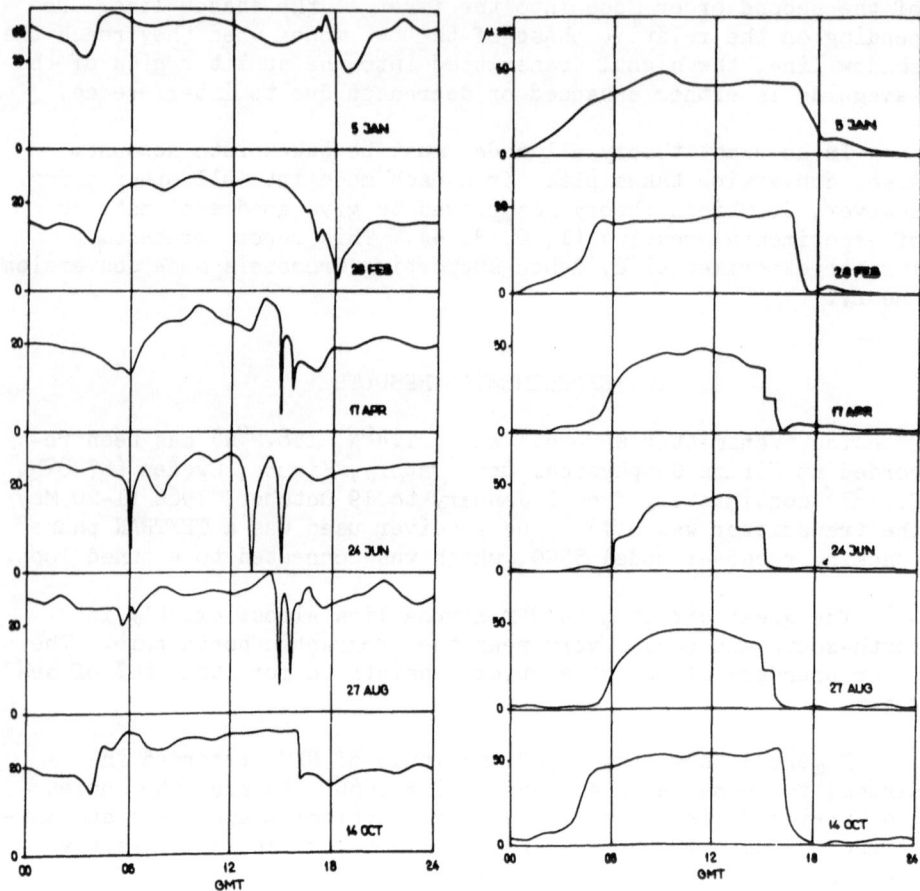

Figure 1.   Amplitude and phase of NPM/26.1 kHz recorded in Kiruna,
            Sweden for some days covering the recording period.   On
            24 June the phase has been corrected for cycle slippages
            in the evening.

even then they were mostly quite weak.   The evening data were much
better and most often it was not difficult to define the times of
phase steps to within a minute.   The curves in Figure 2 show sunset respective sunrise at the following (Table 1) latitudes along
the path NPM–Kiruna and for solar zenith angles X.

        The curves in Figure 2 with parameters according to the table
below are the ones that fit the experimental points best.   Figure
3 shows how the latitude and zenith angle were found for the evening steps.   The horizontal broken lines show the times, which
were taken from the experimental data on 21 June and 10 September.

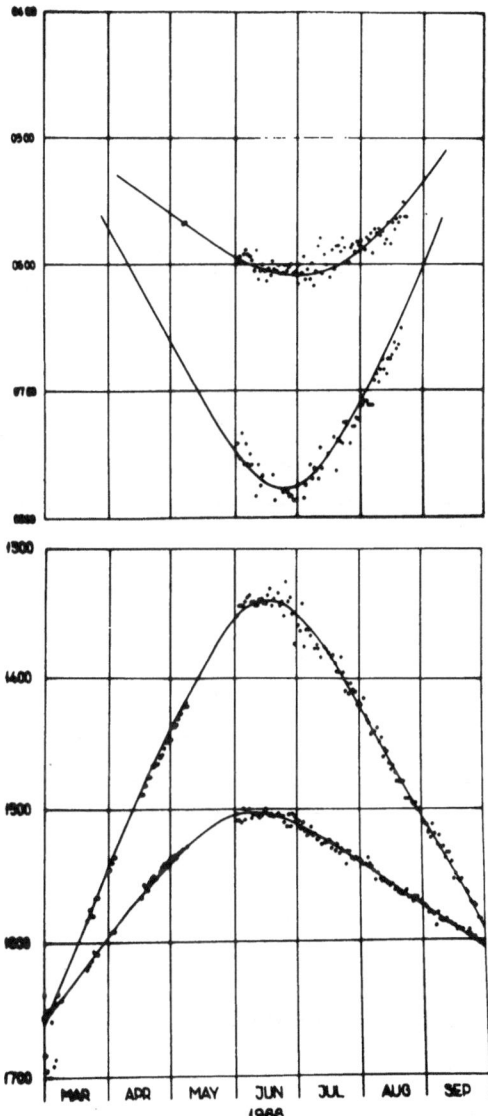

Figure 2.   The times in GMT when phase steps occur as function of
            day of the year. The curves show sunrise at certain
            points along the path and certain solar zenith angles
            according to Table 1 in the text.

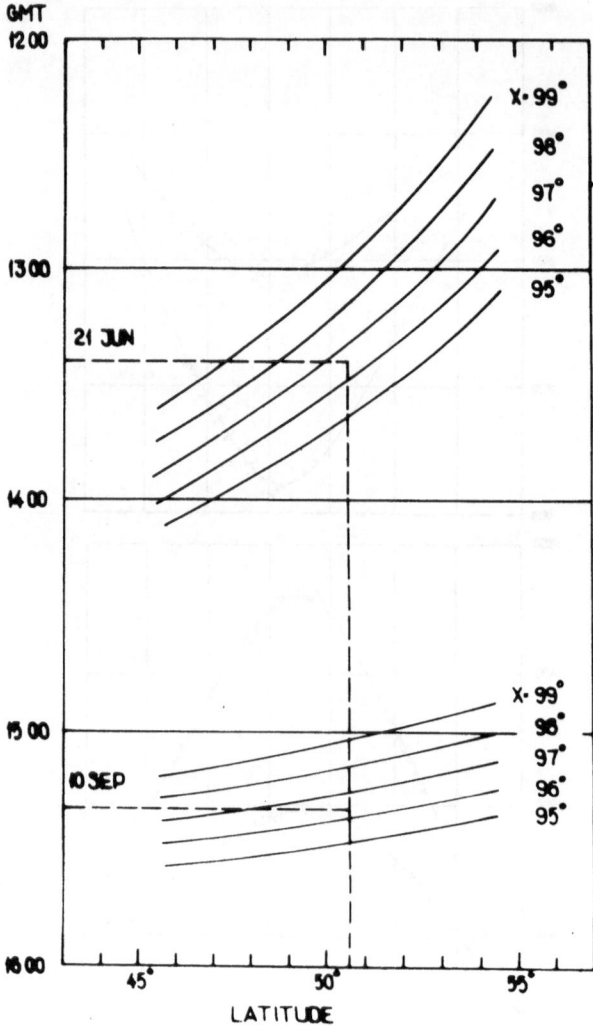

Figure 3. Interpolation to find the sunrise curve, which fits
the times of the first evening step.

	Lat.	X	
First morning step	30.3°	96.0°	Sunset line moves
Second morning step	50.6°	97.5°	away from NPM
First evening step	50.6°	96.5°	Sunrise line moves
Second evening step	30.3°	96.0°	towards NPM

Table 1

Only one pair of values of the solar zenith angle and the latitude angle fits the experimental data. Figure 3 also gives an idea of the accuracy, with which these angles can be determined. For the morning steps the number of points was not sufficient to allow a similar procedure and the latitudes were therefore chosen the same as for the evening steps. The zenith angles, which fitted the data best, were then determined.

The amplitude records were not given much attention. That is because varying duty cycle of the transmitter produced changes in measured amplitude of up to 10 db. Also the Textran receiver is not suitable for amplitude recording during fast phase changes, because unbalance in the tracking servo will produce a too low amplitude indication. It may happen that the receiver switches to "carrier off" if the phase change is to rapid even if the true amplitude is not so small that this should have happened if the phase was steady. In spite of these effects the following features of the amplitude are clear. The morning dips are strongest in the summer when they are of a magnitude of 2.8 db, the first one usually a little stronger than the second one. Approximately two months before and after the summer solstice they do not show up at all. The magnitude of the evening dips (GMT) is about 10 db in the summer. The dips have an increased amplitude both before and after midsummer with maximum in the end of April and in the end of August when the dips mostly are of such magnitude that the signal goes below the threshold of the receiver. Nearer the equinoxes the dips again become smaller until they finally disappear in beginning of March and October respectively. The magnitudes of the two evening dips are always approximately the same. Starting in March evening phase behaviour is normal, in the middle of April a cycle slippage happens sometimes and in June there is almost always a cycle slippage on both phase steps. In the end of August, the normal phase shift becomes frequent again and in September cycle slippage is unusual and disappears completely before the end of the month. Morning phase shift is always normal.

3. COMPARISON WITH THEORY

According to Crombie the signal behaviour at sunset and sunrise is the result of an interference between the first order mode and the converted second order mode. The magnitude of an amplitude dip is a measure of the relative strength of the first and second order modes at the point along the path where the shadow line crosses the path at the time of the dip. If the conversion factor were known the relative strength of the two modes could be calculated.

The magnitude of the amplitude dip, as measured on the record by comparing with the two maxima on both sides of it, can be expressed as

$$ L = \frac{|H_1| - |H_2||\Lambda_{21}|}{|H_1| + |H_2||\Lambda_{21}|} $$

where H_1 and H_2 are the fieldstrengths of the first modes when they arrive at the shadowline from the transmitter, which is in darkness. Λ_{12} is the conversion factor from the second order mode into the first. Using a nighttime model of the waveguide after Wait and Spies (1964) with $h = 90$ km, $\beta = 0.5$ km^{-1}, and infinite ground conductivity, it is obtained for 26.1 kHz:

$$ \text{mod } \Lambda_1 = -21.3 \text{ db} $$
$$ \text{mod } \Lambda_2 = 1.5 \text{ db} $$
$$ \alpha_1 = 2.4 \text{ db/1000 km} $$
$$ \alpha_2 = 2.4 \text{ db/1000 km} $$

The two excitation factors express that the second order mode is 22.8 db stronger than the first at the transmitter. Since the attenuation constants are equal, this is the case at any distance. This is also supported by the experimental results since the evening dips are always about equal. The formula for L then becomes (approx.)

$$ L = \frac{1 - 14 \cdot |\Lambda_{21}|}{1 + 14 \cdot |\Lambda_{21}|} $$

From this the magnitude of the excitation factor can be determined. A cycle slippage corresponds to a negative L which corresponds to a conversion factor larger then than 1/14.

The conversion factor determined from this expression is in summer -15 to -10 db in the end of April and end of August it is about -25 db and decreases to still lower values near the equinoxes. Those values are for the evening phase steps i.e. when sunrise line approaches NPM. For the morning steps the conversion factor is -30 to -40 db in summer.

The phase difference between the two modes should change by one wavelength over the distance D, between the two geographical points, which correspond to the evening steps, see Table above. The difference Δv, in phase velocity then becomes

$$\frac{\Delta v}{c} = \frac{v_2 - v_1}{c} = c/Df$$

D = 2250 km is obtained from the latitude difference and the resulting phase velocity difference is 0.51%. This is the same value as that given by Wait and Spies (1964) for an ionospheric height of 90 km. However, the theoretical difference is not very sensitive to small variations in reflection height.

If the two excitation factors and the conversion factor were purely real the second evening phase step should correspond to a point lying D/2 = 1130 km from the transmitter. The actual distance as obtained from the latitude values of the transmitter and the second evening step (cf. Table 1) is 990 km. The difference could be explained as due to complex excitation factors, Λ_1 and Λ_2, and a complex conversion factor Λ_{21}. According to Wait and Spies

$$\arg \Lambda_1 = 24^{\circ}$$
$$\arg \Lambda_2 = 5.5^{\circ}$$

for a reflection height of 90 km and infinite ground conductivity. With these values the phase of Λ_{21} becomes approximately

$$\arg \Lambda_{21} = 43^{\circ}$$

4. SUMMARY OF RESULTS

1.   The attenuation over sea water is the same by night for the first and second order modes at 26.1 kHz. This is in agreement with Wait and Spies (1964) for nighttime 90 km.

2.   The difference in phase velocity for the first two modes is 0.51%, which agrees with Wait and Spies.

3.   If the theoretical values of Wait and Spies are used for the excitation factors the mode conversion factor is, when the sunrise line approaches NPM and has an angle to the path of 35-40° (summer) about -10 to -15 db. When the angle between the path and sunrise line is 20-30° the conversion factor is

about -25 db. The phase angle of the conversion factor is around 45° which is a mean value for the whole period.

4.   When the sunset line moves away from NPM the conversion factor is -30 to -40 db in summer and less in spring and fall.

No confidence limits have been given to the values quoted above. That depends on that it is difficult to say how good the model is with only two modes and conversion only from second to first mode.

Acknowledgement: The equipment used in this experiment was supplied by the Institute for Exploratory Research, Deal, N.J., USA.

5.   REFERENCES

1.   D.D. Crombie, Periodic fading of VLF signals received over long paths during sunrise and sunset. Radio Science 68D, 27, 1964.

2.   D.D. Crombie, Further observations of sunrise and sunset fading of very-low-frequency signals. Radio Science 1 (New series), 47, 1966.

3.   A.B. Kaiser, VLF propagation over long paths, J. Atm. Terr. Phys. 29, 73, 1967.

4.   D. Walker, Phase steps and amplitude fading of VLF signals at dawn and dusk, Radio Science 69D, 1435, 1965.

DISCUSSIONS OF VLF RECORDS OBTAINED DURING POLAR CAP ABSORPTION

AND AURORAL DISTURBANCES

Alv Egeland

The Norwegian Institute of Cosmic Physics

Oslo, Norway

Abstract: Very low frequency (VLF) radio waves are attractive for long distance transmissions when high reliability and accuracy are important. However, when the phase of the VLF waves changes the signal will alter its frequency. In a navigation system based on VLF transmissions these changes in the propagation time must be taken into account.

In this paper the influence of high latitude disturbances on VLF propagation, based on measurements carried out at Spitzbergen (78° geographic north) from 1965 to 67, are discussed. The principal findings may be summarized as follows:

The correlation between anomalous VLF propagation conditions and disturbed ionospheric activities is high. For VLF waves propagated over distances shorter than say 3000 km the influence of aurora and magnetic activity is particularly marked. Furthermore, radio waves at e.g. 11 kHz are much more sensitive to ionospheric disturbances than waves at 16 kHz.

During two weak PCA events which were hardly detected by HF riometers in Northern Scandinavia, strong VLF phase and amplitude anomalies were recorded for several days. Contradictory to earlier data the major VLF effects were found during day-light periods.

1. INTRODUCTION

The energy of the VLF waves is guided between the earth and the lower ionosphere (D-region) over very great distances. The propagation of these very low frequency radio waves is characte-

117

rized by the very stable phase properties and the relatively low
path attenuation. The VLF propagation is markedly less affected
by ionospheric disturbances than the HF waves are, and the fading
is very slow. These facts together make the VLF radio waves very
attractive for long distance transmissions when high reliability
and accuracy are important.

However, the frequency-widths which can be used for transmissions in the VLF band are very limited and the enormously large
transmitter installations required are the most severe drawbacks
as compared to HF propagation, where the terminal equipment can be
relatively simple.

The theoretical solution of VLF-propagation was first discussed in 1951 by Budden, and his suggested mode theory has been
further developed especially during the recent years (5), (23).
Fairly good agreement between experimental results and this theory
is obtained if a sharply bounded spherical wave-guide model is
used. Further knowledge of the lower ionosphere is very important
if an accurate navigation system is to be based on VLF transmissions. More recently VLF waves have also proved to be a useful
tool for detection and investigation of the disturbed D-region.

During polar cap absorptions and strong auroral disturbances,
when the whole frequency range used for long distance transmissions is interrupted, the use of VLF radio waves is probably the
most reliable communication method. However, during such strong
absorptions, the propagation conditions for VLF waves may also be
seriously disturbed. In this paper concentration will mainly be
on the influence of high-latitude disturbances on VLF propagation.

2. VLF RECEIVING EQUIPMENT

While the amplitude receptions of VLF-waves were initiated
some 50 years ago, it is only during the last 10 years that the
phase variations of VLF waves propagated over long distances have
been measured (3).

At distances greater than a few hundred km it is not practically feasible to record reference signals from the transmitters. Using a highly stable oscillator at the receiver-station,
the phase difference between the received signal and the local
standard can be recorded. Frequency standards with a long term
stability of about one part in 10^{11}/day, are now normally used
(20).

Phase tracking at our VLF station is achieved with an equivalent noise bandwidth of 0.006 Hz (i.e. a time constant of 50 sec).
To provide image-rejection and increase the frequency selectivity,

narrow-band filters are used. The phase is recorded directly in
degrees and can be scaled to an accuracy of approximately 1 degree.

3. VLF PROPAGATION DURING QUIET IONOSPHERIC CONDITIONS

In order to draw attention to the influence of high-latitude
disturbances, a brief summary of some of the main characteristics
of VLF propagation during quiet conditions will be given. For
further informations cf. e. g. Blackband, 1964.

When the VLF path changes from daylight to darkness, rather
regular and systematic variations in the VLF amplitude and phase
are observed. These variations, closely related to the solar illumination, can be explained by changes in reflection heights from
85 to 90 km at night to say 70 to 75 km at day. The reflection
height varies significantly with seasons. During sunrise and sunset the phase and amplitude change rapidly, while during day-time
and night-time conditions the VLF properties are very stable.

For propagation distances shorter than say 1000 km the concept of geometrical ray theory may be used. Due to the interference between the sky-waves and the ground-waves the VLF signals
propagated over short distances are highly variable. The signal
strength varies also with changes in the reflection height and
with the reflection coefficients.

At propagation distances beyond 4-5000 km the higher modes
are highly damped so that the simpler one-mode theory can to a
first approximation be used. In the transition region between say
1000 km and 4000 km it is not possible to make any useful simplification in the general solution (23).

The magnitude and the shape of the phase will depend on the
respective lengths of the paths in darkness and daylight. The
diurnal phase curve shows an approximately trapezoidal shape for
an east-west propagation. For a north-south path the corresponding curve shows a more rectangular form. The magnitude of the
diurnal phase change is largest for short distances and it decreases to a minimum around 2000 km. For greater distances the
phase change increases as the length of the path going from day-
light to dark increases (3). These phase variations depend, however, both on the frequency of the waves as well as on time of the
year.

A typical attenuation rate of VLF waves is about 3 db/1000 km.
The attenuation is higher for land paths than over seas, and it
increases with decreasing wavelengths. The propagation conditions
for VLF waves are normally best during the night. However, the
signal ratio between day/night varies both with distance, fre-

quency and direction of propagation.

In this paper the term phase advance will be used for a decrease in the phase length of the path; i.e. phase advance corresponds to a decrease in the reflection height. The term lag means an increase of the phase length of the path.

4.   ANOMALOUS VLF RECORDS OBTAINED DURING PCAs

Certain solar flares which produce PCA events, also give rise to very large VLF anomalies. Here it should only be pointed out that most of the absorption is located in a layer below the normal D-region (17). Furthermore, PCA events are usually most intense during daylight periods. The day-to-night ratio varies between 3 and 7 with a medium value of 5.3 (14). Satellites often observe proton fluxes so weak that with existing techniques no effect can be measured from the ground. Thus, there is still an uncertainty as to what shall be called a PCA event.

4.1 VLF Measurements at Spitzbergen during PCA Events

The major PCA effects on VLF measurements have hitherto been found when the lower ionosphere is in darkness. VLF observations at Isfjord Radio, Spitzbergen (geographic coordinates 78⁰4' N, 13⁰38' E) during two small PCA events show the major effects during day-time.

A solar flare (coordinates 18⁰N and 27⁰W) of importance 3 was observed between 0225-0414 UT, with max. at 0242 on March 24, 1966. The NBA (24.0 kHz) phase and amplitude curves for the period March 23-27 are shown in Figure 1. A few minutes after the

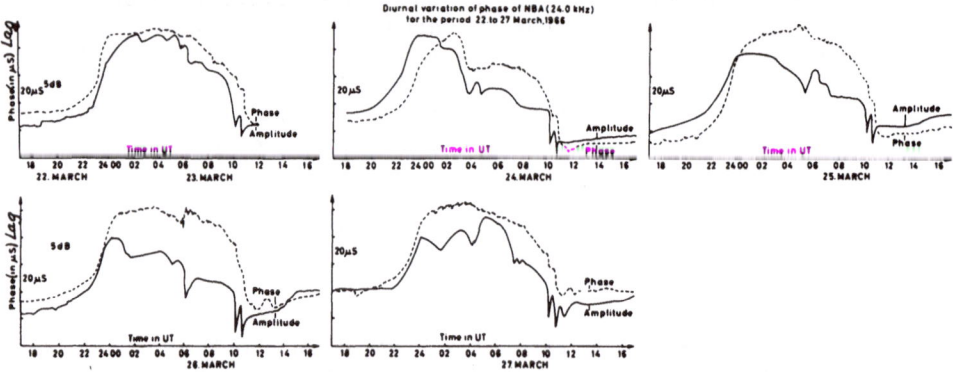

Figure 1. Diurnal phase and amplitude variation of NBA (24.0 kHz)
           for the period March 22-27, 1966.

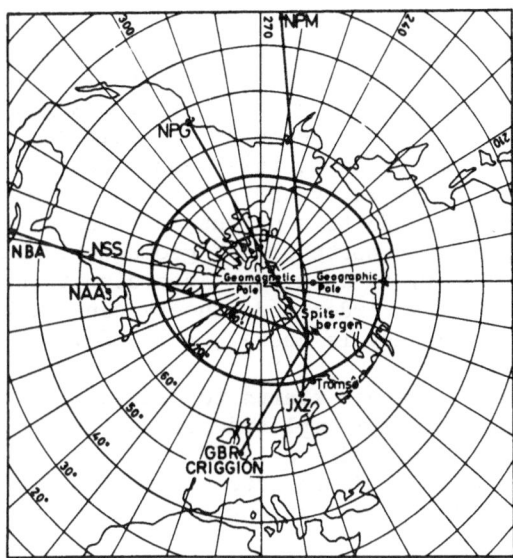

Figure 2.   A map showing some of the different paths monitored in
            a geomagnetic coordinate system. The fully drawn oval
            around the pole represents the centre of the northern
            auroral zone determined from Feldstein's (1960) obser-
            vations.

occurrence of this flare the phase started to change and one hour
later a very large smoothed phase advance of some 28 μs (equivalent to 232 degrees) was observed (Figure 1). The GBZ (19.6 kHz)
transmissions which were recorded simultaneously show a very similar behaviour.

     Weak absorption (less than 1 db at 27.6 MHz) was recorded
both at Tromsø and Kiruna between 0330 and 1430 UT on March 24.
Thus the VLF effects were observed before the HF absorption started. The locations of the transmitters and our receiver station
are shown on the map in Figure 2. No absorption could be seen on
the two following days. At Lycksele, which is 600 km south of
Tromsø, no absorption was found. The HF absorption at the receiver site shows a maximum absorption of 3.2 db at 27.6 MHz,
while the corresponding values for riometers operating at 20 and
40 MHz were 6.2 db and 1.4 db, respectively. The absorption lasted from 0300 to 2200 UT on March 24 and about 0.2 db at 27.6 MHz
absorption was also observed the next day. Normally, a magnetic
storm starts some 20 hours after a PCA flare. But the magnetic
data for March 25, at Tromsø and Isfjord Radio only show moderate

magnetic activity with average K-values less than 3. The VLF measurements shown in Figure 1 strongly indicate that a pronounced PCA event started 0230 UT on March 24, 1966.

The distance between NBA and Isfjord Radio is 9050 km. Assuming that the phase change is due only to changes in the VLF reflection height north of 65° geomagnetic latitude, which is reasonable from the riometer data, this means that about 3000 km of the NBA path are influenced by the PCA event. The observed phase change of 232° corresponds to a height decrease of approximately 30 km for the whole path north of 65° geomagnetic latitude if a first order mode theory is assumed. According to Watt (private communication) the first order mode theory can not be used for this path if the frequency is above say 15 kHz.

The NBA signal strength starts to decrease at the same time as the phase, and a minimum amplitude is observed at 0338 UT (Figure 1) when the amplitude is 13 db below the normal value. The characteristic amplitude effects around sunrise and sunset do not disappear, but the signal strength variation is less than normal.

The most important finding is the observed average phase advance of 25 μs during the day-time of the March 24 and some 10 μs on March 25. On the other hand, the night-values of the phase for the two days following the PCA event, compared to normal values are almost the same. Thus, instead of reducing the diurnal phase variation, this PCA caused an increase in the phase difference from day to night. The VLF phase records for this event show that VLF phase anomalies following this PCA event lasted for 7 days. However, during the last 3-4 days the VLF records are only slightly disturbed.

Both the night-time and the day-time signal strengths of March 24 and 25 are significantly reduced but the diurnal variation is only reduced by approximately 1 db for the two days (Figure 1).

On the GBZ records, a smooth phase advance of 17 μs (equivalent to 123°) was observed between 0253 and 0400 UT on March 24. Also the GBZ signal strength showed a pronounced minimum at 0356 UT on March 24. The GBZ transmitter did not work satisfactory after 0500 UT on this day.

The MPM (26.1 kHz) records of March 24 at Kiruna are shown in Figure 3a together with the average phase curve for the period March 19-26. The path between MPM and Kiruna, which is about 9700 km long, lies close to the geomagnetic pole (Figure 2). A phase advance of only 10 μs, which lasted for about 10 hours, coincided with the start of the PCA event. A small anomalous phase shift

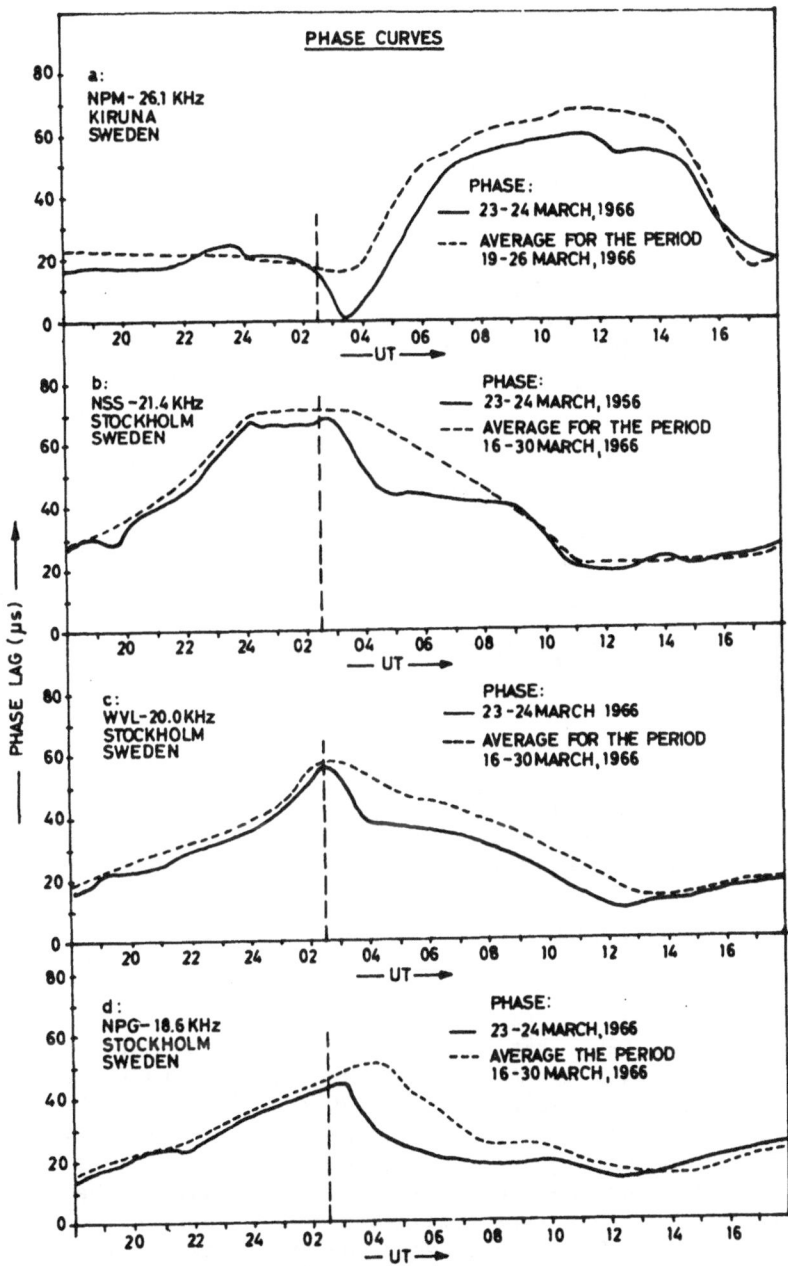

Figure 3. Diurnal phase traces of NPM (26.1 kHz) at Kiruna and of
NSS (21.4 kHz), WVL (20.0 kHz) and NPG (18.6 kHz) at
Stockholm for the March, 1966 PCA event. The dotted
curves represent the average diurnal phase for the
later half of March.

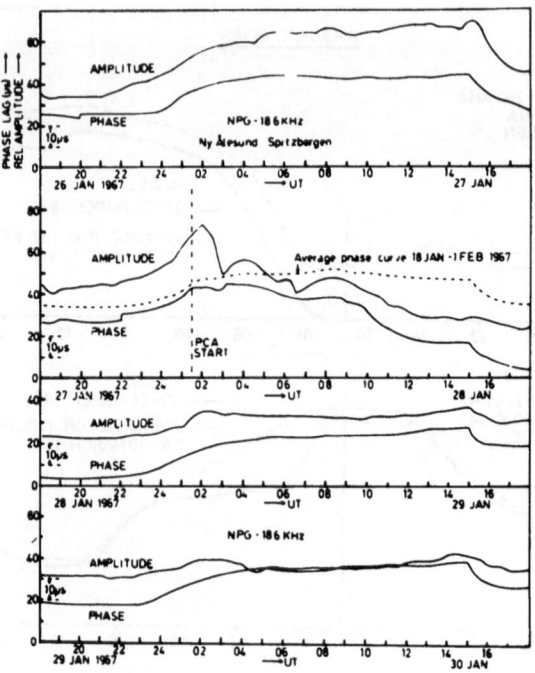

Figure 4.   Diurnal phase and amplitude curves of NPG (18.6 kHz)
            for the period January 26-30, 1967. The average phase
            curve of NPG for the last half of this month is shown
            in the dotted curve.

seems also to have occurred on March 25, but this may not be significant.

     The VLF phase records at Stockholm (about 60° N) of NSS
(21.4 kHz), WWVL (20.0 kHz) and NPG (18.6 kHz) for March 24 are
drawn in Figure 3, b-d. Also for these transmitters the average
phase curves for the period March 16-30 are presented. The largest phase anomalies are observed on NPG, but the NSS as well as
the WWVL records show greater phase changes than those found for
MPM at Kiruna. The NPG and the WWVL records indicate that the
propagation conditions were somewhat disturbed during the day-
time of March 25. The path between NSS and Stockholm reaches a
maximum geomagnetic latitude of 73° N. Thus the observed NSS ano-
lies, although lasting only about 5 hours, indicate that the PCA
layer during the short period of maximum disturbance must have
penetrated below 70° geomagnetic latitude.

From the VLF records at Spitzbergen on January 28, 1967, it
looks as if another PCA event occurred (Figure 4). The HF absorption for this event is even less than found for the March period
presented above. The Spitzbergen records of NAA (17.8 kHz) on
this day show that within an interval of 90 min. the phase is
advanced by 17 μs (or approximately 115°), while the NAA amplitude is depressed by approximately 18 db. However, the NAA phase
continues to advance after this first rapid change, although more
slowly, and a phase shift of more than 20 μs is recorded during
the main daylight period of January 28. The anomalous VLF records
associated with this PCA lasted throughout the night between
January 28 and 29 (10 μs phase changes in average) and during
the day of January 29 the phase started to fluctuate much more
than normally. This is certainly due to the arrival of the more
low energy plasma stream some 24 hrs after the proton event.

The Spitzbergen records of NPG (18.6 kHz) phase for the same
day start to change about 30 min. later than NAA (Figure 4). The
phase advance then goes very rapidly for about one hour. After
this period a gradual phase shift occurred which lasted for several hours. A maximum phase advance of 30 μs (equivalent to 200 degrees) was found ten hours after the start of this PCA event.
Some much weaker phase anomalies are also observed during the next
day. Assuming, which is reasonable from the HF riometer records,
that this precipitation event also is confined to the zone north
of 65° geomagnetic latitude, a change in the VLF reflection height
of about 25 km will explain the observations if a first order mode
theory is used. As the NPG path is located closer to the pole
than the NAA path and has a greater part of its path within the
polar region, this may explain the difference in the change of
the waveguide width for the two stations.

The data presented in the last two sections will be further
discussed in Section 7.

5. ANOMALOUS VLF PROPAGATION DURING AURORA AND MAGNETIC DISTUR-
BANCES

5.1 Brief Summary of Earlier Observations

The VLF effects associated with magnetic activity and auroral
disturbances are most pronounced on high latitude paths (i.e.
paths within or close to the auroral zones). During the sudden
commencement (SC) of a magnetic storm phase advances are often observed. Great phase fluctuations may accompany strong magnetic
bays and storms. These effects, which are recorded as rapid and
deep fading are mainly found during night-time and particularly
during the main phase of the storm. Furthermore, VLF waves with
steep incidence on the ionosphere, (i.e. paths over short distan-

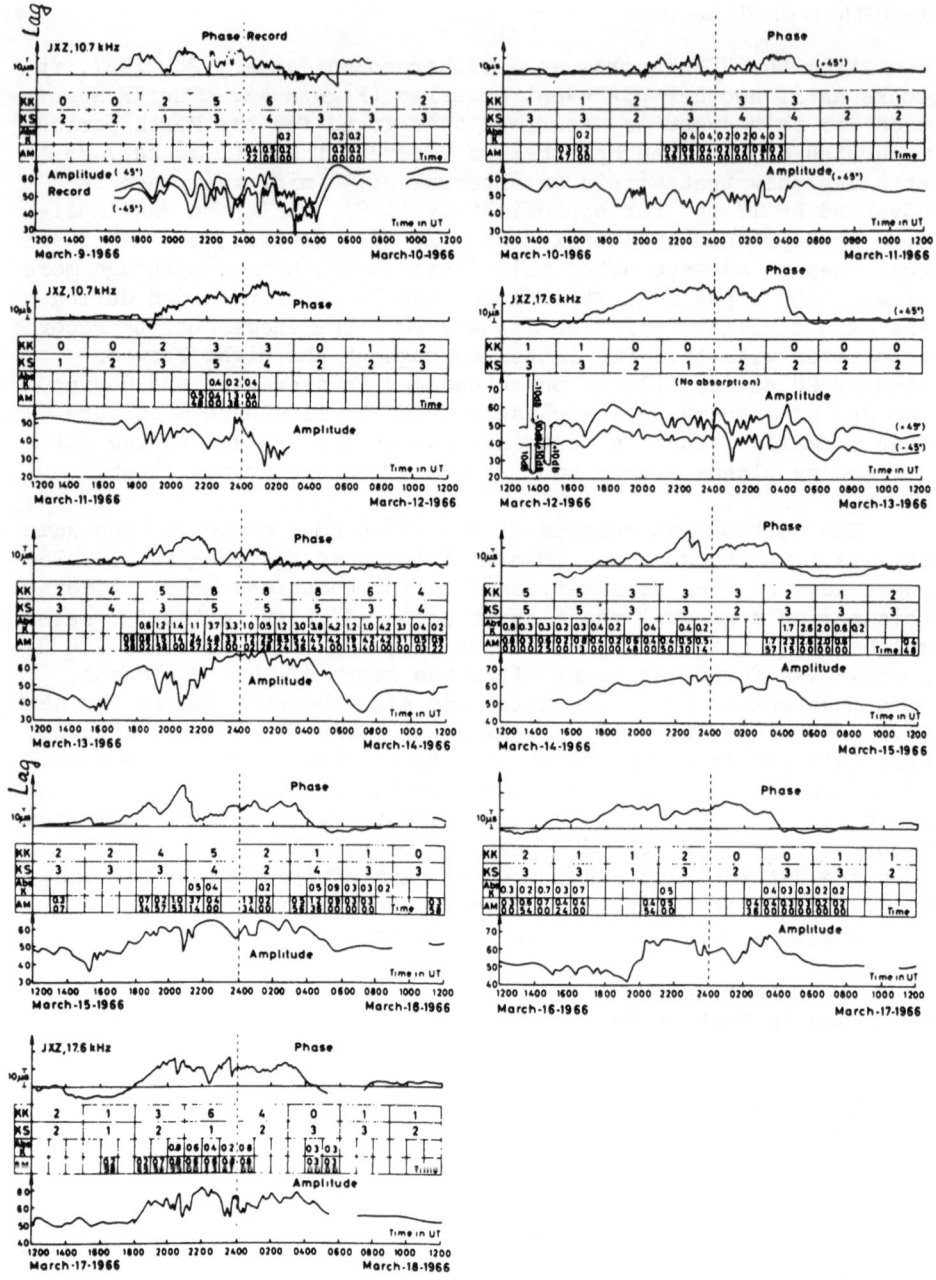

Figure 5. Diurnal phase and amplitude patterns of JXZ at 10.7 kHz
between March 9 and 12, and at 17.6 kHz from March 12 to 18, 1965.
The figures listed have the following meaning: KK = K-values at
Kiruna, KS = K-values at Spitzbergen, Abs K = absorption in db on
27.6 MHz at Kiruna on each whole hour, and Am = maximum absorption
within each hour together with the corresponding time in min.

ces) are particularly susceptible to storm effects. Anomalous VLF effects of this kind may often last several days longer than found on the magnetograms.

Also during day-time bays anomalous VLF variations have been observed. Although the phase may fluctuate much during magnetic disturbances, average height variations of 5 - 10 km over an appreciable part of the path are not unusual. However, the changes in phase and magnetic variations are seldom recorded simultaneously. Time delays of several minutes are often observed. For further informations cf. Kitchen et al., 1953, Bracewell et al., 1951, Egeland et al., 1964, Landmark, 1964, and Burges, 1964.

5.2 The Influence of Ionospheric Disturbances on Phase and Amplitude at 10.7 and 17.6 kHz over a Distance of 1100 km

Some VLF recordings at 10.7 and 17.6 kHz of the Norwegian transmitter JXZ at Novik (geographic coordinates 67°N, 14°E) will be discussed. The centre of the northern auroral zone, determined from Feldstein's (1960) observations, is somewhat south of the midpoint of this path (Figure 2).

The diurnal phase and amplitude variations of JXZ for the period March 9-18, 1966 are seen in Figure 5. In order to check the influence of high-latitude disturbances on VLF propagation, both the absorption values on 27.6 MHz at Kiruna and the magnetic K-values at Spitzbergen and Kiruna are listed (Figure 5).

The K-values show that except for the nights March 12-13 and 16-17, there were more or less continuous magnetic disturbances during the whole period. The night between March 13 and 14 was strongly disturbed. Very strong absorption (up to 8 db) at 27.6 MHz occurred that night, and there were also some weaker HF absorptions on the other days.

By comparing the curves given in Figure 5 with the magnetic and riometer data the following observations should be mentioned:

Quiet ionospheric conditions: The diurnal phase changes at 17.6 and 10.7 kHz vary between 17 - 19 μs and 20 - 22 μs, respectively. If it is assumed that the effect of mode interference is approximately equal on both frequencies, the average diurnal change in reflection height is markedly greater on the lowest frequency (Section 7).

The propagation conditions on 10.7 kHz are best during daytime (cf. the three first days in Figure 5), while on 17.6 kHz the opposite is true. In average the day-time signal strength on 10.7 kHz exceeded the night-time level by approximately 2 db, while on

17.6 kHz the night-time signal level exceeded the day-time level by about 3 db.

The phase change at sunrise is much more rapid than at sunrise and sunset are most pronounced at the highest frequency.

Moderate disturbances: (K-values < 5, absorption at 27.6 MHz < 1.5 db). Moderate ionospheric disturbances do not seem to influence the phase measurements much at 17.6 kHz. However, for a few cases on 17.6 kHz (Figure 5 on March 14, between 2200 and 2300 UT and March 15 between 2000 and 2100 UT) the phase lag as compared to normal day-time values is even greater than during quiet conditions.

The effects of such moderate disturbances are very pronounced on the 10.7 kHz phase. On March 10 between 0200 and 0600 UT, as well as on the night between March 10 and 11 (Figure 5) there were periods when the phase lag relative to the corresponding day-time values was approximately zero. By comparing these observations with calculated and measured electron density profiles at nighttime, a reflection height of 70 to 75 km at 10.7 kHz seems probable even for moderate disturbances (Section 7).

The signal strength on both frequencies decreases during moderate disturbances, but this effect is also far more marked on 10.7 kHz (cf. e.g. the amplitude curves on March 10 between 0200 and 0500 UT, on March 13 between 1400 and 1600 UT and between 2000 and 2200 UT, and on March 15 between 2000 and 2200 UT).

The main conclusion is that the lowest frequency is much more sensitive to moderate disturbances than is the higher one. Similar findings will be shown in next section.

Strong ionospheric disturbances: (K-values $\geq$ 5, absorption $\geq$ 1.5 db). During very strong auroral absorption events the phase of 17.6 kHz radio waves may be advanced up to 20 μs relative to normal values (cf. Figure 5 on March 14 between 0100 and 0600 UT).

The amplitude at 17.6 kHz is also very much influenced during very strong ionospheric disturbances. However, the important point here is that the signal strength increased substantially (e.g. by more than 8 db above normal levels during the early morning of March 14).

5.3 Diurnal Phase and Amplitude Variations at 11.6 and 16.0 kHz over a High Latitude Distance of 2800 km

In this section we will deal with the reception of GBR and Criggion (CRIG). The centre of the auroral zone is located close to the midpoint of the path (Figure 2).

Relatively strong magnetic disturbances K-values between 3 and 6) accompanied by auroral absorption (between 1 and 4 db) occurred between July 27 and 29, and on August 20 and 21, 1965. The 16.0 kHz observations during these days show a change-over from a night-time maximum in signal strength to a maximum during the day-light hours (Section 7).

The average diurnal phase change for these disturbed periods was only 9 μs as compared to 20 μs during quiet conditions. Also for a few periods during day-light when strong absorption occurred phase advances of 2 to 8 μs have been observed.

For the period September 1-16, 1965, the average diurnal phase change of GBR (16.0 kHz) and CRIG (11.6 kHz) propagated over the same path, is 25 μs and 30 μs, respectively. Thus the phase change is again markedly greater on the lowest frequency.

The night between September 27 and 28, 1965 was severely disturbed in Northern Scandinavia, with K-values equal to 7. At Spitzbergen the highest K-values was 5. All auroral absorption values at 27.6 MHz after 1900 UT on September 27 exceeded 1 db, and between 0000 and 0300 UT on September 28 more than 3 db absorption occurred. Both the GBR phase and especially the amplitude for that night are very disturbed (Figure 6). For example, the phase difference between day and night varies between 6 and

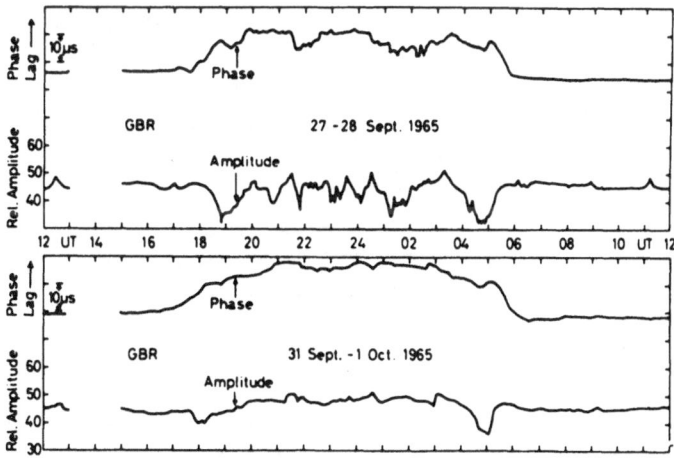

Figure 6. Diurnal phase and amplitude traces of GBR (16.0 kHz) for a disturbed day (September 27-28, 1965) and a quiet day (September 31 - October 1, 1965).

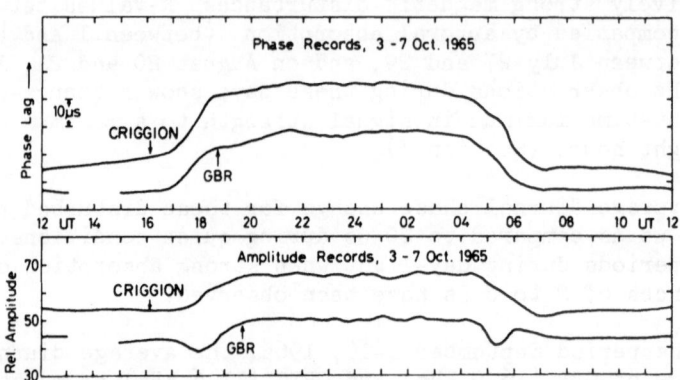

Figure 7. Average diurnal phase and amplitude curves of GBR and
          CRIG for the period October 3-7, 1965.

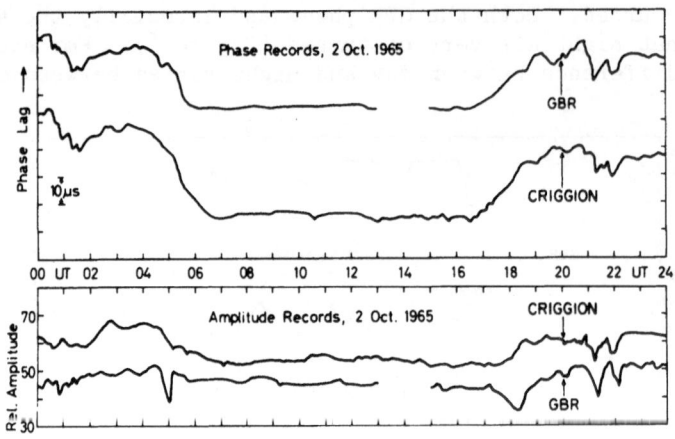

Figure 8 Phase and amplitude curves of GBR and CRIG for the
          disturbed day on October 2, 1965.

10 µs as compared to 18 - 20 µs for the quiet night between September 30 and October 1. Furthermore, the night-time amplitude decreased by max. 4 db below normal values. Another interesting effect is the strong oscillations in the night-time amplitude. Smaller oscillations are also seen on the phase curve.

Figure 7 shows that for the quiet period October 3 - 7, 1965, the average diurnal phase changes at 16.0 and 11.6 kHz are approximately 20 µs and 28 µs, respectively. The day-time signal strengths on GBR and CRIG exceed the night-time value by 1 and 4 db, respectively. These curves should be compared with those in Figure 8 for October 2, which was much disturbed with K-values between 2 and 6, and HF auroral absorption between 0.3 and 3.6 db. The maximum disturbances occurred between 0000 and 0400 and between 2000 and 2300 UT. During these disturbed periods strong irregular variations are seen on all four curves (Figure 8). The amplitude curves (specially on 11.6 kHz) are most irregular, and the average diurnal variation is much reduced. Short time phase variations during the early morning and late evening of more than 15 µs were found.

On the basis of our observations it may be concluded that there is a high correlation between ionospheric activity and anomalous VLF records, and that the correlation coefficient increases when the frequency is lowered. Furthermore, moderate ionospheric disturbances may influence the VLF amplitude records more than the phase. But when more high-energetic particles bombard our ionosphere, resulting in stronger riometer absorption, marked changes of both phase and amplitude are observed for the whole VLF band.

6. THE PRACTICAL USE OF VLF TRANSMISSION

Due to high phase stability found for VLF waves, it was suggested about 10 years ago that VLF transmissions might be used for navigational purposes as well as for frequency- and time-comparisons (18). Quite a few of the VLF transmitters are now stabilized to a very high frequency accuracy.

However, when the phase changes the signal will alter its frequency. (The naturally occurring VLF noise, which is very strong and irregular at high latitudes, once seemed to be an important limitation for the use of transmitted VLF waves. With the use of narrow bandwidths and powerful VLF transmitters, the atmospherics will not seriously disturb VLF propagation). For a navigation system based on VLF transmissions it is necessary, therefore, to correct for changes in the time of propagation in order to obtain the greatest possible accuracy. Various methods for such corrections appear possible, e.g. by introducing large tables which give interpolated and corrected phase values. However, the

phase of VLF waves depends on: 1) frequencies, 2) length and direction of paths, 3) propagation over land and/or seas, and 4) diurnal, seasonal and yearly variations. The method thus will be
very time-consuming and therefore not suitable where a rapid decision is necessary, e.g. in civil and military aeroplanes. A more
rapid but less accurate method has been proposed by Hampton and
Hill (13). Much more data are needed, specially at high latitudes,
before any VLF navigation systems can be made quite reliable.

Furthermore, VLF transmissions are seriously disturbed during
strong ionospheric activity. During such disturbances, when the
width of the wave-guide is very much changed over long-time periods, the problem of correction of VLF data is a serious one.

The changed phase velocity or the produced frequency variation can in principle be smoothed out by using narrow bandwidths
and correspondingly long intergration times. Theoretically the
precision increases as the observation time increases. However,
variations of several of the upper atmospheric parameters normally have periods of 24 hrs and these diurnal changes may be significant as the observation time increases. The precision will
furthermore depend on the long term variations of the ionosphere
(7).

After removing the drift of the frequency standard, frequency
comparison up to 10^{11} for a 24 hrs period may be obtained also
during moderately disturbed periods, but probably not during
strong high-latitude disturbances.

7. SUMMARY AND DISCUSSION

Before we discuss the VLF observations further, a very brief
summary concerning some general facts of disturbances and electron
density profiles at high latitude will be given.

The two major types of high latitude disturbances are auroral
and polar cap absorption. Auroral absorption (AZA) is related to
magnetic disturbances and visual aurora and occurs most often near
the zones of maximum auroral activity. Auroral disturbances are
caused by the precipitation of low energetic particles into the
earth's atmosphere where they interact with the atmospheric con-
stituents. The incoming electrons primarily lose their energy
through various ionization and excitation processes. However, a
fraction of the energy is may-be converted into bremsstrahlung,
which prenetrates further down. PCA events, which are not related
to magnetic and auroral activity in the same way, occur over most
of the polar cap simultaneously (cf. Holt's paper in these Proceedings). Since the particles producing the auroral and polar
cap absorption events are excluded by the magnetic field from the

earth's ionosphere at lower latitudes only VLF paths close to or crossing the pole will be affected by AZA and PCA.

The auroral absorption is mainly caused by electrons above 20 keV, while PCA is primarily due to protons (above 0.5 MeV). Disturbances in the earth's magnetic field are probably produced by the lower tail of the electron spectrum. Roughly speaking it depends on the particle's energy how far down in the ionosphere they are able to penetrate.

The PCA layer is normally stable over periods of many hours and even up to several days. The auroral absorption events, on the other hand, are caused by minor irregular clouds (order of a few hundreds of km) and they normally show very great time fluctuations.

The main VLF effects of these disturbances are an appreciable increase in the contribution of the higher order modes in the depressed wave-guide and a marked change in the reflection coefficient. It should be pointed out, however, that there is a wide variation in the individual aspects of AZAs and PCAs, and the intensities of the ionospheric disturbances do not always agree with the observed VLF anomalies.

On long distance circuits the observed phases and the amplitudes are significantly affected only if the reflection height is lower over a great part of the path. Such conditions occur almost exclusively during PCAs and for a few SID's. From a few riometers it is not possible to obtain information about the extent of an absorption event. For detection of high latitude disturbances VLF measurements are more sensitive than riometers. Rough estimates of the extension of the disturbed region can also be made if several circuits are recorded simultaneously.

An important question is why the strong and long-living VLF anomalies observed during PCAs often correlate so badly with the absorption measurements. The reason is that an increase of a few tens to a hundred free el./cm^3 between 50 and 70 km may explain the VLF observations, but their contribution to the high frequency absorption (as measured with riometers) will be very small (less than 0.1 db). To maintain a peak electron density of say 100 per cm^3 (at high latitude) a flux of about one hundred protons per cm^2 per second is required (10). The absorption produced by this small proton flux, however, can not be measured with the riometer. Thus a weak PCA event may produce very little HF absorption, but still cause great VLF phase and amplitude anomalies when the paths cross the polar cap.

Measurements of the absorption profiles or the electron density distribution in the lowermost ionosphere have been made by e.g. Belrose (2) and Thrane (22) (partial reflections and cross modulations) and by Parthasarathy et al. (17) (multifrequency riometer technique). Furthermore, quite a few profiles have been obtained by rockets. But still the knowledge about electron density and the chemistry in the lowermost ionosphere is poor.

From high latitude rocket measurements of electron profiles between 50 and 100 km a few general points should be mentioned. (For further details cf. Jespersen and Landmark in these Proceedings).

Day-time profiles both during quiet and moderately disturbed periods show no sharp gradient, i.e. the electron density between 50 and 90 km increases at an approximately constant rate. During very strong auroral absorption events, however, a relatively sharp gradient in the day-time electron density below say 70 km, similar to a PCA event, may occur.

During a quiet night the VLF waves at high latitudes are probably reflected from the bottom of the E-layer. The number of free electrons increases much more rapidly with height during the night than during a moderately disturbed day. During stronger ionospheric disturbances the number of free electrons moves downwards, but its shape does not much change. Up to 10^4 el./cm^3 may be observed below 70 km during a magnetic storm.

Simultaneously with our VLF measurements, absorption measurements at multiple frequencies were carried out at Isfjord Radio. On the basis of these measurements electron-density profiles of the lower ionosphere have been derived by the method presented by Parthasarathy el al. (17): The profile most capable of producing the observed maximum absorption values for the PCA event in March 1966, shows that more than 100 el./cm^3 occurred below 55 km. Although this method does not give a unique solution it seems likely that the VLF-waves during the maximum disturbances were reflected from a height below 55 km. This result agrees with what was found by using the first order mode theory (Section 4). However, the contributions from higher order modes may be important. The lack of fine structure in the VLF curves, confirms that the PCA layer must be homogeneous.

The fact that the VLF waves during the PCA events show the greatest phase anomalies during day-time agrees with Reid's (21) conclusion concerning the electron density profile during a PCA, namely: The day-time gradient in the electron density profile between 50 and 70 km is markedly steeper than between 50 and 90 km at night. The phase difference observed at Stockholm, Kiruna and Spitzbergen can be explained if the normal day-night ratio of about 4 for the PCA layer is taken into account.

Some of the data presented in Section 5 will now be discussed. Increases in the ionization above say 70 km would not produce any measurable VLF effect during day-light. However, additional ionization in the 70-80 km region during periods when the VLF path lies in darkness, may cause very pronounced VLF effects. During weak to moderate AZA, it is likely that the main part of the additional ionization is located above say 80 km. The observed data clearly indicate that the VLF effects during magnetic activity and AZA are found mainly during night-time, confirming the fact mentioned above. To observe VLF effects during sunlit atmosphere, an increase in the ionization of the order of say one hundred el./cm^3 below 70 km is needed. Such increases may occur for short periods during SIDs, and very strong AZA over limited areas.

On the basis of VLF phase measurements together with the observed electron-density profiles during quiet conditions, it seems likely that waves between 10 and 30 kHz are reflected from approximately the same height at night. During day-time, however, it is quite likely that the 10 to 12 kHz waves are reflected at significantly lower altitudes than waves at say 20 kHz. An average diurnal difference in reflection height of say 5 km will probably explain the observed phase differences. Thus, VLF measurements may be used to obtain information about the electron density profile.

During moderate disturbances at day-time the whole VLF band is probably reflected from approximately the same heights as during quiet conditions. Greater height changes occur at night when the boundary of the ionosphere is moved considerably downwards even for an absorption event of 1.5 db only. The reduced diurnal phase changes observed at the lowest frequencies during such events agree with this picture. But as auroral disturbances are irregular and limited in extension, this effect is of importance only over short distances. The fact that the higher frequencies are much less influenced during these weakly disturbed events is probably due to increase in free electrons below normal reflection heights, not being sufficient to reflect these waves. However, much is left to be done on this point. The cloud structure of auroral ionization may well explain both the phase and amplitude fluctuations as well as the reduced signal strength found during such events.

During very strong AZAs the number of free electrons is markedly increased at heights below normal, especially on the nightside of the earth. Combining this with the phase records it seems likely that the phase reflection heights may be as much as 25 km below normal during the most disturbed periods. But as strong auroral absorptions are also relatively limited in extension, they will cause only very great phase fluctuations over short distances. The strong phase and amplitude fluctuations observed may be explained by the time and space fluctuation in the structure of the ionization clouds.

From a first order approximation of the mode theory, the change-over from a night-time maximum in signal strength to a maximum during day-time is interpreted as a decrease of the reflection height of say 10 km both day and night.

It is hoped that in the near future the VLF- and particle measurements may be carried out simultaneously at several places. If this is achieved, several of the uncertainties mentioned in this paper could probably be explained and VLF waves would be very useful for investigation of the lower ionosphere. Furthermore, simultaneous observations of several different transmitters, located close together, in the frequency range 10 to 30 kHz, may give valuable information of the electron density profile.

Acknowledgement: I am greatly obliged to Dr. F. Reder at the Institute for Exploratory Research (IER), Deal, New Jersey for using their computer service for the reduction of the phase measurements. The VLF receiving equipment was supplied through IER.

My thanks are specially due to Mr. E. Naustvik for operation of the measuring equipment and analyzing of the VLF data at Spitzbergen. The author is also greatly obliged to Mr. S. Westerlund, Kiruna and Mr. C. Åbom, Stockholm for providing the VLF data presented in Figure 4.

8. REFERENCES

1. P.R. Albee and H.F. Bates, Plan. Space. Sci. 13, 175, 1965.

2. J.S. Belrose, L.R. Bode and L.W. Hewitt, in: J. Frihagen, Electron Density Profiles in Ionosphere and Exosphere, North-Holland Publishing Company, Amsterdam, 1966, p 37 and p 48.

3. W.T. Blackband, Propagation of Radio Waves at Frequencies below 300 kc/s, AGARDograph 74, Pergamon Press, 1964.

4. R.N. Bracewell, K.G. Budden, J.A. Ratcliffe, T.W. Straker and K. Weeks, Proc. I.R.E., 98, 221, 1951.

5. K.G. Budden, The Waveguide Mode Theory of Wave Propagation, Prentice-Hall, New York, 1962.

6.   B. Burgess, in: E. Thrane, Electron Density Distribution in
     Ionosphere and Exosphere, North-Holland Publishing Company,
     Amsterdam, 1964, p 65.

7.   C.J. Chilton, D.D. Crombie and A.G. Jean, in: W.T. Blackband,
     Propagation of Radio Waves at Frequencies below 300 kc/s,
     AGARDograph 74, Pergamon Press, 1964, p 257.

8.   D.D. Crombie, in: J. Frihagen, Electron Density Profiles in
     Ionosphere and Exosphere, North-Holland Publishing Company,
     Amsterdam, 1966, p 118.

9.   A. Egeland, B. Hultquist and J. Ortner, in: G.J. Gassman, The
     Effect of Disturbances of Solar Origin on Communications,
     AGARDograph 59, Pergamon Press, p 79.

10. A. Egeland and W. Riedler, J. Atmos. Terr. Phys. 26, 351,
     1964.

11. A. Egeland and E. Naustvik, Radio Science 2, 1967.

12. Y.I. Feldstein, The Academy Science of the USSR, Moskow, 4,
     61, 1960.

13. D. Hampton and A.S. Hill, in: W.T. Blackband, Propagation of
     Radio Waves at Frequencies below 300 kc/s, AGARDograph, Per-
     gamon Press, 1964. p 231.

14. B. Hultquist, Space Science Reviews, V, 711, 1966.

15. F.A. Kitchen, B.G. Pressey and K.W. Tremellen, Proc. I.R.E.
     100, 1953, p 100.

16. B. Landmark, The Effect of Disturbances of Solar Origin on
     Communications (ed. Gassmann). AGARDograph 59, Pergamon
     Press, 1964, p.65.

17. R. Parthasarathy, G.M. Lerfald and C.G. Little, J. Geophys.
     Res. 68, 1963, p 358.

18. J.A. Pierce, G.M.R. Winkler and R.L. Corke, Nature, 187,
     1960, p 914.

19. E.T. Pierce and H.R. Arnold, Final Report 1081, Stanford Res.
     Inst., Calif., USA, 1963.

20. F.H. Reder and D.M. Viccione, Manual of INT-VLF, Institute
     for Exploratory Research, Deal, N.J., USA, 1966.

21. G.C. Reid, J. Geophys. Res. 66, 1961, p 4071.

22. E.V. Thrane, NDRE Report No. 54, 1967.

23. J.R. Wait, Electromagnetic Waves in Stratified Media, Perga-
     mon Press, London, 1962.

DISCUSSION

The discussion on long wave propagation was mainly concerned with the description and interpretation of anomalies in the amplitude and phase of the sky wave. The latter were particularly important for the proper use of these frequencies for navigational aids. A highlight of the discussion was an elegant simplification of the main features of mode theory by Dr. Watt which clarified the interpretation of many of the anomalies.

The current interest in using long wave propagation when HF fails during PCA events, was reflected in an extensive discussion on the characteristics of these events. The main points stressed were the smooth depression of the phase height in these events, the long duration of the anomaly and its extension to surprisingly low latitudes. Examples were given of large PCA anomalies in I.Q.S.Y. even at 61° geomagnetic latitude in Europe.

The large attenuation of long waves traversing the Greenland icecap was then discussed in terms of mode theory. At night when the second order mode was dominant there was little attachment of the electric field lines in the wave guide to the lower boundary and therefore relatively little attenuation. In the daytime the first order mode was dominant. For this mode strong fields could exist near the lower boundary which would interact with the ice and cause large attenuation. Measurements are needed to find out what thickness of ice is necessary to give serious attenuation when the VLF propagation is dominantly first order mode.

As a whole the part of the discussion concerned with the scientific aspects left the following impressions:

(a) Many scientists were mainly interested in the detail of particular events or of average characteristics which could be used to measure the electron density and collision frequency in the lower ionosphere. Such data were directly of interest to the engineer for short range propagation problems but needed to be supplemented by measurements of variability.

(b) The mode theory was a powerful tool for collecting and summarizing phenomena at longer ranges.

(c) There was a lack of systematic analysis from the engineer's point of view, particularly in describing the range of conditions over which given types of phenomena should be expected and the probable variability of phase and field strength.

(d) Considerable quantities of data exist which could most probably be summarized in forms useful to the practical man but there does not seem to be any way of doing this efficiently at present - the scientists are interested in other problems, yet some specialized knowledge is essential for a useful analysis to be produced.

Design considerations for the establishment of VLF and LF circuits were reviewed. The following factors must be considered:

Radiated Power
Path Attenuation
Median Background Noise
Receiver Performance Factor
Data Rate
Statistics of Signal and Noise
Circuit Protection Factor

In addition time and frequency stability of the propagation path affects the performance of many systems and is an added item that must be considered.

VLF receiver design factors to reduce the effects of impulsive noise were discussed. The technique of using a wide bandwidth intermediate frequency followed by a limiter, followed by a narrow band IF was recommended to system designers. Pr. Piggott pointed out that articles on this aspect of receiver design appeared in the literature as early as in 1939.

Scatter Communication

SURVEY OF IONOSPHERIC AND METEOR SCATTER COMMUNICATIONS

P. J. Bartholomé

SHAPE Technical Centre, The Hague, Netherlands

Abstract: This paper surveys techniques for ionospheric and meteor scatter communication, with particular emphasis on the ARQ
intermittent system. Results of comparing various systems in
terms of power saving and signal duty cycle are given. Experience
obtained with systems based on ionoscatter and meteor-bursts,
developed and tested by STC, is described. A discussion of the
usability of the ARQ principle for operation in arctic regions is
included.

1. INTRODUCTION

Radio propagation by ionospheric scatter and meteor-trail reflections has been studied fairly intensively since the early
1950's. Besides the interest of such studies from a geophysical
point of view, a strong motivation has been to make use of these
two modes of propagation for communicating over the horizon at VHF
frequencies normally unsuitable beyond optical and diffraction
range. However, interest in the possible applications in the communications field has decreased recently, perhaps because of the
somewhat disappointing performance obtained from the first systems
built. The SHAPE Technical Centre has, however, developed two
systems, one using ionoscatter and the other meteor scatter.
These systems have been tested on a 1000-km path between Holland
and South of France at frequencies around 40 MHz.

2.   PRINCIPLE OF ARQ INTERMITTENT SYSTEMS

The principle of operation is the same for both systems. It
is based on the use of a technique called ARQ (Automatic request
for repetition). The information to be sent is encoded using an
error-detecting code and provision is made to repeat any character
which has not been received correctly. Transmission takes place
in the two directions and control signals are exchanged between
the two terminals, each path serving as a feed-back channel for
the other (Figure 1).

Figure 2 shows a time diagram which illustrates how an ARQ
system operates. The detection of an error at Station 1 causes a

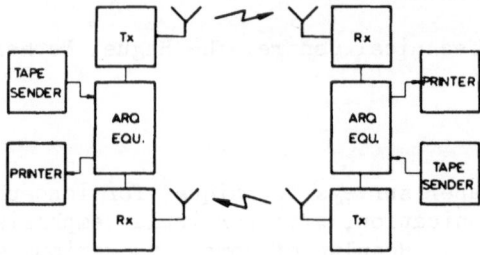

Figure 1. Block diagram of an ARQ system.

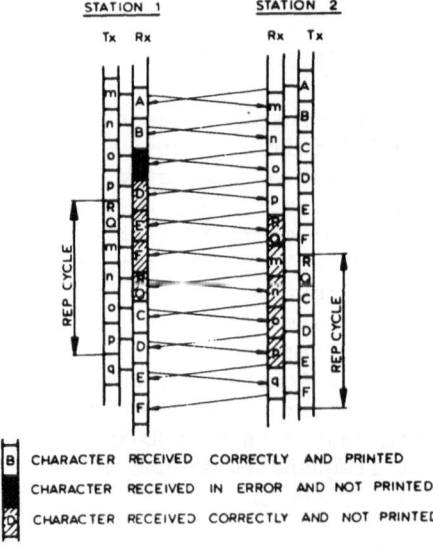

Figure 2. Time diagram of a repetition cycle.

request (RQ) to be sent for the repetition of character C which
was received in error. In practice, because of time delays in the
propagation path and in the equipment, it is necessary for Station
2 to repeat more than one character. In the case shown in Figure
2, four characters are repeated. For reasons of symmetry, Station
1 also repeats four characters. If the radio path is interrupted,
the two terminals continue repeating. As soon as the link is re-
established and after the current repetition cycle has been completed, communication is restored in both directions. A detailed
description can be found in Ref. 1.

Whereas the ARQ technique was originally devised as an optio-
nal feature for HF circuits to safeguard information against signal fades, it has been introduced here as the backbone of intermittent communication systems.

The advantages of an intermittent system over a continuous
one for an ionoscatter link can be shown with the help of Figure 3
where the short-term statistical distribution of the signal received at 40 MHz on an ionoscatter path is represented.

In the no-diversity case (curve III) the distribution function is of the Rayleigh type in the lower part of the amplitude
range where the signal is essentially due to ionoscatter. In the
upper part, the signal is constituted by meteor reflections and
the curve bends upwards as shown. In a conventional system where

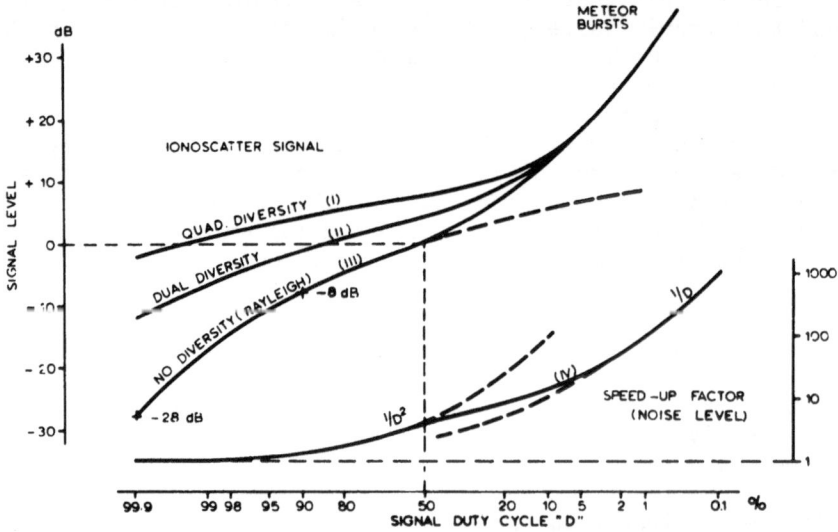

Figure 3. Signal level distribution and speed-up factor in
           intermittent scatter communications

transmission is continuous, the signal is required to be received all the time, or in practice say 99.9% of the time. A fading margin of about 28 db between the median level and the minimum acceptable level must therefore be added to the transmission loss when calculating the system parameters.

The use of ARQ introduces the possibility of intermittent operation and thereby enables the fading margin to be reduced. A reduction of the fading margin from 28 to 8 db brings the signal duty cycle down from 99.9% to 90%, i.e. by a relatively small amount because the fades are of short duration. If the average capacity of traffic is to be kept constant, the transmission rate must be increased by a factor equal to the inverse of the circuit duty cycle. This, however, requires widening the channel bandwidth and the resulting increase in receiver noise level offsets part of the saving made on the fading margin. Since there is no correlation between the signal fluctuations on the two opposite paths of an ionoscatter link because of the difference in frequency, the circuit duty cycle can be taken to be the square of the signal duty cycle D and the speed-up factor is equal to $1/D^2$. For a signal duty cycle of 50% for instance, the circuit duty cycle is 25% and the speed-up factor is 4. Curve IV in Figure 3 shows how the transmission rate must be raised in order to keep the traffic capacity constant when the duty cycle decreases. The vertical scale is such that the same curve can also be used to read the increase in receiver noise level in db.

It can be seen that, as one begins to reduce the fading margin and to transmit faster, the saving increases more steeply than the loss until a maximum is reached for a signal duty cycle of about 50%. As one carries on beyond that point, one gradually enters the region where meteor bursts predominate over the ionoscatter signal and the distribution deviates from the Rayleigh curve. Moreover, as meteor reflections begin to prevail, the signal bursts tend to become correlated on the two opposite paths and the speed-up factor need not be increased faster than $1/D$. This leads to another optimum situation at very low values of duty cycles.

Distribution functions for dual and quadruple diversity are also given in Figure 3 (Curve I and II). When diversity is applied at the reception, one starts with a substantial advantage over the no-diversity case and the signal distribution follows a less steep curve.

Figure 4 illustrates the power saving as a function of duty cycle. The curves shown have been derived from the previous figure by determining the difference in db between curves I, II and III on the one hand and curve IV on the other. Starting from the left end of the diagram (continuous transmission), a first

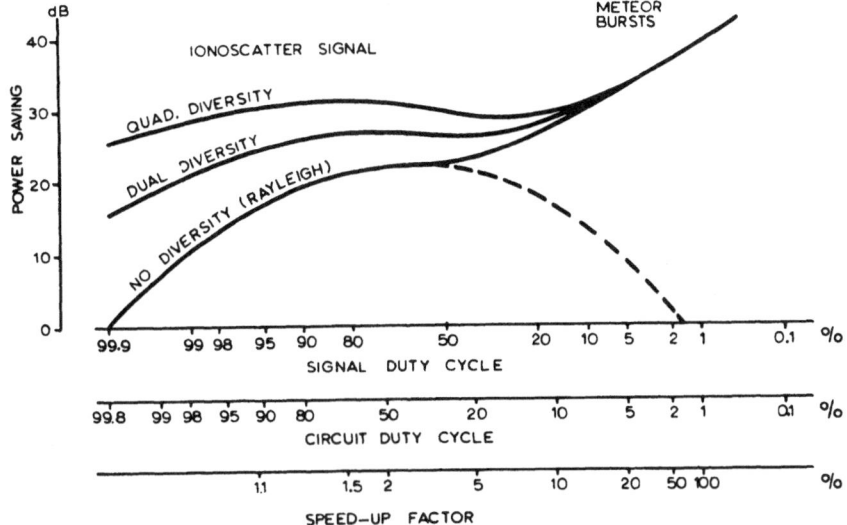

Figure 4. Power saving in intermittent scatter communications

maximum is found for a circuit duty cycle of about 40% in the no-diversity case. It corresponds to a power saving of 22 db. When diversity is used, this maximum occurs earlier and is not so pronounced. Continuing towards lower duty cycle values, the curve bends upwards and rises steadily. For a duty cycle of 1%, the gain over a quasi-continuous system using quadruple diversity is about 12 db. One is prevented from going much beyond the 1% point by practical considerations such as the necessity to keep the transmission rate within reasonable limits or to avoid exceedingly long delays in the transmission of messages.

3. DESCRIPTION OF STC SYSTEMS

The STC ionoscatter system has a speed-up factor of about 1.1 and therefore needs a circuit duty cycle of 90%. It uses eight-fold diversity and the conditions in which it works are approximately those of the first maximum of Figure 4. The meteor-burst system has a speed-up factor of 40 and requires a minimum duty cycle of 2.5%. It works in conditions which are represented by a point at the right end of Figure 4.

Figure 5 shows a picture of the antennas which have been used for both systems. The two masts at the foreground carry the four receiving antennas which provide space and height diversity. The

Figure 5. Photograph of the STC antenna system.

transmitting antenna is visible in the background. The modulation
is frequency-shift keying (FSK) with a 6-kHz separation between
mark and space frequencies. By detecting these separately, a
frequency diversity gain is obtained.

(a) <u>The ionoscatter system</u>. The ionoscatter system has a nominal
      capacity of four 50-baud telegraph channels in time-division
      multiplex. Figure 6 shows the cumulative distribution of the
      signal-to-noise ratio (SNR) measured over an 11-month period
      of test. With 5 kW of transmitter power, the SNR obtained
      exceeded the minimum required of 10 db for about 99.9% of the
      time. As for the error rate, it was not higher than 1 in
      3000 characters when conditions were marginal. Figure 7
      shows the dependence of the signal level received with the
      length of the path. The upper curve holds for high-gain an-
      tennas such as corner reflectors. The useful range of dis-
      tances is approximately from 500 to 2000 km. The lower curve
      holds for medium-gain Yagi antennas. The upper limit is low-
      er in this case, say 1600 km.

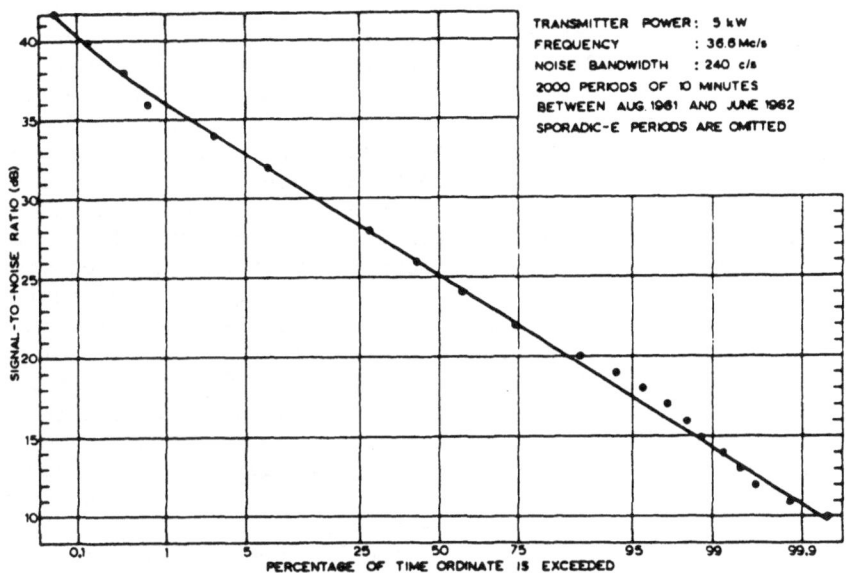

Figure 6.   Cumulative distribution of signal-to-noise ratio on the STC ionoscatter link.

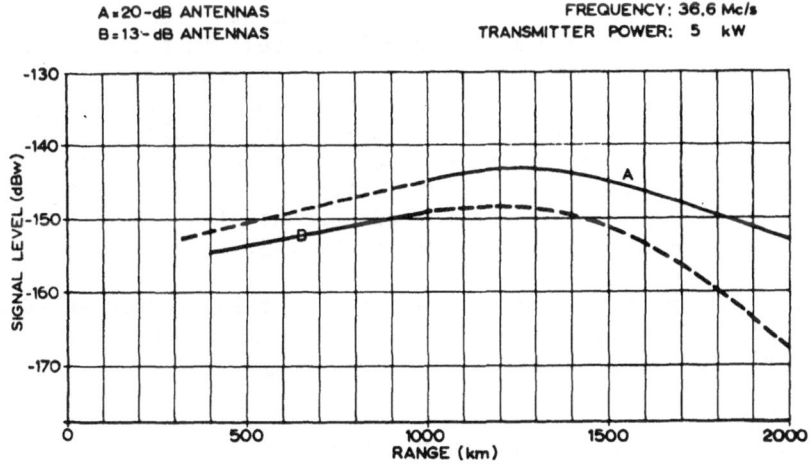

Figure 7.   Distance dependence of signal level on an ionoscatter path.

(b) <u>The meteor-burst system</u>. The meteor-burst system uses an ARQ
     technique which is in principle the same as in the ionoscat-
     ter system but certain refinements have been introduced to
     take care of phasing problems raised by the high rate of
     transmission (2000 bauds) and by the continuous changes in
     the length of propagation paths created by meteor reflec-
     tions. Unlike most other meteor-burst systems, the STC sys-
     tem uses a gating technique which is not based on an assess-
     ment of the SNR by reference to a preset threshold. With the
     adoption of an ARQ technique, the threshold concept has been
     abandoned and the only criterion used for controlling the
     flow of information is the conformity of the received signal
     to the code. Figure 8 shows a block diagram of the equipment
     contained in a terminal station. A complete description of
     the system is given in Ref. 2. The average capacity of the
     meteor-burst system is equivalent to one 50-baud channel when
     the duty cycle is 2.5% and varies in proportion with it. The
     results of 12 months of test, using 200 W of transmitter
     power, have shown that the hourly average capacity exceeded
     one channel in 90% of cases, the average over 24 hours being
     equivalent to about 3 channels. Figure 9 shows the values of
     the hourly capacity on 4 consecutive days in December 1965,
     during one of the quietest periods of the year as far as the
     meteor activity is concerned. Figure 10 shows the time de-
     lays measured in the transmission of a message of 150 charac-
     ters (25 words) for the same month and for the best and the
     worst times of the day. These delays include electrical
     transmission time and waiting times between bursts. In the
     early morning, the time delay was in most cases less than
     20 seconds and seldom exceeded 1 minute. In the late after-
     noon, it varied between zero and 3 or 4 minutes, the average
     being about one minute. On a continuous system working at 50
     bauds, the transmission of such a message would take 22.5 se-
     conds. Owing to the use of ARQ, the error rate is in 90% of
     cases less than 1 in 3000 characters. The peak worst value
     is about 1 in 1000. A detailed account of the performance of
     the system at 40 MHz is given in Ref. 3.

     The meteor-burst system is at present being tested at 40 and
     at 100 MHz to determine by comparative measurements the
     variation of performance with frequency.

     4. APPLICATIONS FOR COMMUNICATIONS IN THE ARCTIC

     Certain aspects of ionoscatter and meteor-burst propagation
will now be discussed briefly in relation to the particular case
of communications in the arctic.

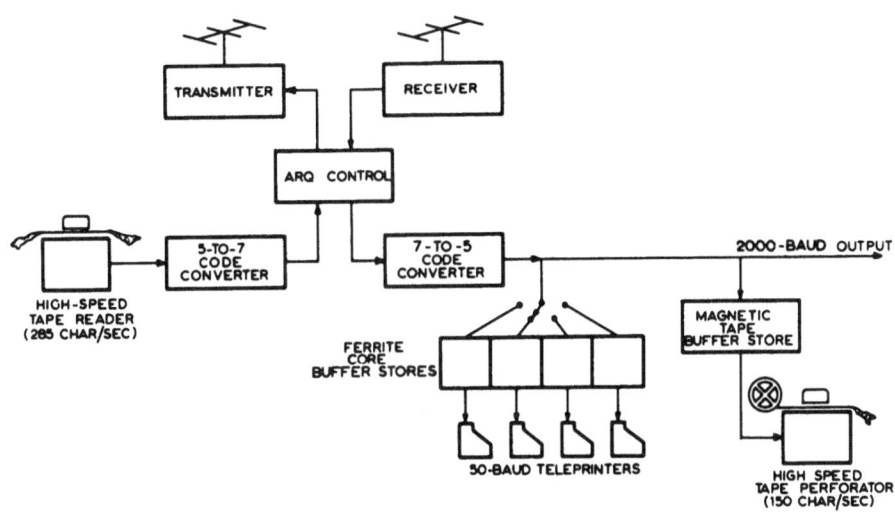

Figure 8. Block diagram of meteor-burst terminal equipment.

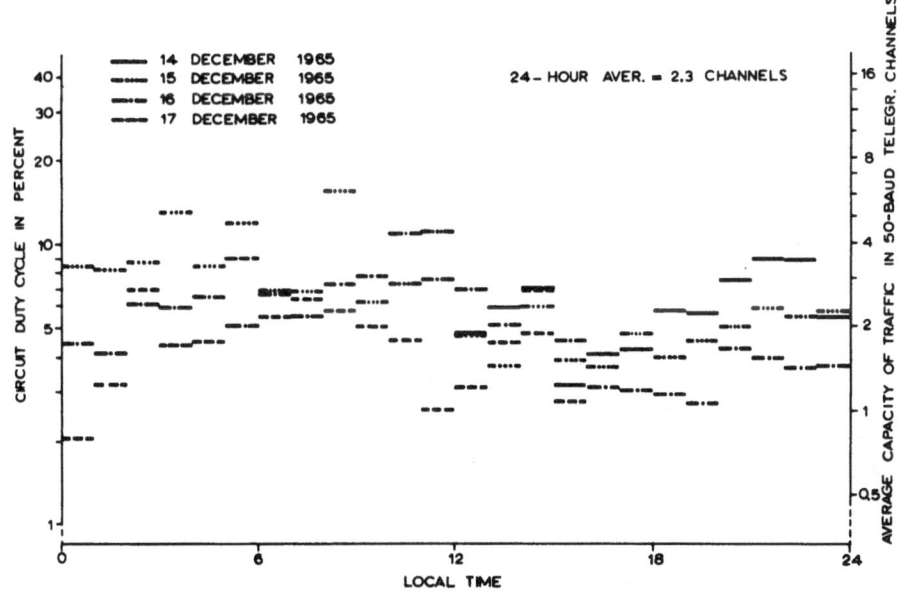

Figure 9. Diurnal variation of traffic capacity (December 1965).

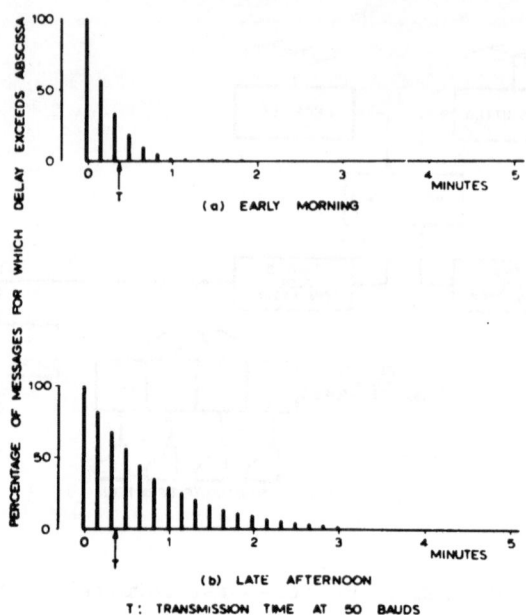

Figure 10.   Transmission delays for a 150-character message.

     The transmission loss on an ionoscatter path increases with
frequency while the level of the cosmic noise decreases. With
antennas of equal gain, the SNR varies as f^{-5} so that the useful
frequency range is practically limited to about 60 MHz. In the
case of meteor signals, both the peak amplitude of the burst and
their rate of decay are frequency dependent and the power requirements vary approximately as $f^{-2.5}$. The useful range of frequencies extends to 100 MHz at least.

     During ionospheric disturbances, meteor reflections and iono-
scatter signals may be attenuated, as are HF signals reflected by
the ionosphere, since most absorption takes place in the D region.
The absorption is, however, much smaller, its value in db being
roughly proportional to f^{-2}. Moreover the noise, which is of ex-
tra-terrestrial origin, is also attenuated and this helps to pre-
serve the SNR.

     In the case of ionoscatter, there is experimental evidence
that an increase in ionization also intensifies the scattering
process. The net result depends on the frequency used and on the
height at which abnormal ionization is created. In some cases,
absorption occurs below the scattering region and its effect pre-

dominates at frequencies below say 45 MHz whereas above that frequency a signal increase is observed (4). In other cases the signal is enhanced at all frequencies, especially at the lower end of the VHF band (5).

The effect of abnormal absorption on a meteor-burst link is to reduce the duty cycle. The weaker reflections are attenuated below the noise level and the stronger ones are reduced in duration. The way the duty cycle is affected by an increase in transmission loss is a function of frequency. At 30 MHz, an additional loss of 10 db reduces the duty cycle by a factor of 10 to 1, whereas at 100 MHz the reduction due to the same loss of 10 db may be of 3 or 4 to 1 only.

For both types of system the higher frequencies are least affected and they should be preferred when reliability is an essential requirement. The choice of an optimum frequency for a link in arctic latitudes would therefore be a matter of compromise between low power requirements and high percentage of reliability.

At 60 MHz, an ionoscatter system such as the STC system would probably achieve with 10 kW of power a reliability of 90% to 95% with an error rate of 1 in 3000 characters or better. The tests with the meteor-burst system at 100 MHz have not progressed sufficiently to enable firm conclusions to be drawn but it is already obvious that the power needed will be of the order of 5 kW and that a higher rate of transmission, e.g. 10,000 bauds instead of 2000, would be indicated.

5.   CONCLUSIONS

In arctic regions, ARQ intermittent systems using ionoscatter or meteor-scatter may be of particular interest in view of their relative ability to operate in spite of ionospheric disturbances. If propagation conditions degrade, the main result is a temporary reduction of the traffic capacity but the quality of the transmission channel remains unaffected. High-speed meteor-burst systems should be given special consideration since they are more economical in power and offer a higher probability of retaining their minimum performance in all circumstances.

6. REFERENCES

1. P.J. Bartholomé and I. Vogt, Ionoscatter Communications: New design concepts and experimental results, SHAPE Technical Centre Technical Report TR-53, April 1965.

2. P.J. Bartholomé, The STC Meteor-Burst System, SHAPE Technical Centre, Technical Memorandum TM-156, February 1967.

3. P.J. Bartholomé, Results obtained with the STC Meteor-Burst Communications System, SHAPE Technical Centre, Technical Memorandum TM-165 (to be published).

4. J.C. Blair, R.M. Davis and R.C. Kirby, Frequency dependence of D-Region scattering at VHF, Radio Sci. J. of Res. of the NBS/USNC-URSI 65D, No 5, 1961.

5. C. Collins and L.A. Maynard, Simultaneous VHF Riometer and Forward-Scatter Observations of the Disturbed Lower Ionosphere, These Proceedings.

SIMULTANEOUS VHF RIOMETER AND FORWARD-SCATTER OBSERVATIONS OF THE

DISTURBED LOWER IONOSPHERE

C. Collins and L. A. Maynard

Defence Research Telecommunications Establishment

Ottawa, Canada

Abstract: Continuous observations have been made over a two-year period of the propagation modes of 41 and 104 MHz signals transmitted over a 1400 km circuit close to the auroral zone in northeastern Canada. A 30 MHz riometer located on one side of the path with a broad-beam antenna directed towards the mid-point of the forward-scatter circuit was used for continuous measurement of absorption on the path. Illumination of the common volume by the two systems thus provided simultaneous observations of the ionization changes in the lower ionosphere during magnetically disturbed periods. It has been found that variations in the intensity of the signal propagated by the scattering regions near the path midpoint are strongly correlated over long periods with variations in the intensity of the cosmic noise as seen by the riometer. Temporal variations of the ratio of the median scatter-signal intensity to absorption over the two-year period provide qualitative information about the characteristics of the precipitating particle flux and the disturbed ionosphere below about 90 km.

1. INTRODUCTION

The nature and cause of the irregularities in the ionization of the lower ionosphere which are responsible for the reflection and scattering of very-high frequency radio waves in and near the auroral zone is perhaps one of the most intriguing problems in auroral physics, and although there have been numerous reports in the literature on various aspects of the problem, it is probably safe to say that we are still some way from achieving a full understanding of the physical processes involved in the phenomena. In

this paper we shall present the results of some measurements of
the intensities of 41 and 104 MHz signals obliquely propagated by
these ionic inhomogeneities which occur in the lower ionosphere
during magnetically-disturbed periods. The measurements were made
on a 1400 km circuit in north-eastern Canada close to the auroral
zone and they are part of an extensive study of meteoric and other
modes of VHF propagation.

For this part of the study a 30 MHz polar riometer was used
to illuminate the mid-point of the path in order to obtain simultaneous measurements of the cosmic noise absorption in or near the
scattering region. It is perhaps interesting to note that although
riometers have been in use for about 15 years and have contributed
greatly to our knowledge of auroral-zone phenomena there are few
accounts of their application in this way to the study of oblique-
ly-propagated signals. This is not due to any inherent difficulties in the use of the riometer but almost certainly to the mere
logistic problems involved in siting the equipment so that the absorption measurement can be made on the region responsible for
reflecting or scattering the oblique signal. Ekre and Landmark
(1) used an HF ionosonde to relate the occurrence of absorption at
the mid-point of a VHF forward-scatter path to the intensities of
the received signal, but the limited dynamic range of the ionosonde restricts its usefulness during disturbed conditions, and to
date, most of the studies relating absorption and oblique propagation have been of a statistical nature and have not been based on
simultaneous observations.

2. CIRCUIT DESCRIPTION

The location of the forward-scatter circuit and the riometer
are shown in Figure 1. Two CW transmitters operating on 41.3 and
104.1 MHz with powers of 140 and 1400 watts respectively were installed at Goose Bay, Labrador. The receiving terminal was at
DRTE, Ottawa. The antennas at both ends were 5-element Yagis
whose main lobes intersected at the mid-point of the path at a
height of 90 km. The 30 MHz riometer was located at Mont Joli on
the south shore of the St. Lawrence River. The antenna was a
crossed pair of 5-element Yagis pointing at the pole star. This
had the dual advantage of illuminating the mid-point of the scatter circuit and eliminating the diurnal variation in the cosmic
noise. Signal intensities at both Ottawa and Mont Joli were recorded logarithmically on conventional chart recorders. The
hatched portion shows the common area at 90 km illuminated by the
transmitting and receiving antennas. The double-hatched portion
has been drawn from the 3 db points of the riometer antenna. The
path mid-point geomagnetic latitude is ~61° which corresponds to
an L value of about 4.

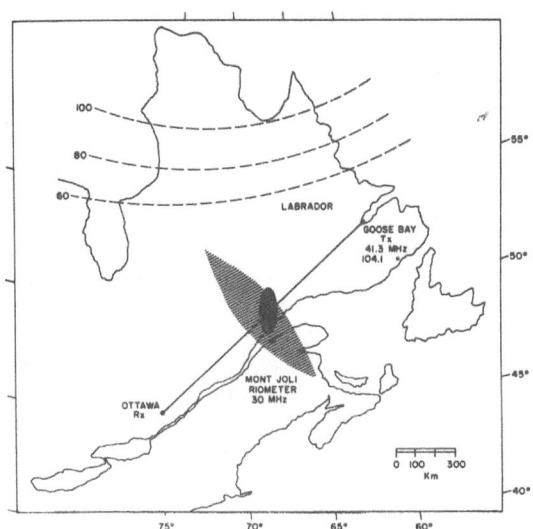

Figure 1. Location of forward-scatter circuit and the riometer.

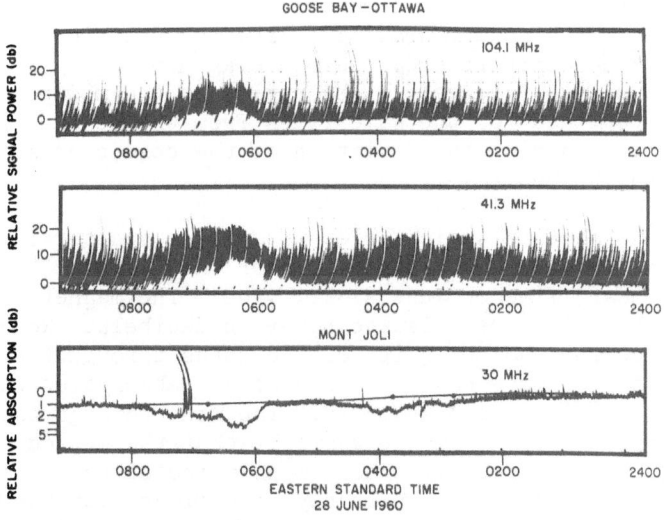

Figure 2. Sample of A3-type signal.

3. SIGNAL CHARACTERISTICS

VHF radio waves can be propagated over this circuit by at
least 5 different modes but in this paper we shall discuss only
one of these. This is the signal designated by Collins and Forsyth
as A3 (2). Considerable information on the statistical properties of this type of signal is available from work done at DRTE
and at the Universities of Saskatchewan and Western Ontario. It
is observed only during periods of magnetic disturbance. Angle-
of-arrival measurements show a spread of up to 20 degrees either
side of the great-cicle bearing. It is usually a long-enduring
signal with a slow fading rate (~3 cps) and of sufficient intensity to produce a distinctive trace on the records and partially obscure the ever-present meteor signals. It is believed that the A3
signal is scattered from irregularities near the centre of the
path at heights below 100 km. Figure 2 shows a typical recording
of this type of signal at 0200 to 0400 hours on 41.3 MHz and again
at 0600 to 0800 hours on both frequencies.

The equipment was operated continuously from April 1960 to
July 1962 during which time the A3 signal enhancements were observed 55 times for intervals varying from approximately 30 minutes to 6 hours. The signal was usually observed on both frequencies but was 10 to 15 db weaker at 104 MHz. The magnetic K index, as given by the Agincourt Observatory near Toronto, varied
from 3 to 6 for these periods. Since no integration was used in
the receiving and recording circuits, an "average" received-signal
intensity was obtained by drawing a line through the middle of the
A3 trace on the chart record. This average intensity was then
scaled at 5-minute intervals. Absorption at the mid-point of the
path was determined for the same times from the 30 MHz riometer at
Mont Joli by measuring the deviation of the cosmic noise signal
from the undisturbed level.

Figure 3 shows a typical plot of average scatter-signal intensity against cosmic noise absorption for the period 2345 to 0200
Eastern Standard Time on June 21-22, 1960. The magnetic K index
was 4. Both scales are relative power in decibels. No correction
has been made for the obliquity of the paths through the ionosphere and no estimate has been made of any absorption which the
scatter signal might have suffered. The line through the points
is only a visual best fit. It was felt that the method of scaling
the records did not justify a more sophisticated treatment. The
straightness of the line is probably significant but the most interesting feature of this plot, bearing in mind the duration of
the signal, is the small amount of scatter in the points.

Figure 4 shows three more examples of these plots. They are
similar to that of Figure 3 except that the slope of the line
through the points is different for each period. It should be

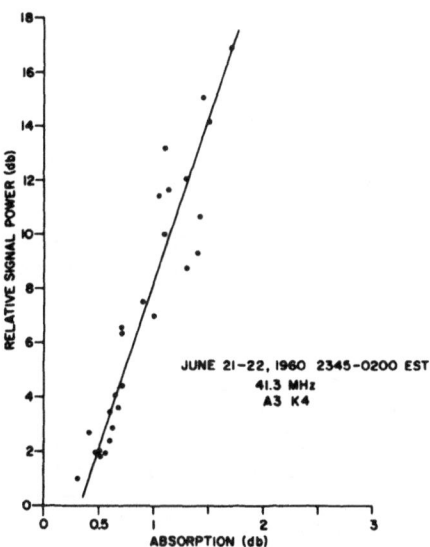

Figure 3. Average scatter-signal intensity vs cosmic noise
absorption.

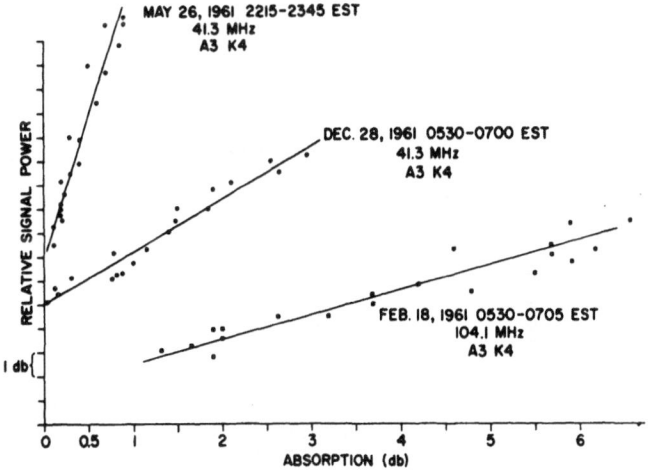

Figure 4. Scatter-signal intensities vs cosmic noise
absorption.

noted that their relative vertical position has no significance. On May 26, there was a relatively large variation in the scatter signal with very little change in the absorption while on February 18, these conditions were reversed. The latter case cannot be attributed to absorption since the 104 MHz transmission would experience only about one half the absorption measured by the riometer, assuming that it passed through the entire absorbing region. In some cases, it was found that a straight line did not fit the points and two examples of these are shown in Figure 5. However, there were very few of these. Most of the plots were similar to those shown in Figures 3 and 4 and in all cases the most striking feature is the remarkably small amount of scatter in the points.

The interpretation of these results is made somewhat more difficult because we do not have any direct measurements of the height of the region or regions involved, nor do we have any information about the radiation responsible for the ionization enhancement. However, there is some indirect information on both these points which permits us to make at least some qualitative comments on the foregoing plots. Extensive measurements have been made by several Canadian groups (2, 3) on VHF forward-scatter signals propagated over paths varying in length from 860 km to 1910 km. These studies showed that the A3 type of signal was not observed on the two longest paths (1910 and 1880 km) although it was frequently observed on shorter paths in the same general area. This absence of the signal on the long paths was almost certainly

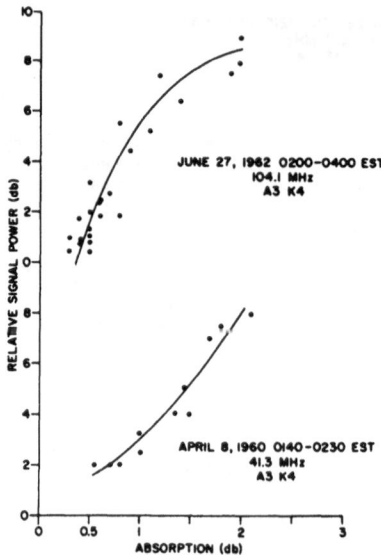

Figure 5. Measurements where curved lines fit observational points.

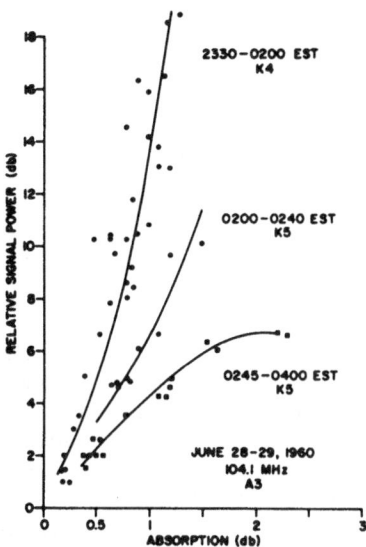

Figure 6. Scatter-signal power vs absorption for three consecu-
           tive periods during the same night.

due to the "horizon cut-off" of the antennas and indicates that
the A3 signal is scattered from a height less than 90 km. Although the height of the absorbing region cannot be determined
from these measurements, there is, in spite of the conflicting reports in the literature, considerable evidence to show that it too
is in the lower ionosphere and probably extends well below 90 km
(4).

     As for the radiation responsible for the electron density en-
hancements, it is reasonable to suppose, since we are considering
only magnetically disturbed periods and the propagation path was
close to the auroral zone, that the enhancements were due to
streams of precipitating electrons with energies greater than 40
keV. If this were so then the different slopes in the plots of
scatter-signal power vs absorption in Figures 3 and 4 could be
interpreted as being due to variations in the spectra of the incoming particles. This is seen more clearly in Figure 6 which
shows similar plots of scatter-signal power vs absorption for
three consecutive periods during the same night. It can be seen
that the slope of the plots decreases with time although once
again there is very little scatter in the points for each period.
This suggests that the spectrum remains constant for appreciable
periods then changes abruptly, becoming harder towards the morning hours, driving the absorbing-scattering region lower or, if

there are two regions, increasing the electron density in the lower absorbing region relative to the scattering region. Such an interpretation would be consistent with the observations reported by Johansen of NDRE (5) who made simultaneous measurements of auroral absorption and visible auroral emissions.

Perhaps the simplest conclusion that one might draw from the correlation of the variations in the scatter-signal power and the absorption is that both the scattering and the absorption occur in the same region, or that the absorbing region lies immediately below the scattering region. In either case, it is believed that the pertinent height range is 70 to 85 km. It is generally accepted that the non-deviative absorption measured by the riometer at 30 MHz is proportional to the electron density. However, one cannot easily make the same statement about the intensity of the scatter signal which is a function of the irregularities in the electron density. The theories of weak scattering which one would expect to be applicable here have been discussed extensively in the literature by Booker (6) and others. These have shown that the scatter-signal power is proportional to the mean square deviation of the electron density in the irregularities and to the size of the irregularities. Both of these parameters might be expected to change with variations in the intensity of the ionizing radiation although they will almost certainly be dependent upon the nature of the mechanisms producing the irregularities.

It is not the purpose of this paper to discuss the various mechanisms which have been proposed to account for the irregularities in the electron density but there is one other observation which can be made from the results presented here which is pertinent to any such discussion. The simultaneous occurrence of the absorption and the scatter-signal enhancement indicates that the irregularities are due either to structure in the incoming particle stream or to long-enduring inhomogeneities in the atmosphere. The first possibility seems rather remote since any structure in the particle flux would tend to be smeared out as a result of the numerous collisions which the incoming electrons make before they are thermalized in the D-region. The second possibility, that the inhomogeneities exist in the atmosphere, is much more tenable. There is ample evidence from the visual and radio observations of meteor trails for the existence of turbulence at these heights. Such turbulence would produce density variations in the neutral gas which might be "mirrored" in the ionization produced by the incoming particles or, irregularities in the velocities of the neutral particles might be transferred by collisions to the ionization and give rise to irregularities in the electron distribution.

Regardless of the mechanisms involved, if the neutral atmosphere plays a significant role in the scatter phenomenon, one

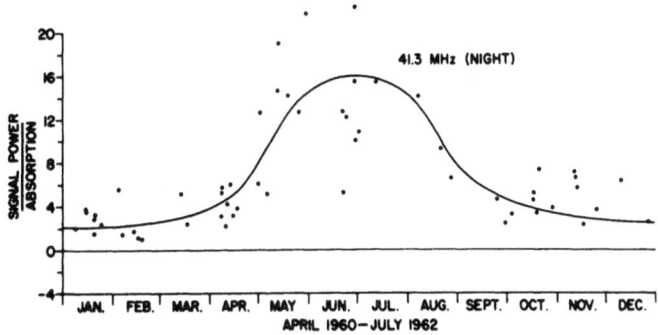

Figure 7. Daily plot of the ratio of scatter-signal intensity to
           absorption.

might expect to see some evidence of this in the data presented
here. With this possibility in mind, reference is made to Figure
7. This is a daily plot of the ratio of scatter-signal intensity
to absorption, which is just the slope of the other plots. The
solid line through the points is again only a visual aid and not a
best-fit curve. These measurements were made during the night on
41 MHz but the same kind of distribution was obtained for the measurements made during the day and night on both frequencies. It
is immediately obvious that this is quite unlike the seasonal
variation of most auroral phenomena believed to be caused by particle precipitation. There are no equinoctial maxima but there
is, in spite of the scatter, a significant increase in the ratio
during the summer months. This seasonal variation might be related to the magnetospheric changes responsible for the precipitation of the charged particles but, and we have no evidence as yet
to support this view, it seems more plausible to attribute this
result to seasonal changes in the neutral atmosphere which are reflected either in the amount of turbulence present or in a change
of height of the absorbing-scattering regions.

4. CONCLUSION

In conclusion, the qualitative picture of the lower ionosphere one gets from the strong correlation between absorption and
scatter-signal intensities is of a common or closely associated
absorbing-scattering region in the height regime of 70 to 85 km
where the electron density, enhanced by precipitating particles,
has irregularities produced in it, probably by turbulence in the
neutral gas; irregularities whose temporal variations may be attributed to changes both in the neutral atmosphere and the spectra
of the incoming particles.

We need hardly say that this study is not complete. It has raised several interesting questions and we hope to be able to pursue them further. We would like to examine the events which extend through the sunrise and sunset periods, the scatter signals during SID's and PCA when there can be no structure in the particle stream, and the properties of the neutral atmosphere for a proper understanding of the seasonal variation. At the moment, we are not sure whether we should be consulting the plasma physicists for more information about the instabilities in particle streams or the meteorologists regarding instabilities in the neutral atmosphere. We shall probably have to do both.

5. REFERENCES

1. H. Ekre and B. Landmark, Norwegian Defence Research Establishment Report No. 27, Pt. II, 1958.

2. C. Collins and P.A. Forsyth, Journal of Atmospheric and Terrestrial Physics, 13, 315, 1959.

3. F.D. Green, Patterns in the behaviour of VHF bistatic radio reflections, PhD Thesis, University of Saskatchewan, 1961.

4. B. Hultqvist, Planetary and Space Science, 12, 579, 1964.

5. O.E. Johansen, Planetary and Space Science, 13, 225, 1965.

6. H.G. Booker, Journal of Geophysical Research, 64, 2164, 1959.

METEOR BURST COMMUNICATIONS IN THE ARCTIC

L. A. Maynard

Defence Research Telecommunications Establishment

Ottawa, Canada

Abstract: Absorption and multipath propagation are important effects to be considered in operating meteor-burst systems at high
latitudes. The greater part of this lecture is devoted to a description of multi-path modes as observed on two Canadian test
circuits. Observed effects of an intense PCA event are discussed.

1. INTRODUCTION

Two obvious and pertinent propagation factors influencing
meteor burst communications at high latitudes are absorption and
multi-path propagation. In general, absorption results in a
lowering of the overall average field strength or alternately a
reduction in the system duty cycle of a meteor burst communication
system. On the other hand, multi-path propagation causes intersymbol interference and results in high error rates. Several
techniques have been developed which greatly improve system operation under multi-path conditions, including rapid frequency step-
ping and error detection and correction techniques. However, in
order to optimize system designs incorporating such techniques, it
is desirable to have a good description, at least in a statistical
sense, of the intensity, delay times, dispersion, and fading characteristics of the multi-path modes.

2. CIRCUIT LOCATION AND TEST PROGRAM

The results discussed in this paper are derived from experimental measurements over two Canadian test circuits between Goose
Bay Labrador and Ottawa, in Eastern Canada, and Edmonton and

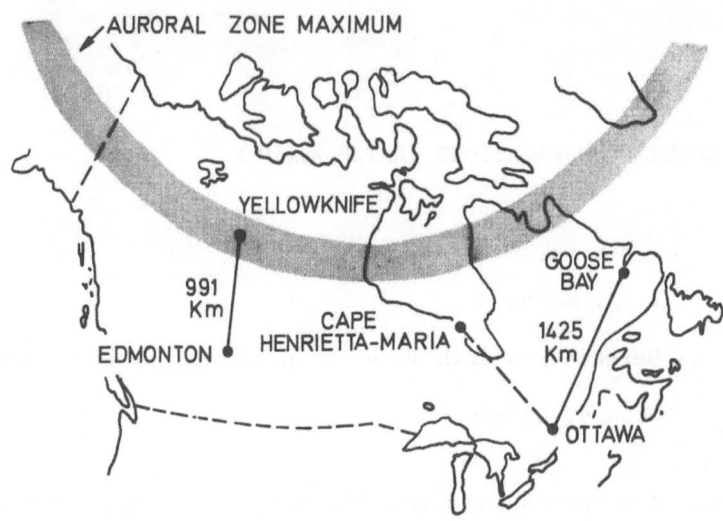

Figure 1. Location of test circuits.

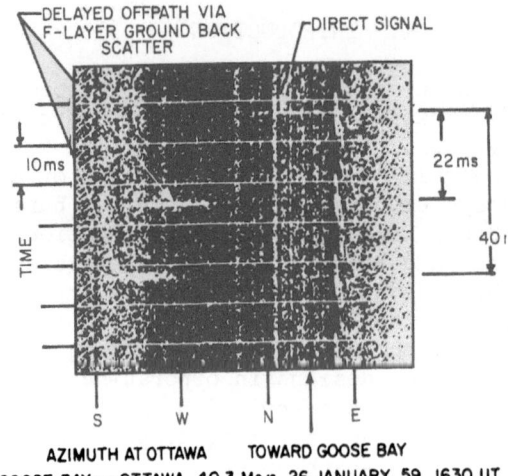

Figure 2. Records showing F_2-region supported ground-scatter
modes.

Yellowknife in West-Central Canada, with midpaths at about 61° and 65° Geomagnetic Latitude respectively. Most of these measurements were undertaken over the last sunspot number maximum in the years 1958 to 1961. Figure 1 shows the location of the circuits.

The facilities and tests described in this paper are as follows:

1. JANET B equipments operating between Edmonton and Yellowknife at frequencies near 40 MHz and near 50 MHz.

2. C.W. tests between Goose Bay and Ottawa at 41 and 104 MHz.

3. Pulse tests between Goose Bay and Ottawa at 40 MHz.

4. Polar Riometer measurements at mid path on the Goose Bay - Ottawa circuit.

The bistatic pulse system operating between Goose Bay and Ottawa had the facility to measure both the azimuthal direction of arrival and travel time of the pulse transmissions originating from the Goose Bay site. The receiver antenna system employed a continuously rotating Yagi.

3. SIGNAL BEHAVIOUR

The direct E-region scatter and meteor burst signals were used as a reference delay by which multi-path delays were measured. Mode delays ranging from 1 to 50 milli-seconds in excess of direct E-region modes were observed. Most delayed signal transmissions observed are attributed to three types of propagation. These included F_2 region supported ground back scatter, direct off-path auroral reflection and a combination of auroral reflection and subsequent F-region propagation. Figure 2 illustrates typical records showing the presence of F_2-region supported ground-back-scatter modes. Delay time marker lines running horizontally are separated by 10 milli-seconds in time with delay time increasing downwards. The abscissa on the slide represents direction. The transmitter located at Goose Bay, is in a northeasterly direction. The direct signal propagated by meteor bursts is seen arriving from the Goose Bay direction near the top of the figure. Another signal delayed by 22 milli-seconds is noted arriving from the west. This delayed signal is attributed to F_2 supported ground back scatter. In this figure, a second hop ground back scatter mode delayed by 40 milli-seconds is also present. The two hop mode is not often observed, however. Figure 3 illustrates another commonly observed delayed mode. Again the direct E-region signal is observed arriving from the transmitter direction. Another signal arriving 2 milli-seconds later is seen arriving from a northly direction. The delayed mode shown here is most often observed at

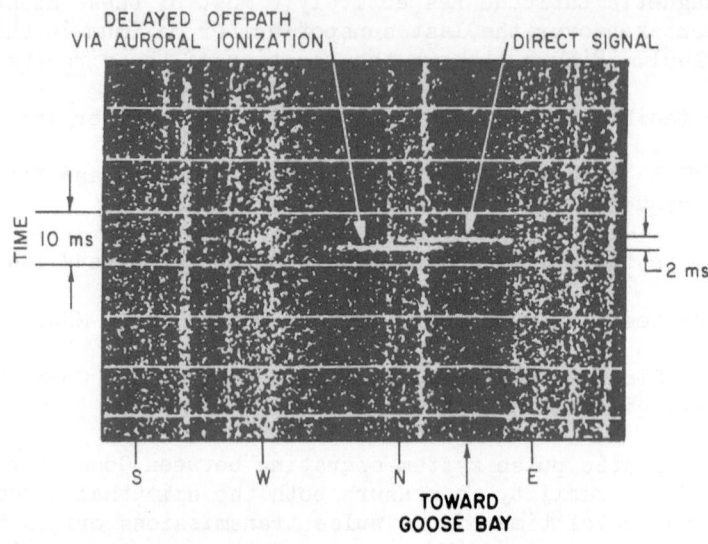

Figure 3. Record with off-path auroral reflection.

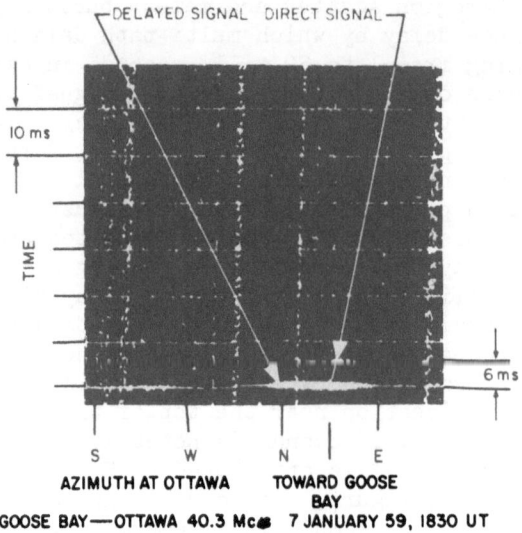

Figure 4. Example of delayed signal possibly associated with F_2-
region propagation and auroral ionization.

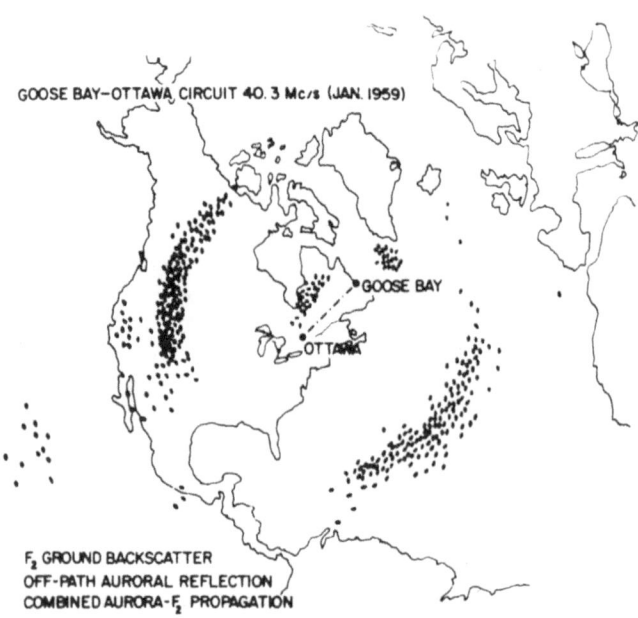

Figure 5.   Location of echoing regions for the month of
            January 1959.

night and is attributed to off-path auroral reflections. Figure
4 illustrates the third class of delayed signals which have been
observed. The delayed mode is seen to arrive from the transmitter
direction in this case, but with a delay some 6 milli-seconds
greater than the direct E-region mode. This delayed signal seems
to be associated with both F_2-region propagation and auroral ionization. Figure 5 is a plot of echoing regions for the month of
January 1959. Each dot shown here represents 30 minutes of signal.
The dot positions represent the apparent center of the scattering
region as determined by the travel time and direction of arrival
of the pulse transmissions originating from Goose Bay, Labrador.
Areas remote from the transmission path represent regions of reflection of F_2 layer supported back scatter. Reflections associated with dots to the north of the path are attributed to off-path
reflection by radio aurora. Most of these echoes occur during the
night and show strong correlation with magnetic activity. The
area shown to the northeast of Goose Bay represents reflection region points which have been associated with combined field-aligned
ionization and F_2 region forward propagation. A closer examination of the echoing pattern of F_2 ground back scatter shows that
the scatter regions possess a certain degree of symmetry about the
transmission path. In general, it may be seen that scatter re-

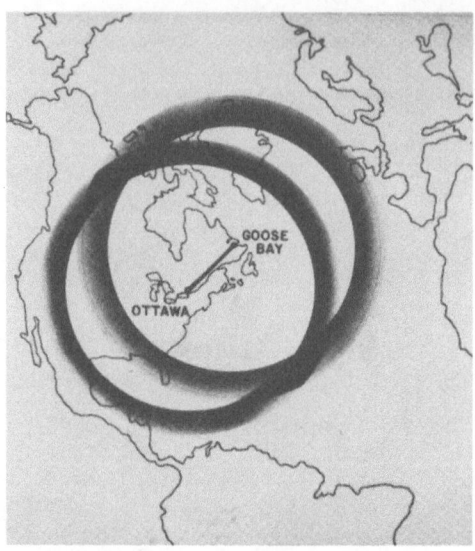

Figure 6.   Display of areas illuminated by antennas at Goose Bay
            and Ottawa.

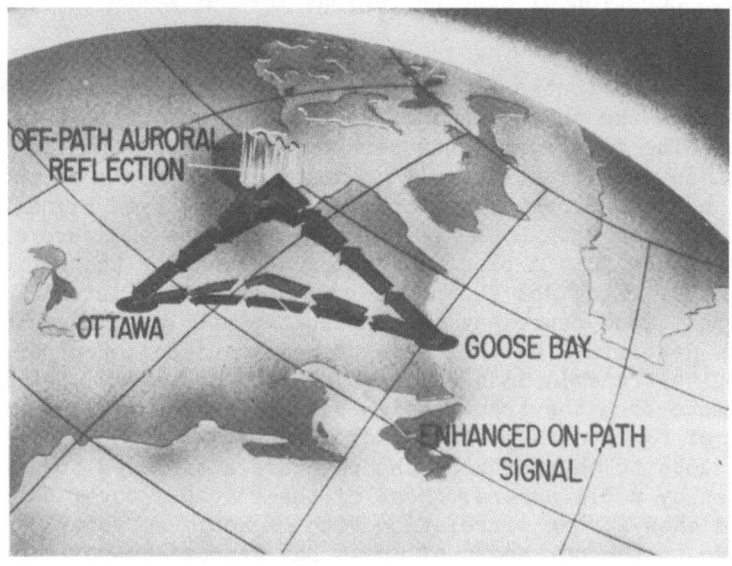

Figure 7.   Pictorial view of off-path auroral reflection.

gions are located primarily off to the sides of the transmission
path and disticnt nulls in the ground scatter region exist off the
ends of the path. This echo pattern is seen to be slightly offset toward the south. A possible explanation as to why the regions to the sides of the transmission paths seem to be favourable
for this delayed mode is illustrated in figure 6. Areas illuminated by a signal of F_2 by the antenna at Goose Bay are shown in the
ring symmetrical about Goose Bay. At the same time, the area
symmetrical about Ottawa represents the area of sensitivity of the
Ottawa antenna for a one-hop F_2 propagation mode. These two regions coincide in directions to the sides of the transmission path
but towards the end of the path the receiving antenna is unable to
receive energy from the same area that the transmitting antenna is
illuminating. It might be pointed out here that this effect
should be observed only on relatively long paths. Over shorter
paths, illuminated areas behind each terminal would not differ in
location enough to make the effect observable and ground-back-
scatter should be observable in all directions. F_2 region back
scatter was observed as much as 90% of the mid-day hours in winter
months during the sunspot maximum period. Figure 7 represents a
pictorial view of off-path auroral reflection. For radio energy
originating at Goose Bay to be reflected from field-aligned auroral ionization in this region ray paths come within about 5 degrees of the specular reflection condition assuming field aligned
ionization. Invariably, a strong direct path signal accompanied
any off-path auroral signal observed during these times. The direct path signal was generally greater in amplitude than the off-
path auroral signal associated with it. The dots northeast of
Goose Bay in the preceding slide represented reflection regions
attributed to a combined auroral reflection and F_2 supported propagation mode. This mode manifests itself as a pulse arriving
from the transmitter direction approximately 6 ms later than a
direct meteor burst signal. Figure 8 shows the proposed ray path
for this type of propagation. Here pulses originating at Goose
Bay travel to field aligned ionization where they are reflected
forward and travel to Ottawa via F_2 region propagation. This mode
is observed only at mid-day and in most cases no direct E-region
activity was observed other than the meteor burst mode. Propagation modes involving auroral ionization were observed for 1515
hours at 41 MHz in a test period of 18 months between May 1960 and
October 1961 over the Goose Bay to Ottawa path. Over this same
path at 104 MHz auroral supported multi-path modes were observed
for 85 hours in the same test period. The rapid decrease in occurrence with increasing frequency, however, is offset by the ac-
companying rapid decrease in system duty cycle at the higher operating frequencies.

During times when multi-path conditions existed, high error
rates were often observed on the JANET B test circuit between
Edmonton and Yellowknife. Error rates approached 5 to 10% during

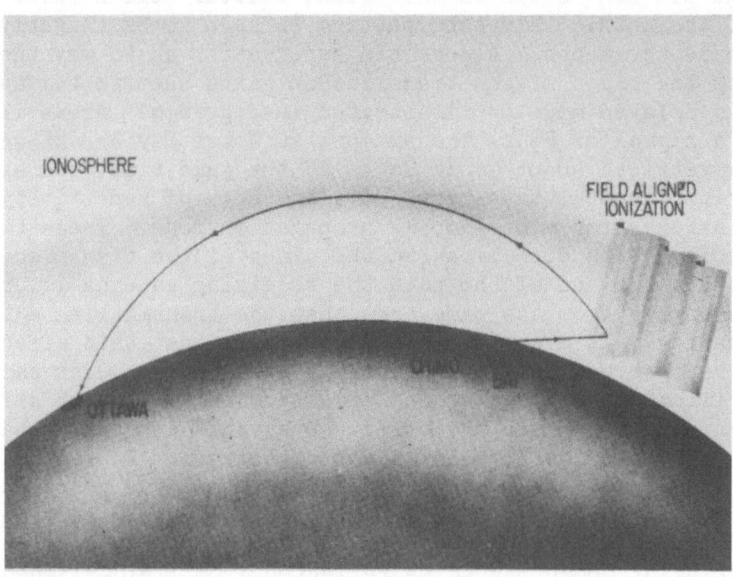

Figure 8. Proposed propagation for off-path signal.

times of high geomagnetic activity. Furthermore, error rates up
to 5% were observed during times that F_2 ground back scatter oc-
cured. In addition to the problem of multi-path modes which frequently occur on high latitude scatter circuits, both polar cap
and auroral absorption also has profound effects on the reliability of communications systems operating at high latitudes. Polar
cap absorption caused an extended series of blackouts during the
initial Edmonton to Yellowknife JANET B system tests in July 1958.

     Figure 9 illustrates the effect of an intense PCA event on
meteor burst propagation at 41 and 104 MHz. The upper portion of
the frame shows 30 MHz riometer absorption as a function of time
for mid-path on the Goose Bay to Ottawa Ontario path. The time
shown in this slide is universal time. The center portion of the
slide shows a plot of 104.1 MHz meteor signal duty cycle over this
path. The shaded region on the graph represents the normal limits
of variation in duty cycle due to the diurnal variation in meteoric activity and to the random nature of the meteoric signal.
Signal duty cycle is seen to drop below its normal range for three
intervals during the four days plotted. These periods of signal
blackout coincide fairly well with peaks in absorption as measured
by the riometer at mid-path. These periods are the only times in
the test period of over 18 months that the 104 MHz meteor burst
signals have been significantly affected by absorption. Gaps in

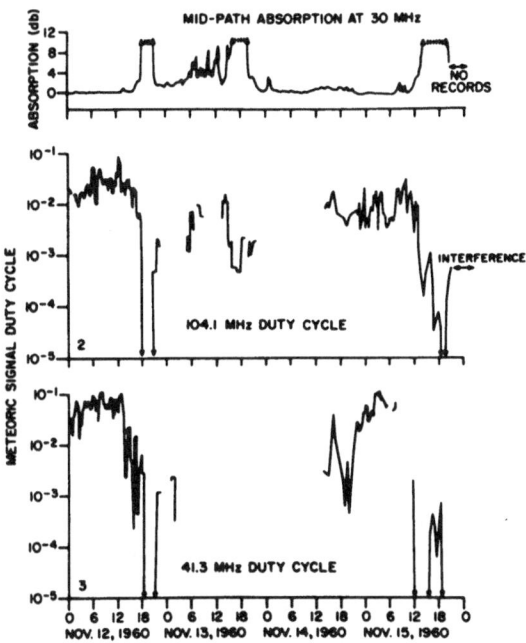

Figure 9. Effects of intense PCA event on meteor burst propaga-
           tion at 41 and 104 MHz.

the duty cycle plot in this figure and the lower curve are due
mainly to cluttering of the meteoric signal by auroral type signals. The 41 MHz signal duty cycle is shown in the lower portion
of the figure and is affected to a greater extent by the absorption. Significant changes in duty cycle occur as early as 1400 UT
on November 12, about 3/4 hour after the solar flair at 1325 UT.
During the period shown, the 104 MHz duty cycle deviated by a
ratio greater than 1000 to 1 from normal causing essentially a
meteoric signal blackout at times. It can be seen then that even
at frequencies in excess of 100 MHz, polar cap absorption can have
an extreme effect on the reliable operation of scatter circuits
operation at high latitudes.

     In a statistical study over the previous sunspot maximum pe-
riod, absorption caused effective signal blackouts at 41 MHz for a
total of 204 hours during the previous 18 month test period discussed above. At 104 MHz, absorption of intensity sufficient to
cause signal blackout, occurred for a period of 7 hours in the
same operational period.

VHF BISTATIC AURORAL BACKSCATTER COMMUNICATION

G. Lange-Hesse

Max-Planck-Institute for Aeronomy, Institute for

Ionospheric Physics, Lindau, West Germany

Abstract: In the first sections a short review is given about the
basic characteristics of the VHF-backscatter-phenomenon: Aspect
sensitivity control by the geomagnetic field, back-scatter mechanism, and wavelength dependence. Then the geometry of the propagation path of VHF bistatic aurora backscatter communication is
computed according to a theoretical model and compared with experimental observations. The computed curves in space along which
the backscattering centres must be located (backscatter curves)
coincide at least at one point with the simultaneous visual aurora
display responsible for the observed backscatter communication.
The computed maximum-possible-distance-curves for this kind of
communication are in reasonably good agreement with the observations. Observed deviations from these curves at greater distances
can be explained mainly by magnetic dip angle variations during
geomagnetic storms. A short review about time variations (diurnal
and seasonal) of bistatic auroral backscatter communication, as
well as the control of the frequency of occurrence by the geomagnetic activity in different latitudes, is given in another section. In a final section an explanation of some of the features
of the auroral backscatter phenomenon is given using the new plasma acoustic wave theory.

1. INTRODUCTION

The fact that aurora influences radio wave propagation was
known before World War II. Reflections of radio waves from ionization associated with aurora - called: radio aurora - were first
investigated in 1938 by Harang and Stoffregen at Tromsø (1940) in
the lower VHF range. At about the same time amateur radio opera-

tors in North America had discovered that VHF radio wave propagation via auroral ionization was possible (Tilton, 1944; Moore,
1951). Ionosonde echoes in the HF range associated with aurora
were reported at Tromsø/Norway in 1933 by Appleton et al (1937).
After World War II and particularly during the last fifteen years
radio aurora has been studied by a number of researchers in North
America, the U.K., in Norway, Sweden, Finland, Germany, Russia,
New Zealand, and the Antarctic.

Most studies of radio aurora[*] have been made by means of radar, i.e. with transmitter and receiver at the same place. Several summaries of the results of these radar auroral experiments
have been published by numerous authors e.g. Little et al (1956),
Lange-Hesse (1957), Birfeld (1960), Peterson (1960), Bagaryatsky
(1960), Booker (1960), Chamberlain (1961), Hultqvist et al (1964),
and Leadabrand (1965a).

2. ASPECT SENSITIVITY CONTROL BY THE GEOMAGNETIC FIELD

The reduction of VHF radar echoes from aurora has shown that
these echoes can only be obtained from a very restricted strip of
sky corresponding to the region where the line of sight from the
radar location to the aurora intersects the local geomagnetic
field lines at right angles. Figure 1 shows the geometry of the
VHF auroral backscatter problem. On the condition that perpendi-
cularity is necessary, auroral displays at points A and B in Figure 1 used to give backscatter-echoes at the radar site, but not
at the points C and D. Perfect 90° intersection, however, is not
required, because of the finite length of the auroral scatters.
Depending upon the frequency, peak power and sensitivity of the
radar, echoes can be obtained at intersection angles that differ
from 90° by as much as 10°. The deviation of the intersection
angle from 90° is called the off-perpendicular angle.

In many of the cases the observed deviation from normal incidence is not a true off-perpendicular angle, since the geomagnetic
field lines can change their orientation during a magnetic storm
which accompanies aurora displays especially at lower latitudes.
Størmer (1926) describes an analysis of the movement of the radi-
ant point of coronal forms. This point was traced through a movement of as much as three to four degrees in inclination during the
course of the great auroral display of March 22-23, 1920. The auroral rays forming the corona are aligned with the geomagnetic

---

[*] In this paper the following expressions are mostly used as
synonyms without alluding to different physical mechanisms:
radio aurora, auroral echoes, auroral backscatter, and
auroral reflections.

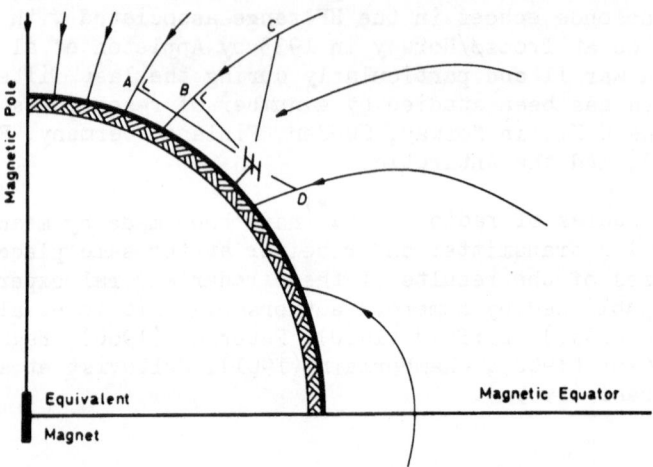

Figure 1. VHF radio wave auroral backscatters are only possible
           when the direction of radio-wave-propagation and the
           direction of the lines of force of the earth's magnetic
           field are perpendicular at the reflection point. Under
           this assumption auroral displays at points A and B used
           to give backscatter-echoes, but not displays at points
           C and D.

field, and their motion reflects a similar distortion of the local
magnetic field, probably at 200-300 km heights. As the disturbance current system during a geomagnetic storm is usually considered
to flow in the E-layer, one would expect any field distortions to
be as great or greater there than those occurring at more considerable heights. This means that in many of the cases the true
off-perpendicular angle is not the observed one for an undistorted
geomagnetic field especially during strong geomagnetic storms
(Kp = 8 and 9).

3. BACKSCATTER MECHANISM

From the numerous models which have been proposed for the
backscattering of VHF radio waves from aurora, the theory of auroral radar scattering as developed by Booker (1956), based upon
the observed wavelength dependence and aspect sensitivity (controlled by the geomagnetic field geometry), appears to give the most
satisfactory explanation of the observations, except for occasions
in the HF and lower VHF range when auroral ionization becomes over-
dense. The Booker theory involves the concept of scattering by
nonisotropic irregularities in electron densities. These irregula-

rities provide discontinuities in the dielectric constant which
are short compared to the radar wavelength and which scatter back
only a small fraction of the incident energy towards the radar
(reflection coefficient ≈ 10^{-6}).

To produce the observed aspect sensitivity these irregularities must be in the form of columns of ionization with their long
axes parallel to the earth's magnetic field. The length of the
irregularities then determines the scattering aspect sensitivity
of auroral reflections. The discontinuity of electron density and
the dimension of the irregularity transverse to the earth's magnetic field determines the wavelength dependence of the auroral
backscatter echoes.

4. WAVELENGTH DEPENDENCE

The wavelength dependence of the auroral echoes is well pronounced. The echo power decreases strongly with increasing frequency. Wavelength dependence together with aspect sensitivity
have been measured by a number of workers, e.g. Presnell et al
(1959), Blevis et al (1963), Flood (1960), Stone et al (1959).
The results obtained by Leadabrand (1962) and Leadabrand et al
(1965) with elaborate auroral equipment located at Fraserburgh,
Scotland, however, could be interpreted most easily, primarily
because of the narrow beam involved. The radar was specifically
designed for this purpose having identical beamwidths (1.2°) at
400 and 800 MHz. Thus the identical volume of aurora was illuminated simultaneously at both frequencies. Wave length dependence
observations carried out with this equipment resulted in a power
law dependence of λ^{7} and aspect sensitivity observations in an
energy decrease of 10 db/per degree off-perpendicular. The interpretation of these results according to the booker scattering
theory indicates that the size of the backscattering irregularities is 45 to 90 m along the magnetic field lines and 0.7 metres
across the field lines.

5. BISTATIC AURORA BACKSCATTER COMMUNICATION

As well as by radar studies, radio aurora have also been
studied by investigating oblique - or bistatic - auroral reflections, i.e. with transmitter and receiver at different places.
Observations of this kind were carried out in northern Scandinavia
by recording the transmissions from several FM broadcast stations
between 87 and 100 MHz at one place in case of aurora backscatter
(Egeland et al, 1961; Egeland, 1962a, b; Oksman 1964, 1966a, b).
Observations of bistatic radio reflections in the lower VHF range
were carried out in Canada (Collins et al, 1959, Green, 1961) and
in USA (Dyce, 1955b).

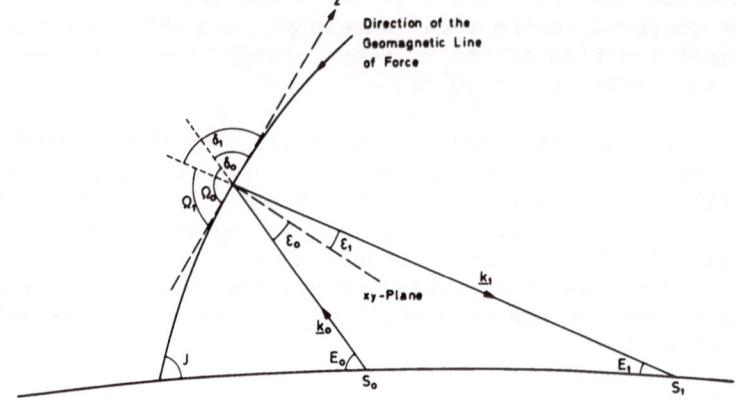

Figure 2.   Cross-sectional view of the earth with the geometry of
            the propagation path in the vertical direction for VHF
            bistatic auroral backscatter propagation between the
            two points So and S_1
            $\underline{k}_o$, $\underline{k}_1$ vector of the wave normal of the incident and
                     backscattered wave, resp.,
            ε_o, ε_1 angle between k_o and k_1, resp. and the xy-plane.
            J magnetic dip angle.
            Ω_o, Ω_1 propagation angle = angle between the direction
                     of radio wave propagation and the magnetic li-
                     nes of force.
            E_o, E_1 elevation angle above the horizontal.

        Investigations of VHF bistatic auroral reflections have also
been made by means of a geographically extended net of amateur
radio stations. Observations of this kind were first carried out
in North America during the years before the IGY (Moore, 1951;
Dyce, 1955a; Gerson, 1955 a, b). During the IGY and later years
some observations were carried out in Europe especially in the
U.K. (e.g. Stone, 1960; 1965; Smith-Rose, 1960; Newton, 1966) and
Germany (Lange-Hesse, 1962, 1963a, b; 1964a, b; Lange-Hesse et al,
1965, 1966).

        The Booker scattering theory as described before has been
worked out in considerable detail for the radar backscatter case.
This theory has been extended to oblique - or bistatic - auroral
backscatter propagation by Egeland (1962b). Figure 2 shows a representation of the geometry of the propagation path in the vertical direction in the case of bistatic auroral backscatter communication between the points S_o and S_1, the geometry of the propagation path in the horizontal direction is shown in Figure 4. The
z-axis of a three-dimensional xyz-coordinate system in Figure 2

is tangential to the geomagnetic line of force at the point of the backscattering centre or irregularity. The xy-plane is perpendicular to the geomagnetic line of force; $\bar{k}_o$ and $\bar{k}_1$ are the vectors of the wave normal of the incident and backscattered wave, resp.; Ω_o and Ω_1 are the "propagation angles" = the angles between the direction of radio wave propagation and the magnetic lines of force; ε_o and ε_1 are the angles between $\bar{k}_o$ and $\bar{k}_1$, resp. and the xy-plane. According to the theory (Egeland, 1962b) optimum conditions are given for the possibility of VHF aurora backscatter propagation, if the relation $\cos \Omega_o + \cos \Omega_1 = 0$ is fulfilled at the point of the backscattering centre. This means that the vectors $\bar{k}_o$ and $\bar{k}_1$ describe the same angle with the xy-plane, in this case is $\varepsilon_o = \varepsilon_1$ and the vector $(\bar{k}_o - \bar{k}_1)$ is perpendicular to the xy-plane. Optimum conditions of that kind will be referred to as "ideal backscatter conditions".

If the angles ε_o, ε_1 or Ω_o, Ω_1 deviate from the ideal backscatter conditions by only a few degrees (similar to the radar off-perpendicular backscatter case) bistatic backscatter propagation in principle is possible according to the theory. The backscattered power, however, descreases very rapidly with increasing angle deviation from the ideal conditions. One therefore needs very strong transmitters and antennas, with high gain, in order to establish communication, if the angles mentioned before deviate by only a few degrees from the ideal backscatter conditions especially in the UHF and higher VHF range. For particulars see the curves in the papers Egeland (1962b) pp. 198 to 201, and Czechowsky (1966) Figure 14 and A 11.

Azimuth, range, and intensity of the aurora backscattered radiation can be recorded by the pulse radar method. The limitations are that scattering in other directions than directly backwards to the radar location cannot be studied and that the pulse signal is not suitable for investigating the changes of fine structure in the reflection process. This, however, is possible by investigating oblique - or bistatic - auroral backscatter signals transmitted from FM broadcast stations or beacon transmitters.

An aurorally-propagated VHF signal has a characteristic growl or hiss due to a fast fading that is at an audio rate up to several hundred cycles per second. As the carrier frequency is increased to higher VHF frequencies, the growl increases in pitch. Amplitude modulated phone signals are badly garbled although relatively slow CW-telegraphy can get through without difficulty. Unlike E- or F-layer propagation, strongest signals are usually obtained when both stations point their directional antennas northward toward the aurora, regardless of the actual great-circle bearing between the stations. During especially strong auroras the signals may appear to come from a variety of directions spread about north.

Bistatic auroral backscatter propagation is potentially useful for communication purposes, but is of even more importance because of the interference it may cause in communication circuits.
Because the aurora is capable of backscattering over long distances, care must be taken in the assignment of operating frequencies
to avoid interference between transmitters even though they are
widely separated in the conventional sense. This is of great importance in the VHF range where a transmitter power of the order
of 100 watts and low gain directional antennas are sufficient to
obtain auroral backscattered long distance communications up to
1000 km and more. This fact is well proved by the extended observations of radio amateurs, who use the backscatter feature of
the aurora in order to carry out bistatic auroral backscatter communications between two stations in the 144 and 50 MHz amateur
band (a comprehensive list of bistatic auroral backscatter communications carried out by radio amateurs in the 144 MHz-band in
Middle Europe from 1957 to 1962 is published by Lange-Hesse,
1963b). The possibility of interference by auroral backscatter in
the UHF range between transmitters with powers of the order of
100 - 1000 watts and widely separated in the conventional sense is
very low. As mentioned in section 4 the backscattered power decreases strongly with increasing frequency with a power law dependence of λ^7. Therefore, one can neglect in a first approximation
the auroral backscatter in this frequency range for low power
transmitters. This also is confirmed, as far as it is known to
the author, by the fact that no auroral backscatter communication
was reported by radio amateurs in the 435 MHz amateur band. UHF
military radars, however, having high transmitter power and high
antenna gain can be expected to show auroral echoes on their
screens even at locations far from the auroral zone. In these
cases targets will need to be detected among these auroral "clut-
ter" echoes.

6. COMPUTATION OF THE LOCATION OF THE BACKSCATTERING CENTRES IN
    CASE OF BISTATIC AURORAL BACKSCATTER PROPAGATION

On the map of Europe in Figure 3 the solid curves at the left
represent the location where the line of sight from London in different directions intersects the geomagnetic lines of force at
constant angles at the height of 110 km. This is about the mean
observed height of the backscattering centres (see e.g. Unwin
1958, Unwin 1959a, Barber et al, 1962, Leadabrand et al, 1965).
In Figure 3 the curves are shown for an angle of intersection of
88°, 90° and 92°. The curves were computed with the help of an
electronic computer using magnetic dip angle and declination from
the ground (description of the method see Millman, 1959 and Egeland, 1962b). The calculation of the curves was restricted to
elevation angles E_0, E_1 (Figure 2) greater than or equal to zero
for the line of sight to 110 km height. The dotted curves to the

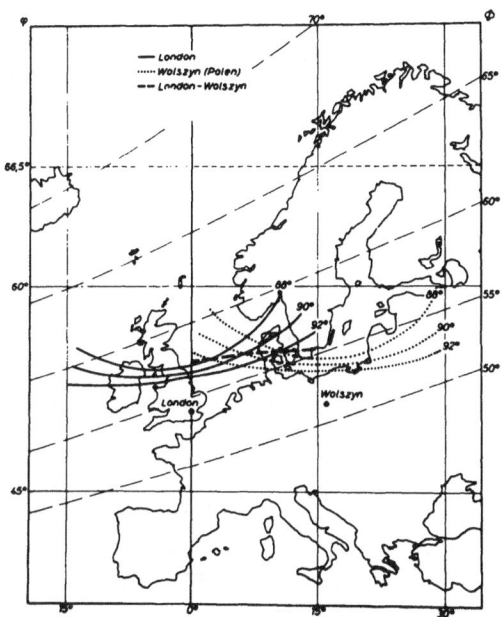

Figure 3. Curves of constant intersection angle between the direction of radiowave-propagation and the magnetic lines
of force at the 110 km height level for the two points
London (___) and Wolszyn (Poland) (......). The
dashed line represents the location at the 110 km
height level where the "ideal backscatter conditions"
$\cos \Omega_0 + \cos \Omega_1 = 0$, or $\varepsilon_1 = \varepsilon_2$ (Figure 2) are fulfilled for bistatic auroral backscatter communications
from London to Wolszyn; ϕ = geographic latitude, Φ =
geomagnetic latitude. After Lange-Hesse et al (1965).

right in Figure 3 show the same as the solid curves to the left
but referred to Wolszyn (Poland). It can be seen in the figure
that the curves intersect each other. For the two locations,
London and Wolszyn, the ideal backscatter condition $\cos \Omega_0 + \Omega_1 =$
0 or $\varepsilon_0 = \varepsilon_1$ (Figure 2) is fulfilled along the dashed line in Figure 3 for elevation angles E_0, E_1 of the wave-normal with the
ground greater than or equal to zero. As one can see in Figure 3
the dashed line connects the points of intersection of a) the 88°
- London curve with the 92° - Wolszyn curve, b) the 92° - London
curve with the 88° - Wolszyn curve, and c) the 90° - London and
Wolszyn curves.

The locations where the ideal backscatter conditions (as
shown in Figure 3) are fulfilled at 110 km height are shown in Fi-

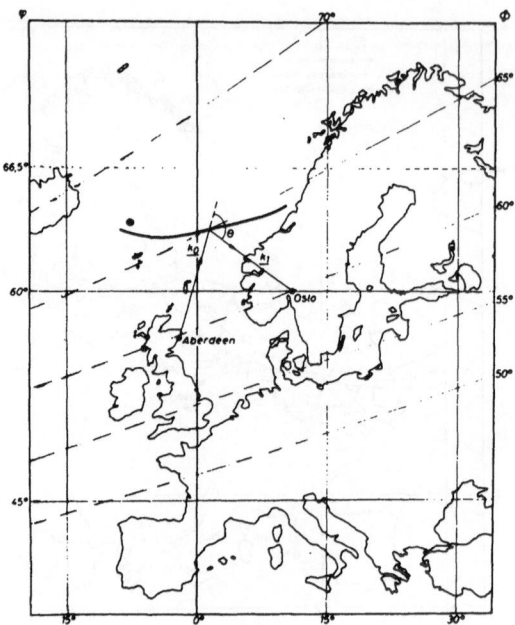

Figure 4. Geometry in the horizontal direction for VHF bistatic
           auroral backscatter communications between the two
           points Aberdeen and Oslo. The solid curve at about
           $\phi \approx 65^{\circ}$ (called "backscatter curve") represents the lo-
           cation at the 110 km height level where the ideal back-
           scatter conditions (Figure 3) are fulfilled, this means
           where aurora (and simultaneously aurora ionization)
           must be located in case of the occurrence of aurora
           backscatter communication between Aberdeen and Oslo;
           Θ angle between the wave normals of the incident and
           backscattered wave, ϕ geographic latitude, Φ geomagne-
           tic latitude. After Lange-Hesse et al (1965).

gure 4 for the two points Oslo and Aberdeen. If auroral backscatter communication occurs between these points an aurora (and simultaneously auroral ionization) must occur at the same time at
one point along the solid curve in the figure. The solid curve in
Fig. 4 leads from the geomagnetic latitude $\phi \approx 68^{\circ}$ in the west to
$\phi \approx 64^{\circ}$ in the east close to the auroral zone ($\phi \approx 67^{\circ}$). This
means that auroral backscatter communication occurs relatively
frequently between these points.

    The backscattered power also depends on the angle Θ between
the two wave normals $\overline{k_0}$, $\overline{k_1}$ of the incident and backscattered wave
(Figure 4). Θ can vary between $\Theta = 180^{\circ}$ (radar case, the two

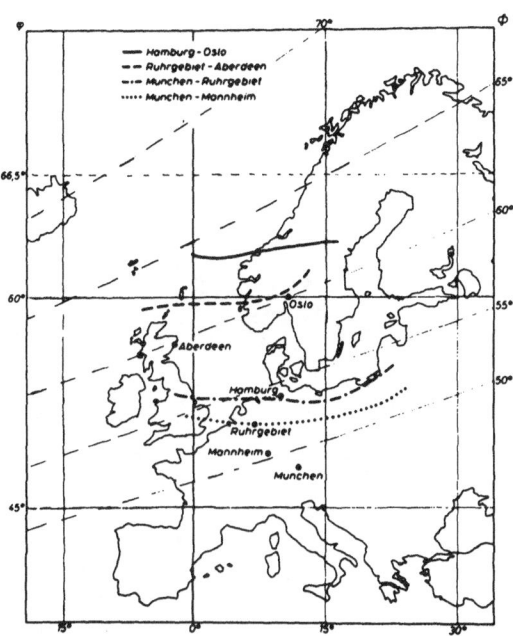

Figure 5. Map of Europe with the location of the backscatter
curves (Figure 4) for VHF bistatic auroral backscatter
communication between the pairs of stations specified
below. Height level = 110 km
a) solid curve   : Hamburg - Oslo
b) dashed curve   : Ruhr District - Aberdeen
c) dashed - dotted curve : Munich - Ruhr District
d) dotted curve   : Munich - Mannheim
φ geographic latitude,   Φ geomagnetic latitude.
After Lange-Hesse et al (1965).

points fit together, Figure 1) and small values of θ(forward-scatter case). The backscattered power varies in these extreme cases
according to the theory in a ratio of about one to two.

The map of Europe in Figure 5 represents four different curves. The ideal backscatter conditions as described before are
fulfilled along these curves for the pairs of stations specified
in the figure text. These curves are referred to as "backscatter
curves" in the following. The most northern solid backscatter
curve refers to communications from Hamburg to Oslo and the most
southern dotted backscatter curve refers to communications from
Munich to Mannheim (sout west Germany).

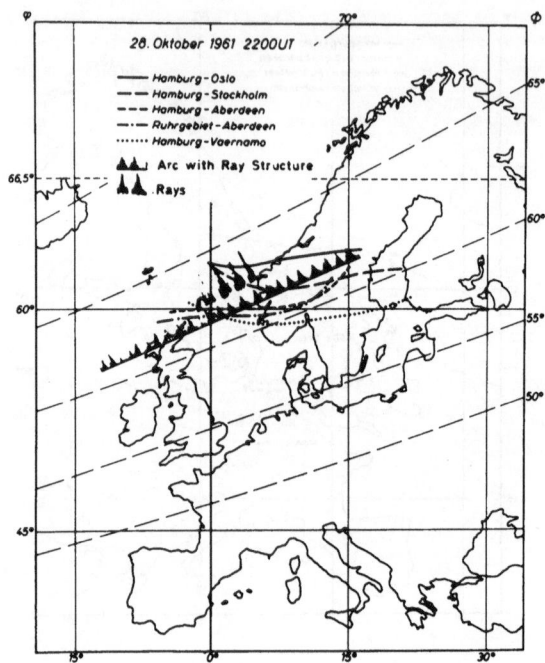

Figure 6. Location of the visual auroral display on October 28,
           1961 at about 2200 UT (Kp = 7+) and the location of the
           backscatter curves for the VHF bistatic auroral back-
           scatter communications carried out simultaneously be-
           tween the pairs of stations specified below.
           ———— Hamburg - Oslo
           — — Hamburg - Stockholm
           – —– Hamburg - Aberdeen
           —.— Ruhr - District - Aberdeen
           ...... Hamburg - Vaernamo
           ϕ,Φ see text Figure 5. After Lange-Hesse et al (1966).

7. RELATION TO THE LOCATION OF VISUAL AURORA DISPLAY

     From the observed data reported by radio amateurs (Lange-
Hesse, 1963b) it follows that e.g. on October 28, 1961 at about
2200 UT, VHF bistatic auroral backscatter communications were possible from Hamburg to Oslo, Stockholm, Aberdeen (Scotland) and
Vaernamo (southern Sweden) as well as from the Ruhr District to
Aberdeen. The computed backscatter curve for the pairs of stations previously mentioned by name are represented in Figure 6.
The location of the auroral display observed simultaneously is also represented in Figure 6. It is an arc with ray structure and

single rays. It can be seen that at least one part of the backscatter curves shown in the figure coincides with the visual aurora display. The backscattering centres at these parts of the
backscatter curves (coinciding with the aurora) have caused the
communication between the pairs of stations specified in the text
of Figure 6. According to this the backscattering centres which
caused the communication from Hamburg to Vaernamo (dotted backscatter curve in Figure 6) were located half way between Scotland
and southern Norway. At the time of day given in Figure 6 stations in southern Germany (e.g. in Munich) tried to get VHF communications via backscatter but without success. It follows, however, from Figure 5 that the backscatter curves for communications
from Munich to Mannheim and to the Ruhr District are located much
more to the south than the auroral display in Figure 6. These two
curves, therefore, did not coincide with the aurora so that it was
impossible to get VHF contacts from Munich via auroral backscatter
to Mannheim and to the Ruhr District.

An example similar to that in Figure 6 is shown in Figure 7
for September 4, 1958 at about 2300 UT. The backscatter curves,
for all auroral backscatter communications observed during that
time between pairs of stations specified in the figure text, are
represented on the map together with the visual auroral display
observed simultaneously. It can be seen (as in Figure 6) that at
least one part of the backscatter curves coincide with the visual
aurora. In the majority of the cases more than one part of the
backscatter curves coincide with the aurora. This is the reason
that the backscattered signals can come from a variety of directions spread about north during strong auroral activity as demonstrated in Figure 7. At the time given in Figure 7 it was impossible to get auroral backscatter communications from Hamburg
to Oslo and Stockholm. The reason for this is the fact that the
relevant backscatter curves (Figure 5 and 6) are located more to
the north than the visual aurora (Figure 7) so that it was not
possible for the aurora and the backscatter curves to intersect.

The geophysical events described in Figure 6 were accompanied
by a degree of geomagnetic activity of Kp = 7+, the events described in Figure 7 by Kp = 8+ and the visual auroral displays are
located at lower latitudes than the displays in Figure 6. This
southward shift of the visual aurora with increasing Kp-degree is
known (see e.g. Lange-Hesse, 1960; Akasofu, 1964). The simultaneous shift of the backscattering centres to the south was investigated by Lange-Hesse (1964b).

According to the results shown in Figure 6 and 7 it is principally possible to reconstruct later, from the observations of an
geographically extended net of observing stations (amateur stations), the location of the visual auroral displays in first
approximation for overcast or foggy days in a region which is

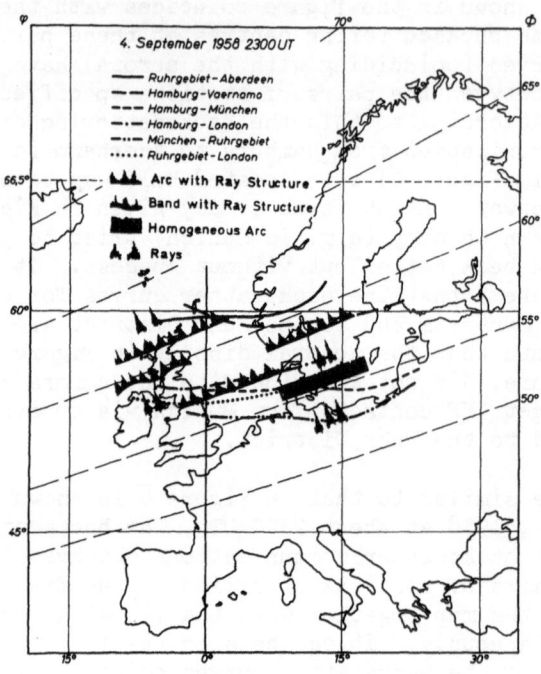

Figure 7.   Location of the visual aurora display on September 4,
            1958 at about 2300 UT (Kp = 8+) and the location of the
            backscatter curves for the VHF bistatic aurora back-
            scatter communications carried out simultaneously be-
            tween the pairs of stations specified below.
            ——————— Ruhr District – Aberdeen
            —— ——   Hamburg – Vaernamo
            — —— —   Hamburg – Munich
            —.——   Hamburg – London
            —..——   Munich – Ruhr District
            ......   Ruhr District – London
            φ, Φ see text Figure 5.   After Lange-Hesse et al (1966).

covered by the backscatter curves belonging to the net of observing stations. Maps with additional backscatter curves (additional to those shown in Figure 5, 6 and 7) are given by Lange-Hesse
et al (1965), Czechowsky (1966), and Lange-Hesse (1967) for middle
European, U.K. and southern Scandinavian regions.

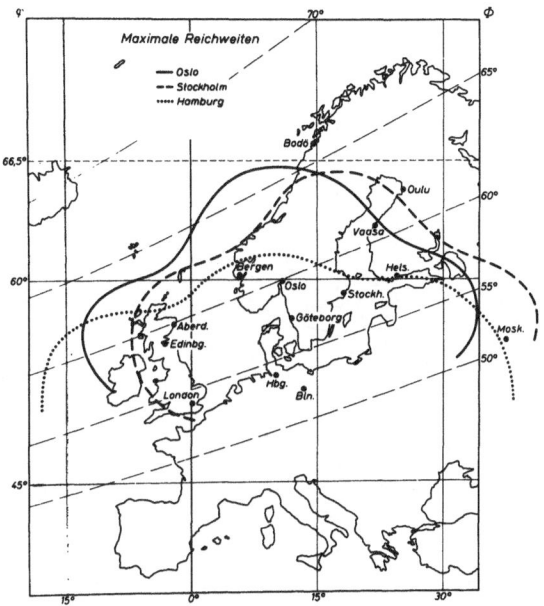

Figure 8.   Computed maximum-distance-curves for VHF bistatic au-
            roral backscatter communications from Oslo (——),
            Stockholm (— — —) and Hamburg (••••••).   Adopted
            height above the ground of the backscattering centres
            h = 110 km. ϕ, Φ see text Figure 5.   After Lange-
            Hesse et al (1966).

8. MAXIMUM DISTANCE FOR BISTATIC AURORAL BACKSCATTER
COMMUNICATIONS

    The three curves shown in Figure 8 represent the computed
maximum distances which can be contacted by VHF bistatic auroral
backscatter communications from Hamburg (dotted curve), Stockholm
(dashed curve), and Oslo (solid curve).   The curves are valid for
a height of the backscattering centres of 110 km above the ground.
For the computation of the curves only those directions of propagation of the backscattered wave are taken into consideration
which fulfil the ideal backscatter condition.   That means
$\cos \Omega_0 + \cos \Omega_1 = 0$ or $\varepsilon_0 = \varepsilon_1$ (Figure 2).   According to theoretical estimations mentioned before (Czechowsky, 1966) the ideal
backscatter condition must be fulfilled for communications with
low power transmitters (e.g. the power used by radio amateurs is
only of the order of 50 to 100 watts) in order to have sufficient
signal strength at the receiving point.

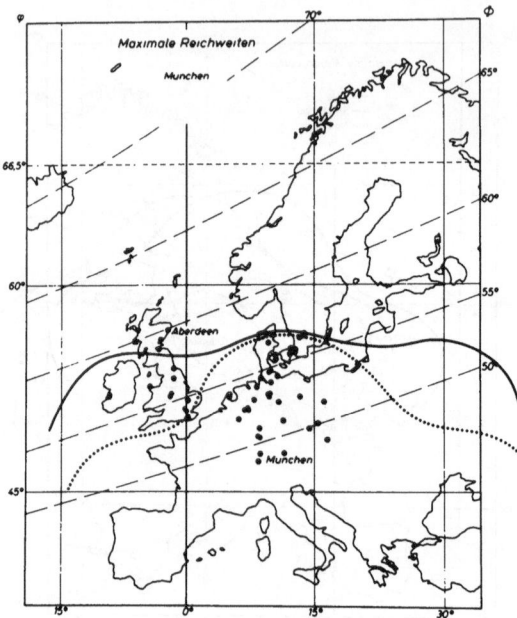

Figure 9.   Similar curves as in Figure 8 but computed for Munich
            for the two adopted heights h = 110 km (———) and h =
            200 km (••••••) of the backscattering centres. The
            dots represent stations which could be contacted from
            Munich via VHF auroral backscatter communications.
            φ, Φ see text Figure 5. After Lange-Hesse et al (1966).

        In a supplement to Figure 8 the curves shown in Figure 9 re-
present the computed maximum-distance-curves for Munich for the
two heights of the backscattering centres of 110 and 200 km. The
dots in Figure 9 represent stations which could be contacted from
Munich via VHF auroral backscatter communication on 144 MHz. A
few of these stations were contacted from Munich more than once.
If these recurring contacts are taken into consideration by corresponding (statistical) weights, only 5% of the total number of
contacts were carried out with stations located outside the maxi-
mum-distance-curve for 110 km height, but 30% of the contacts were
carried out with stations located outside the 200 km curve (dotted
curve in Figure 9). It follows from the results presented in
Figure 9 that the computed maximum distances for a height of the
backscattering centres at 110 km coincide to a first approximation
with the observations.

        Leadabrand et al (1958) have computed similar curves as shown
in Figure 8 and 9 for some points in North America but using the
basic magnetic elements from the geomagnetic dipole field which is

only a first approximation to the true geomagnetic field. The
curves in Figures 8 and 9, however, are computed using the observed magnetic dip angle and the declination from the ground.
Leadabrand et al (1958b) has also computed for each maximum - distance - curve the appertaining "region of useful ionization", a
region of approximate semicircular shape which must be covered by
auroral ionization in order to get communication to every point
within the maximum - distance - curve.

9. PROBABLE REASONS FOR THE MAXIMUM DISTANCE VARIATIONS

9.1 Influence of Height Variations of the Backscattering Centres

     Backscatter curves for communications from Munich to Hamburg
for the two heights 110 km (solid curve) and 200 km (dashed curve)
of the backscattering centres are shown in Figure 10. The backscatter curve for 200 km is located more to the north than the 110
km curve. The maximum - distance - curve (Figure 9), however,
shows smaller distances for the 200 km than for the 110 km scatter
height. The reason for this is the fact that the propagation
angle Ω (Figure 2) increases with increasing height. Therefore,
it becomes more difficult to fulfil the ideal backscatter condition $\cos \Omega_0 + \cos \Omega_1 = 0$ with increasing height for definite regions. This causes the reduction of the maximum distance with increasing height of the backscattering centres. The observed increase of the maximum distance for 5% of the observations carried
out at Munich (Figure 9) therefore, can not be caused by an increase of the height of the scatter centres.

     It is known, that in middle latitudes the frequency of occur-
rence of aurora increases with increasing geomagnetic latitude.
If one considers auroral backscatter communications between two
fixed stations (e.g. Hamburg and Munich in Figure 10) the backscatter curve shifts to the north with increasing height of the
scatter centres (Figure 10). Due to this, the curves move into a
region of higher aurora probability which in turn causes a higher
probability of the occurrence of auroral backscatter communications. The conclusion from this behaviour is the fact, that an
increase of the height of the backscattering centres causes a reduction of the maximum distance and indeed an increase of the probability of occurrence of bistatic auroral bacskcatter communications.

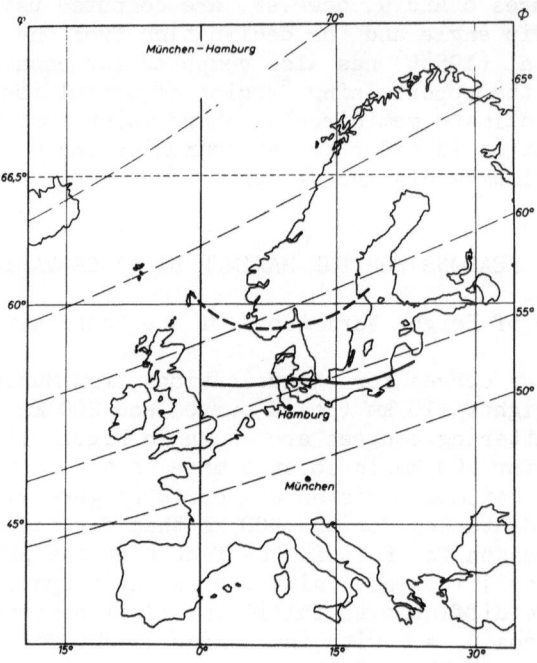

Figure 10.   Location of the backscatter curves (described in
             Figures 3, 4 and 5) for VHF bistatic auroral back-
             scatter communications from Munich to Hamburg for the
             two adopted heights h = 110 km (———) and h = 200 km
             (······) of the backscattering centres. φ, Φ see
             text Figure 5. After Lange-Hesse et al (1966).

9.2 Influence of Magnetic Dip Angle Variations and Radio
     Meteorological Influence

      All of the propagation angles, backscatter and maximum – dis-
tance – curves shown in the figures of this paper are computed
using the assumption of an undisturbed geomagnetic field or for
the "static situation". It is well known that during stronger
geomagnetic storms the geomagnetic lines of force can change their
orientation in space, that means that the dip angle J deviates
from its normal value (Størmer, 1926). In connection with a variation ΔJ of the dip angle the ideal backscatter condition may be
fulfilled for propagation paths which do not normally fulfil this
condition under geomagnetic undisturbed conditions. The propagation path for the communication Munich – Aberdeen deviates from
the ideal backscatter condition by 1.3°. For this special case it
is computed, that a ΔJ of 1° will result in the exact fulfilment

of the ideal backscatter condition (Figure 2), (Czechowsky, 1966;
Lange-Hesse et al, 1966). The auroral backscatter communications
from Munich to stations beyond the 110 km maximum - distance -
curve in Figure 9 were carried out during times of stronger geomagnetic storms which have caused dip angle variations of 1° or
more (Czechowsky, 1966; Lange-Hesse et al, 1966). Dip angle variations in connection with geomagnetic storms therefore can cause
deviations from the maximum - distance - curves up to 300 km and
more. These deviations are larger than those caused by radio meteorological influences (Czechowsky, 1966; Lange-Hesse et al,
1966). The dip angle variations therefore provide an explanation
of the observed communications beyond the computed maximum distances (Figure 9).

10. TIME VARIATIONS OF BISTATIC AURORA BACKSCATTER COMMUNICATION

10.1 Diurnal Variation

The lower diagram in Figure 11 shows the diurnal variation of
the frequency of occurrence of VHF auroral backscatter given by
the number of days with the occurrence of bistatic auroral backscatter communications (aurora contacts) observed and reported
during the time from September 1957 to April 1961 by a net of amateur radio stations in median latitudes ($48^\circ \leq \phi \leq 54^\circ$). Two
strong maxima occur, the first at about 1700 to 1800 hr and the
second after midnight local time. One sharp minimum occurs at
about 2100 hr, a second broad one extends from early morning until
noon. The results shown in the lower diagram of Figure 11 are
confirmed by the upper diagram of this figure which shows the diurnal variation of the frequency of occurrence of VHF auroral radar backscatter carried out at about the same geomagnetic latitude. The geomagnetic latitudes given in Figure 11 give the location of the observing stations. The backscattering centres,
however, are located about 3° to 10° more to the north (see e.g.
Figures 3, 4 and 5).

The times of maximum frequency of occurrence in Figure 11
show a strong dependence on the geomagnetic latitude of the observations. The approximate times of peak backscatter activity for
observations taken at various geomagnetic latitudes have been investigated by Egeland et al (1961) and Unwin (1966). The results
of Unwin are summarized in Figure 12. The solid spirals in Figure
12 represent the various echo maxima for moderate magnetic disturbance. There is a clear indication of three separate spirals
of maximum backscatter activity. The left-hand spiral is hereafter referred to as the E (evening) spiral. The centre spiral is
hereafter referred to as N (night) spiral. The right-hand spiral
of Figure 12 is referred to as M (morning) spiral. The position
and relative prominence of the spirals is sensitive to the level

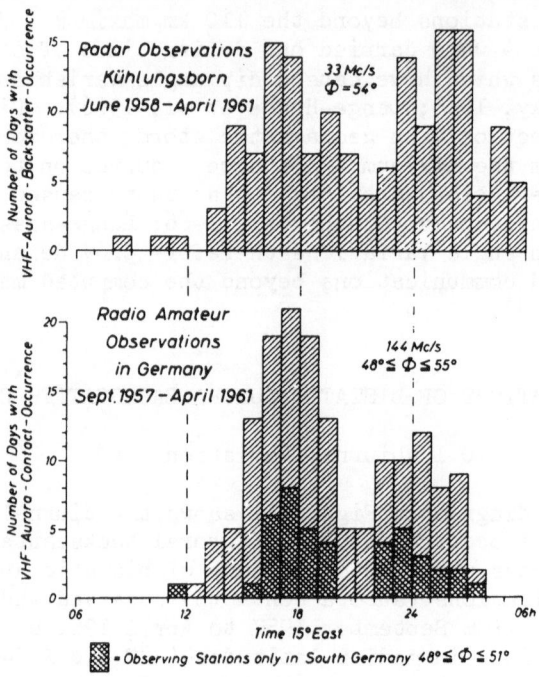

Figure 11. Comparison of the diurnal variation of the frequency
            of occurrence of auroral backscatter echoes according
            to amateur observations (low figure) and to radar ob-
            servations at about the same geomagnetic latitudes but
            on different frequencies. After Lange-Hesse (1962).

of magnetic activity and therefore there is a slightly different
picture than that in Figure 12 for weak magnetic disturbance (Unwin, 1966). The backscatter spirals in Figure 12 are similar to
the (dashed) spirals of maximum magnetic activity derived by Feldstein (1963) and others. If one takes into account that the backscatter spirals in Figure 12 represent the location of the backscattering centres (not the location of the observing stations)
and that the difference between geomagnetic and local time in
middle Europe is not more than about one hour, the results from
Figure 11 (times of peak backscatter activity) coincide with those
given in Figure 12.

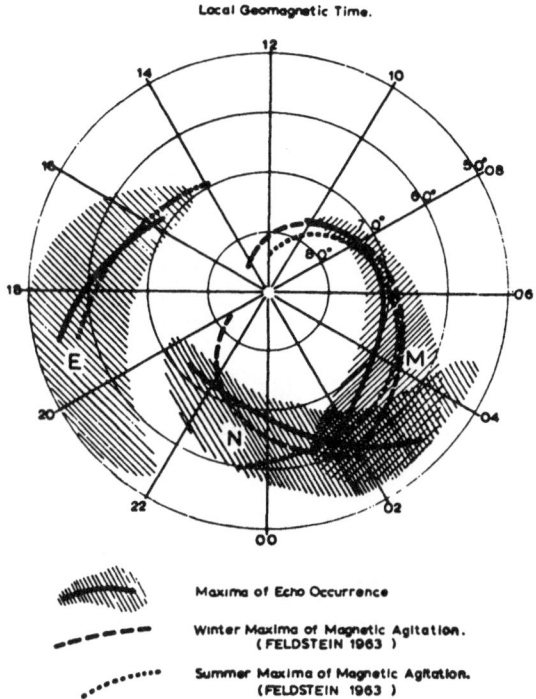

Figure 12.   Geomagnetic times of peak auroral echo activity and
             peak magnetic activity as observed at various geomag-
             netic latitudes. The solid spirals represent the va-
             rious backscatter maxima for moderate magnetic distur-
             bance and the dashed spirals the various maxima of
             geomagnetic activity. After Unwin (1966).

10.2 Seasonal Variation

The diurnal variation of the frequency of occurrence of auroral backscatter echoes in sub-auroral latitudes (Figure 11)
shows only a small seasonal variation: The afternoon maximum is
later in winter than in summer and the early evening minimum is
later and less pronounced in winter than in summer (see e.g.
Lange-Hesse, 1963a). More about seasonal variation of the diurnal
variation as a function of latitude see e.g. Unwin (1966).

The seasonal variation of the frequency of occurrence of VHF
aurora backscatter shows distinct maxima during equinox and the
autumn maximum is higher than the spring maximum (see e.g. Lange-
Hesse, 1963a). This variation has close similarity with the sea-

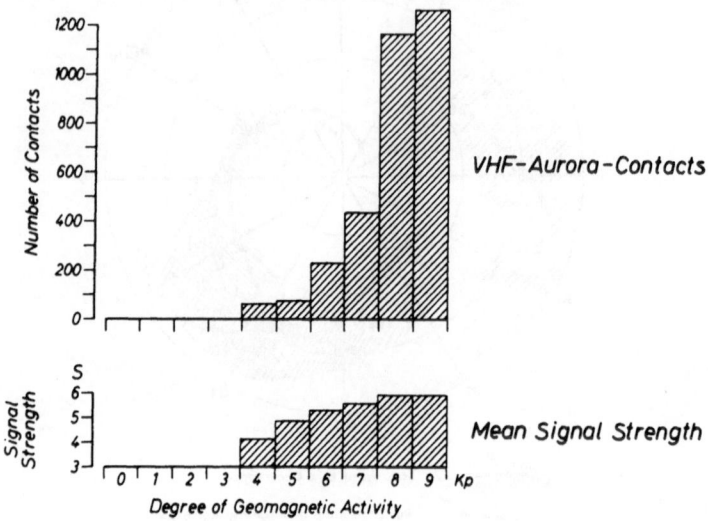

Figure 13. Frequency of occurrence of VHF bistatic auroral communications (auroral contacts) as well as variation of
mean signal strength S of the auroral-backscattered
signal as a function of the degree of the planetary
geomagnetic activity Kp according to amateur observations on 144 MHz in middle European and southern Scandinavian latitudes. After Lange-Hesse (1963a).

sonal variation of the visual auroras (see e.g. Meinel et al,
1954).

11.   OCCURRENCE FREQUENCY AS A FUNCTION OF GEOMAGNETIC LATITUDE
      AND GEOMAGNETIC ACTIVITY DEGREE

     The southward shift of the backscattering centres with in-
creasing Kp-degree was mentioned in section 7 (Lange-Hesse, 1964b).
According to the observations carried out by VHF radio amateurs in
Europe the backscattering centres in average shift down to about
$60°$ to $63°$ geomagnetic latitude ϕ during Kp = 7 (Figure 6), to
about $\phi = 53°$ to $57°$ during Kp = 8 (Figure 7) and down to about
$\phi = 52°$ and less during Kp = 9. In and close to the auroral zone
backscattering centres occur usually during Kp = 2 to 6 (Egeland
et al, 1961). Since low Kp-degrees (2 to 6) are much more frequent than high Kp-degrees (Bartels, 1958) VHF bistatic auroral

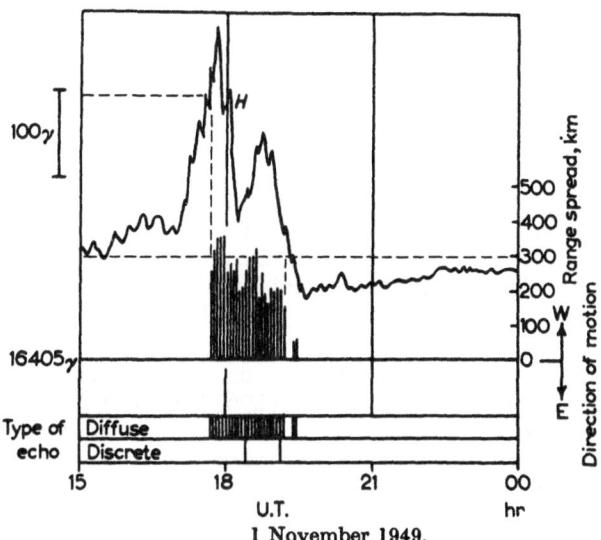

Figure 14. Complete daily trend in geomagnetic disturbance and
radar auroral echo characteristics observed with 72
Mc/s radar aquipment at Jodrell Bank Experimental
Station. After Bullough et al (1957). Note that the
backscattering of radio waves begins only when the
departure of the geomagnetic H-component exceeds some
threshold value (about 200 γ).

communications near the auroral zone are relatively frequent (on
the average every second or third day), at latitudes of about
Φ = 50°, however, they are relatively seldom and then only during
sunspot maximum years (about two to five times a year during sunspot maximum). For more about occurrence frequency in latitudes
near the auroral zone see e.g. Egeland et al (1961), Egeland
(1962a); in subauroral latitudes see e.g. Lange-Hesse (1962,
1963b, 1964a, 1964b).

12. PLASMA ACOUSTIC WAVE THEORY

Figure 14 shows the complete daily trend in geomagnetic and
auroral backscatter activity at Jodrell Bank (near Manshester) on
72 MHz. The vertical lines indicate the occurrence of auroral
backscatter echoes. The magnetogram is the horizontal component
from Eskdalemuir in Scotland. By comparing the time of the beginning of the auroral backscatter phenomenon with the magnetogram of
the horizontal component in Figure 14 it can be recognized that
the backscattering of radio waves takes place only when the depar-

ture of the magnetic variation from the undisturbed value (horizontal dashed line in Figure 14) exceeds a fixed value of about
200 γ, that means that the current strength of the auroral jet
must exceed some threshold value before the field-aligned irregularities (which give rise to the backscatter) occur. The direction of the magnetic variation in Figure 14 reverses in sign at
about 1900 hr (caused by the 180°-reversal of the current flow of
the electrojet). The radar backscatter echoes disappear at about
the same time, that means at a remarkable lower threshold value
than had been observed in the growth phase of the electrojet between 1700 and 1800 hr. This is similar to a hysteresis-like behaviour. The amplitude of the negative departure of the magnetic
variation in Figure 14 at about 1930 hr (which corresponds to the
growth phase of the electrojet in the opposite direction) seems to
be below the threshold value so that no radar echoes occur during
and after this time. Bowles et al (1963) have found similar results at the magnetic equator by comparing the scattered power at
50 MHz from E-region field-aligned equatorial irregularities with
the magnetogram of the horizontal component. The backscattering
of the radio waves at the equator also takes place only when the
current strength of the electrojet along the geomagnetic equator
and, in relation to this, the amplitude of the variation of the
horizontal component H of the earth's magnetic field exceeds some
threshold value. Bowles and colleagues have also found a hysteresis-like behaviour at the geomagnetic equator in the correlation
between the amplitude of the variation of the H component and the
occurrence and the amplitude of the backscattered echo.

The hysteresis-like behaviour at the equator in the approximate proportionality of the echo power, with the fluctuation of
the H trace, causes the proportionality curve followed in the
afternoon to differ from the curve followed in the morning. This
difference has now been shown to be the result of changes in the
magnetic field arising externally to the equatorial electrojet.
The externally caused changes in the magnetic field were estimated
by referring to magnetographs made outside the zone of the equatorial electrojet. When these changes were subtracted from the H
variations at the equator, the hysteresis-like behaviour almost
always collapsed to a single line (Cohen et al, 1963).

There is fairly good evidence that the hysteresis-like behaviour of the auroral backscatter echoes can be similarly explained.
Investigations have not yet been made to check this in the same
manner as was adopted at the equator. Principally there are experimental difficulties to investigate exactly the hysteresis-like
behaviour of the auroral backscatter echoes. Since the auroral
electrojet, contrary to equatorial electrojet, has no fixed location in space, it shifts to lower latitudes with increasing geomagnetic activity, difficulties therefore arise in carrying out
geomagnetic measurements always exactly below the electrojet.

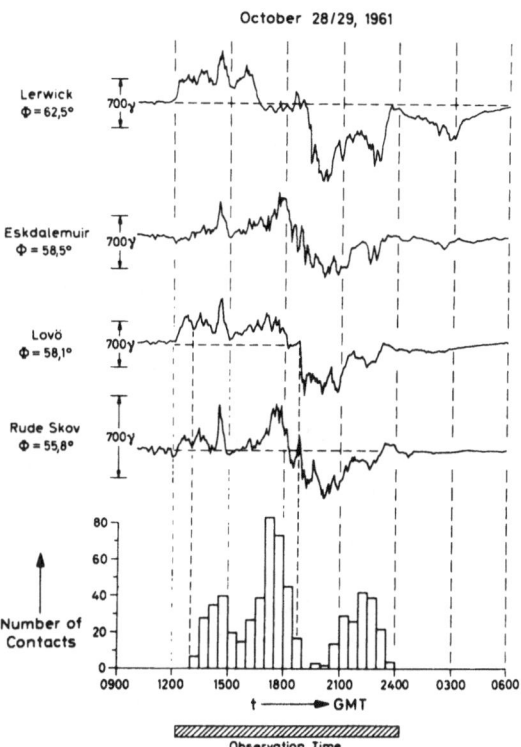

Figure 15.   Complete daily trend in geomagnetic disturbance characteristics (H-component) at four European Observatories: Lerwick (Shetland Islands), Eskdalemuir
(Scotland), Lovö (Sweden), and Rude Skov (Denmark).
The horizontal dashed line in the magnetograms gives
in a first approximation the undisturbed H-value.
The lower diagram shows the simultaneous frequency of
occurrence of VHF bistatic auroral backscatter communications (auroral contacts) in middle Europe. Note
that (as also shown in Figure 11) the backscattering
of radio waves begins at about 1300 UT only when the
departure of the geomagnetic H-component exceeds some
threshold value. The peaks in geomagnetic departures
are accompanied by an simultaneous increase in the occurrence frequency of aurora backscatter communications. The observation time (lowest diagram) is the
time when radio amateurs operate their stations look-
ing for possible aurora backscatter signals from other
amateur stations. Observational data taken from
Lange-Hesse (1963b).

Some years ago Buneman (1963) and Farley (1963) pointed out independently, the possibilities that an ionospheric current system like the equatorial and polar electrojet represents a "two-stream" instability described in modern plasma theory. Bowles et al (1963) applied this concept to explain the existence of field-aligned E-region irregularities at the magnetic equator. Buneman (1963) and Bowles et al (1963) have suggested that this theory is applicable also to explain the field aligned auroral ionization irregularities. The irregularities are spatially and temporally correlated with the electrojet. The threshold suggests that the irregularities are generated by the current stream. According to the theory of Farley or Buneman such a threshold can be interpreted as the onset of a plasma instability that produces "acoustic plasma waves" which are thought to be the immediate cause of the electron density irregularities.

The threshold and the hysteresis-like behaviour, as shown in Figure 14 for the radar case, also occur in the case of bistatic auroral backscatter communication as can be seen in Figure 15. The lower diagram of this figure shows the frequency of occurrence of VHF bistatic auroral backscatter communications (auroral contacts) in Middle Europe on October 28/29, 1961 according to amateur observations on 144 MHz (Lange-Hesse, 1963b). The sharp and deep minimum at about 2000 UT also occurs in the average diurnal frequency variation as shown in Figure 11. The backscatter curves, for most of the communications used for the diagram in Figure 15, are located over southern Scandinavia. Therefore, the magnetogram of the H component from Rude Skov (near Copenhagen) and Lovö (near Stockholm) is representative for the approximate current strength of the electrojet and the threshold, indicating the start of the formation of the irregularities responsible for the observed auroral backscatter communications.

According to the magnetograms in Figure 15 the growth of the electrojet starts at about 1200 hr. The threshold occurs at about 1300 hr. The reversal of the electrojet occurs at about 1900 hr. At about the same time the possibility for auroral backscatter communication disappears at a remarkably lower threshold than in the growth phase at about 1300 hr. This hysteresis-like behaviour can also be observed in Figure 15 during the negative (down) departure of the H-component. The results discussed previously indicate that the plasma acoustic wave theory gives a possible explanation of the early evening minimum in the diurnal occurrence frequency of VHF auroral backscatter. During this minimum the current strength of the electrojet is below the threshold, so that the irregularities disappear. Caused by the hysteresis-like behaviour a time lag occurs between the reversal of the electrojet and the peak of the frequency minimum.

Observations at the equator have shown that the threshold shows a frequency dependence (Bowles et al, 1963). The threshold must be higher for echoes to be observed at 148 MHz than for echoes to be observed at 50 MHz. If one applies this concept to the auroral zone it might be an additional explanation for the observed higher frequency of auroral backscatter echoes on lower frequencies compared to higher frequencies (see e.g. Currie et al, 1953).

Moorcroft (1966) has made a detailed interpretation of the observed frequency dependence of the auroral backscatter phenomenon. He concluded that the acoustic wave theory appears to provide the most reasonable explanation of the auroral backscatter frequency dependence.

Leadabrand (1965b) has made a comparison between 400 and 800 MHz radar auroral backscatter data in order to check the validity of the acoustic wave theory. This comparison indicates that, although many of the key features of the plasma acoustic wave concept can be found in the radar auroral data, the spectral characteristics of the auroral echoes, however, are much too complicated to be simply understood in their present form.

In summary one can say: One of the main result of the new plasma acoustic wave theory is the fact, that to get VHF-aurora-backscatter echoes you need not only an aurora and the fulfilment of the geometrical conditions mentioned before but also an electrojet with a current strength which is beyond the threshold. The occurrence of backscatter echoes indicates, that the plasma in the electrojet has a special fine structure, i.e. the field-aligned backscattering centres.

Within the auroral zone and in subauroral latitudes the visual aurora is normally closely correlated in space with the polar electrojet. One therefore observes in these latitudes a relatively good correlation between visual and radio aurora in directions where the geometrical conditions are fulfilled. Within the polar cap, however, there often occurs aurora during very low geomagnetic activity. There is fairly good evidence that a large number of these aurorae do not give rise to VHF-backscatter echoes, since the electrojet is below the threshold in these regions. The inverted case was also reported from a location in Antarctica, i.e. the occurrence of backscatter echoes from the electrojet without the occurrence of a visual aurora (Shipstone, 1966).

In view of this it would perhaps be reasonable to change the term "radio aurora" to "radio electrojet", since the electrojet is mainly responsible for this phenomenon, and also to see the VHF backscatter phenomena in the auroral zone and at the equator from the same point of view. The new plasma acoustic wave theory of-fers much new incentive to the radio aurora, or using the new term, to the radio electrojet researchers for future work.

Acknowledgements: The author wishes to express his thanks to Dr. J. Paton of the World Data Centre C (visual aurora), Edinburgh, for the copies from the synoptic aurora maps of European sector, further to many European Magnetic Observatories for copies from the magnetograms, and to the Institute for Radio Meteorology and Marine Meteorology at the University of Hamburg for the radio meteorological data. Finally, the author wishes to express his thanks to many European radio amateurs for the careful and extended observations of VHF bistatic aurora backscatter communications and to the RSGB (Radio Society of Great Britain) and the DARC (German Amateur Radio Club) for collecting the comphrehensive observation data. Here again is a good example of amateur radio supplying research information which is difficult and more expensive to obtain in any other way.

13.   REFERENCES

1. S.-I. Akasofu, "The latitudinal shift of the auroral belt", J. Atm. Terr. Phys. 26, 1167-1174, 1964.

2. E.V. Appleton, R. Naismith and L.J. Ingram, "British radio observation during the Second International Polar Year, 1932-33", Phil. Transact. Roy. Soc. A 236, 191-259, 1937.

3. B.A. Bagaryatsky, "Some results of radar studies of polar aurora", Spectral electro-Photometrical and Radar Researches of Aurora and Airglow, No. 2-3, Pub. House of Acad. of Science of USSR, Moscow, 1960, pp 7-14.

4. D. Barber, H.K. Scitcliffe and C.D. Watkins, "Some observations of meteors and aurorae at 300 and 500 Mc/s using a large radio telescope. II. Observations of the aurora borealis", J. Atm. Terr. Phys. 24, 599-609, 1962.

5. J. Bartels, "Planetarische erdmagnetische Aktivität in graphischer Darstellung: tägliche Cp, 1937-1958, dreistündliche Kp, 1937-1939, 1950-1958". Abh. Akad. Wiss. Göttingen, Math.-Phys. Klasse, Beiträge zum Internationalen Geophysikalischen Jahr, Heft 3. Publisher: Verlag Vandenhoeck und Ruprecht, Göttingen (Germany), 1958.

6. J.G. Birfeld, "Radar observations of Polar aurorae", Bull. (Izvestia) Acad. Sci. USSR, Geophys. Ser., No. 12, 1248-57, 1960.

7. B.C. Blevis, J.W.B. Day and O.S. Roscoe, "The occurrence and characteristics of radar aurora echoes at 488 and 944 Mc/s", Can. J. Phys. 41, 1359-1380, 1963.

8. H.G. Booker, "A theory of scattering by nonisotropic irregularities with application to radar reflection from the aurora", J. Atm. Terr. Phys. 8, 204-221, 1956.

9. H.G. Booker, "Radar studies of the aurora", in Physics of the Upper Atmosphere, ed. J.A. Ratcliffe, New York: Academic, 355-375, 1960.

10. K.L. Bowles, B.B. Balsley and R. Cohen, "Field aligned E-region irregularities identified with acoustic plasma waves", J. Geophys. Res. 68, 2485-2501, 1963.

11. K. Bullough, T.W. Davidson, T.R. Kaiser and C.D. Watkins, "Radio reflections from aurorae-III. The association with geomagnetic phenomena", J. Atm. Terr. Phys. 11, 237-254, 1957.

12. O. Buneman, "Excitation of field aligned sound waves by electron streams", Phys. Rev. Letters, 10, 285-287, 1963.

13. J.W. Chamberlain, "Physics of the Aurora and Airglow", New York: Academic, 1960.

14. C. Collins and P.A. Forsyth, "A bistatic radio investigation of auroral ionozation", J. Atm. Terr. Phys., 13, 315-345, 1959.

15. B.W. Currie, P.A. Forsyth and F.E. Vawter, "Radio Reflections from Aurora", J. Geophys. Res. 58, 179-200, 1953.

16. P. Czechowsky, "Analyse von Rückstreubeobachtungen ultrakurzer Wellen an Polarlichtern - Steuerung durch die Richtung der erdmagnetischen Feldlinien und Zusammenhänge mit der geographischen Lage der rückstreuenden sichtbaren Polarlichter und der Mophologie der begleitenden erdmagnetischen Störungen", Diplom-Arbeit (Master's Thesis) University of Göttingen (Germany), 1966.

17. R. Dyce, "More about VHF auroral propagation", QST 39, (Jan. 1955), 11-15, 1955a.

18. R. Dyce, "VHF auroral and sporadic - E propagation from Cedar Rapids, Iowa, to Ithaca, N.Y.", Transact. Inst. Radio Engers, AP-3, 76-80, 1955b.

19.   A. Egeland, J. Ortner and B. Hultqvist, "A study of the sta-
      tistics of VHF oblique auroral reflections", Scientific Re-
      port, No. 7, Kiruna Geophyscial Observatory, 1961.

20.   A. Egeland, "Studies of auroral reflections in the VHF band.
      I. Experimental investigations, with special regard to time
      variations, fading rate, azimuthal distributions, and polari-
      zation characteristics", Arkiv för Geofysik, 4, 6, 103-169,
      1962a.

21.   A. Egeland, "Studies of auroral reflections in the VHF band.
      II. Comparison of experimental results with theoretical mo-
      dels", Arkiv för Geofysik, 4, 7, 171-209, 1962b.

22.   D.T. Farley, Jr., "A plasma instability resulting in field
      aligned irregularities in the ionosphere", J. Geophys. Res.
      68, 6083-6097, 1963b.

23.   Y.I. Feldstein, 1963 Sbornik III, No. 5. "Sbornik" refers to
      the U.S.S.R. results of the IGY, published by Akademii Nauk
      SSSR. Section III deals with Geomagnetism. Section IV with
      Aurorae and Airglow.

24.   W.A. Flood, "Simultaneous VHF auroral backscatter measure-
      ments", J. Geophys. Res. 65, 2261-2268, 1960.

25.   N.C. Gerson, "Diurnal variation in auroral activity", Proc.
      Phys. Soc. 68, 408-414, 1955a.

26.   N.C. Gerson, "Radio observations of the aurora", J. Atm. Terr.
      Phys. 6, 263-267, 1955b.

27.   F.D. Green, "Pattern in the behaviour of VHF bistatic radio
      reflections in the auroral zone", Radio Studies, Report No.
      RS-9, Institute of Upper Atmospheric Physics, Univ. of Sas-
      katchewan, Saskatoon, Canada, 1961.

28.   L. Harang and W. Stoffregen, "Echoversuche auf Ultrakurzwel-
      len", Hochfreq. u. Elektroak. 55, 105-108; and Nature 142,
      832-833, 1940.

29.   B. Hultqvist and A. Egeland, "Radio Aurora", Space Science
      Review, 3, 27-78, 1964.

30.   G. Lange-Hesse, "Rückstrahlung kurzer und ultrakurzer Wellen
      an Polarlichtern", Arckiv d. elektr. Übertragung (AEÜ) 11,
      253-261, 283-288, 1957.

31. G. Lange-Hesse, "Südlichste Ausdehnung von Nordlichtern nach deutschen Land- und Seebeobachtungen im Internationalen Geophysikalischen Jahr", Naturwissenschaften 47, 423-424, 1960.

32. G. Lange-Hesse, "VHF-long-distance-propagation in Middle Europe by aurora-backscatter", Arch. d. elektr. Übertr. (AEÜ) 16, 251-261, 1962.

33. G. Lange-Hesse, "Seasonal influences on VHF-aurora-backscatter in Middle Europe", Z. Geophys. 29, 35-44, 1963a.

34. G. Lange-Hesse, "German Aurora Observations 1957-1962, Part A: Aurora Observations by Means of VHF Radio Waves, Part B: Observations of Visual Aurora", Abhandlungen der Akademie der Wissenschaften in Göttingen, math.-phys. Klasse, Beiträge zum Internationalen Geophysikalischen Jahr, Heft 10, published by Verlag Vandenhoeck and Ruprecht, Göttingen (Germany), 1963b.

35. G. Lange-Hesse, "VHF-long-distance-propagation in middle latitudes by aurora-backscatter - Influence of the geomagnetic-activity-degree on the communication frequency as a function of magnetic dip of the observing stations", Arch. d. elektr. Übertr. (AEÜ) 18, 430-438, 1964a.

36. G. Lange-Hesse, "VHF-bistatic-aurora communications as a function of geomagnetic activity and magnetic latitude", in Arctic Communications, AGARDograph 78, ed. B. Landmark, Oxford: Pergamon Press, 253-262, 1964b.

37. G. Lange-Hesse and P. Czechowsky, "VHF bistatic aurora backscatter communications and the relation to the location of the visual aurora displays", Arch. d. elektr. Übertr. (AEÜ) 19, 511-514, 1965.

38. G. Lange-Hesse and P. Czechowsky, "VHF bistatic aurora backscatter communications - Comparison of the observations with the theory", Arch. d. elektr. Übertr. (AEÜ) 20, 365-375, 1966.

39. G. Lange-Hesse, "Radio Aurora, Part I. Observations, Part II. Comparison of the observation with a theoretical model", in AURORA and AIRGLOW, ed. B. McCormac, Reinhold Publishing Corp. New York, N.Y., 1967.

40. R.L. Leadabrand and I. Yahoff, "The Geometry of auroral communications", IRE Trans. Ant. Propag., January 1958, 80-87.

41. R.L. Leadabrand, "Radio studies of the aurora", J. Phys. Soc. Japan 17, (Supp. A-1), 218-222, 1962.

42. R.L. Leadabrand, J.C. Schlobohm and M.J. Baron, "Simultaneous
     very high frequency and ultra high frequency observations of
     the aurora at Fraserburg, Scotland", J. Geophys. Res. 70,
     4235-4284, 1965.

43. R.L. Leadabrand, "Electromagnetic measurements of aurora", in
     Auroral Phenomen, ed. M. Walt, Stanford: Univ. Press,
     London: Oxford Univ. Press, 99-129, 1965a.

44. R.L. Leadabrand, "A comparison of radar auroral reflection
     data with acoustic wave theory, Radio Science, 69D, 959-964,
     1965b.

45. C.G. Little, W.M. Rayton and R.B. Roof, "Review of ionosphe-
     ric effects at VHF and UHF", Proc. IRE 44, 992, 1956.

46. A.B. Meinel, B.J. Negaard and B.J. Chamberlain, "Statistical
     analysis of low-latitude aurora", J. Geophys. Res. 59, 407-
     413, 1954.

47. G.H. Millman, "The geometry og the earth's magnetic field at
     ionospheric heights", J. Geophys. Res. 64, 717-726, 1959.

48. R.K. Moore, "A VHF propagation phenomena associated with au-
     rora", J. Geophys. Res. 56, 97-106; and "Aurora and magnetic
     storms", QST 35, 15 (June 1951); and QSR 23, 78 (May 1939),
     author unknown, 1951.

49. D.R. Moorcroft, "The interpretation of the frequency depen-
     dence of radio aurora", Planet. Space Sci. 14, 269-275, 1966.

50. C. Newton, "The Society's IGY aurora programme - part I and
     II", RSGB Bulletin (Journal of the Radio Society of Great
     Britain) 42, 289-294, 785-790, 1966.

51. J. Oksman, "On the geometry of auroral reflections", Annales
     Academiae Scientiarum Fennicae A VI Physica 169, 1-16
     (Helsinki), 1964.

52. J. Oksman, "Movement of radio aurora", Geophysia (Helsinki)
     9, 3, 235-250, 1966a.

53. J. Oksman, "Massung von Radionordlichtern in Sodankylä/
     Nordfinnland", Kleinheubacher Berichte, published by Fern-
     meldetechnisches Zentralamt, F. Gr. VC, Darmstadt (Germany)
     10, 1966b.

54. A.M. Peterson, "The aurora and radio wave propagation", The
     Radio Noise Spectrum, ed. Donald H. Menzel, Cambridge, Mass:
     Havard Univ. Press, 1960, pp 7-42.

55. R.I. Presnell, R.L. Leadabrand, A.M. Peterson, R.B. Dyce, J.C. Schlobohm and M.R. Berg, "VHF and UHF radar observations of the aurora at College, Alaska", J. Geophys. Res. 64, 1179-1190, 1959.

56. D.M. Shipstone, University of Sheffield, Phys. Dept. Private communication, August 1966.

57. R.L. Smith-Rose, "Some radio aspects of the IGY RSGB-Bulletin (Journal of the Radio Society of Great Britain) 35, 392-394, 1960.

58. C. Störmer, "Résultat des mesures photogrammétrique des aurores boréales observées dans la norwège méridionale de 1911 à 1922", Geofysiske Publikasjoner 4, Nr. 7, Oslo, 1926.

59. M.L. Stone, R.P. Ingalls, C.H. Dugan and L.P. Rainville, "Simultaneous auroral observations at two ultra-high frequencies", paper presented at URSI meeting, Washington, D.C. 1959.

60. G.M.C. Stone, "Amateur radio participating in the IGY", RSGB-Bulletin (Journal of the Radio Society of Great Britain) 35, 395-397, 1960.

61. G.M.C. Stone, "The Radio Society of Great Britain and IQSY", Interradio. The International Radio Journal, ITU Centenary Edition, Geneva 20, 24-27, 1965.

62. E.P. Tilton, "On the very highs", QST, 28, 41-43, 86, 1944.

63. R.S. Unwin, "The geometry of auroral ionization", J. Geophys. Res. 63, 501-506, 1958.

64. R.S. Unwin, "Studies of the upper atmosphere from Invercargill, New Zealand: Part I - Characteristics of auroral radar echoes at 55 Mc/s", Ann. de Géophys. 15, 377-394, 1959a.

65. R.S. Unwin, "The morphology of the VHF radio aurora at sunspot maximum, Part I - Diurnal and seasonal variations, Part II - The behaviour of different echo types, Part III - Movement of echoes", J. Atm. Terr. Phys. 28, 1167-1181, 1183-1194, 1966.

DISCUSSION

The following points were made in the general discussion which took place after the presentation of the papers in the "Scatter Communication" session:

- In designing meteor-scatter circuits it should be recognized that the channel is capable of supporting very high instantaneous signalling rates. For instance at DRTE, tests have been conducted at 104 MHz with data rates up to 100,000 bits/s over a 1400 km path.

- From measurements made by STC on ionoscatter and meteor-scatter systems using diversity and ARQ, it can be concluded that meteor-scatter requires about five times less power per telegraph channel than ionoscatter.

- Measurements of duty cycle for meteor-scatter systems made at latitudes 30 to 40°N show greater diurnal variations than those made by STC at a latitude of about 45°N. This is to be expected from geometric considerations since the observed meteor rate is highest in the direction of the earth's motion about the sun.

- Although only A3-type signal enhancements were considered in the paper by Collins & Maynard, similar observations would be made with signal enhancements of the types A1 and A2 which are also related to auroral ionization.

- Although the antenna design for meteor-scatter circuits is not generally critical, for best results consideration may be given to so-called meteor hot-spot regions, i.e. regions of high meteor occurrance, which can be on either side of the great circle path between the two stations.

- It was noted that there is no contradiction between the observations made by Blair et al and those made by Collins & Maynard concerning the frequency dependence of signal enhancement and absorption during disturbed periods. The fact that lower frequencies are absorbed in some cases and enhanced in others depends on whether or not most absorption occurs below the scattering region.

- Duty cycle obtainable on auroral backscatter circuits depends on the degree of geomagnetic activity, latitude of the stations, time of day, solar cycle and on the frequency of transmission. No reliable data are available at middle geomagnetic latitudes but tests are being planned in Germany by Max-Plank Institute at Lindau/Harz. Some scattered data obtained from radio amateures are presented in the paper by Dr. Lange-Hesse.

HF Communications

SURVEY OF THE EFFECTS OF IONOSPHERIC PROPAGATION ON HF COMMUNICATION SYSTEMS

W. L. Hatton

Defence Research Telecommunications Establishment

Ottawa, Canada

Abstract: A high signal frequency propagated via the ionosphere is generally exposed to time and frequency dispersion and to the effect of spatial variations across the reflection volume. This lecture deals with the performance of various modulation systems in the presence of dispersion. Methods available for optimizing HF channel behaviour are discussed.

1. INTRODUCTION

The purpose of this paper is to outline, in general terms, the limitations that ionospheric propagation places on high frequency communications and also provide an introduction to how these limitations affect various modulation systems.

Reflections of high frequencies from the ionosphere can provide an economical solution to communications in the Arctic for all requirements except those of high capacity and high reliability. While HF radio can be unreliable part of the time due to the unfavourable behaviour of the ionosphere, it can provide communication over any range with elementary equipment most of the time. Furthermore the equipment is easy to operate and costs very little. On the other hand by the use of the most modern techniques that are now available a greater reliability could be realized than has ever been achieved in the past, (1). HF radio has provided the backbone of long range communications in the Arctic just as it has elsewhere in the world for thirty years.

If all the knowledge that now exists, and which will be revealed during the Study Institute on Ionospheric Radio Communica-

tions in the Arctic, were used HF radio would go a long way to
solving a large part of our requirements for communications in the
Arctic.

While other systems may give higher reliability, the cost of
a long-range low-capacity high frequency circuit is orders of magnitude less than for other systems.

Because of the vital interest in providing economical communications in the Arctic, it is essential that high frequency radio
be used efficiently and to do this it is essential to know something of the characteristics of the medium and also to know how to
communicate effectively through it, (2).

2.   CHARACTERISTICS OF HF RADIO CHANNELS

To take full advantage of HF radio propagation as a means of
communication a complete understanding of the channel characteristics is required. Although high frequencies have been in extensive use for world wide communications since the mid 1920's, until
the oblique sounder (3) came into use in the 1950's only a vague
idea of the characteristics of the HF channel was available. It
wasn't until the late fifties and early sixties with the development of the linear time variable channel mode (4) that the characteristics of the channel could be related to the performance of
various modulation systems. It may be added that it is only in
the last ten years or so that our technology has advanced to the
point where this new knowledge of the channel and its interaction
with communication systems could be exploited.

Before proceeding with a more detailed examination of the
high frequency channel an important characteristic of HF noise,
that is of particular interest to Arctic communication, should be
pointed out. This is the variation with latitude. As shown in
Figure 1, atmospheric noise decreases sharply with latitude (5).
Since most atmospheric noise originates in the tropics there is an
even sharper drop of noise with latitude during periods of ionospheric absorption (6). This decrease may more than compensate
for loss of signal strength during some disturbances. If man made
noise overrides the atmospheric noise this advantage is lost.

A variation exists in HF propagation that is not as important
in other modes. This is the extreme variability of the extent of
the frequency spectrum that propagates via the ionosphere. This
variation has resulted in the need for a flexible approach to frequency usage. Unfortunately the degree of flexibility provided by
international regulations and existing systems was decided on before a clear understanding of the variations of the HF band existed. Figure 2 shows the available band between the maximum ob-

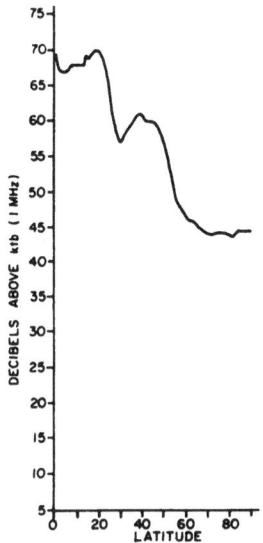

Figure 1. Variation of atmospheric noise with latitude.

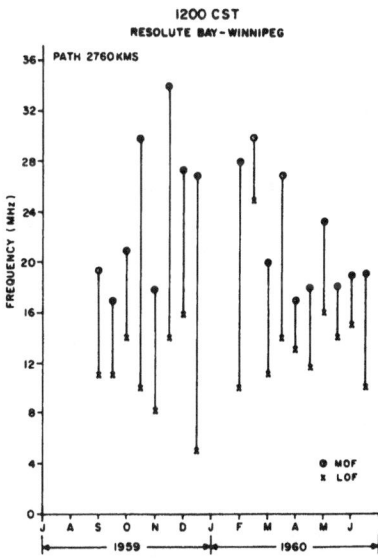

Figure 2. Available HF band-Resolute Bay to Winnipeg.

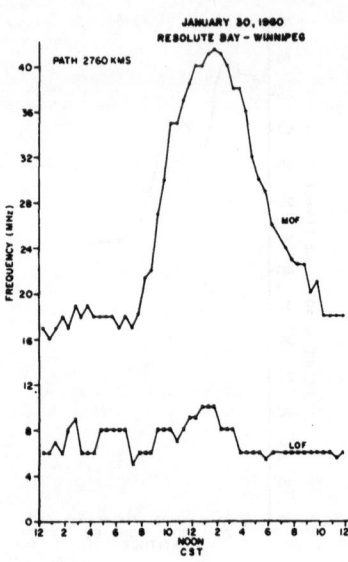

Figure 3.   Variation of MOF & LOF.

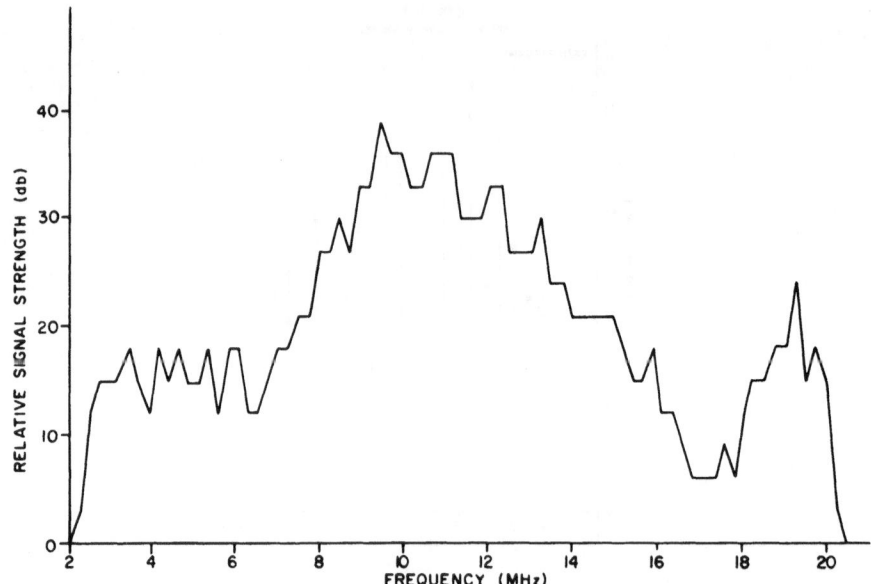

Figure 4.   Signal strength versus frequency.

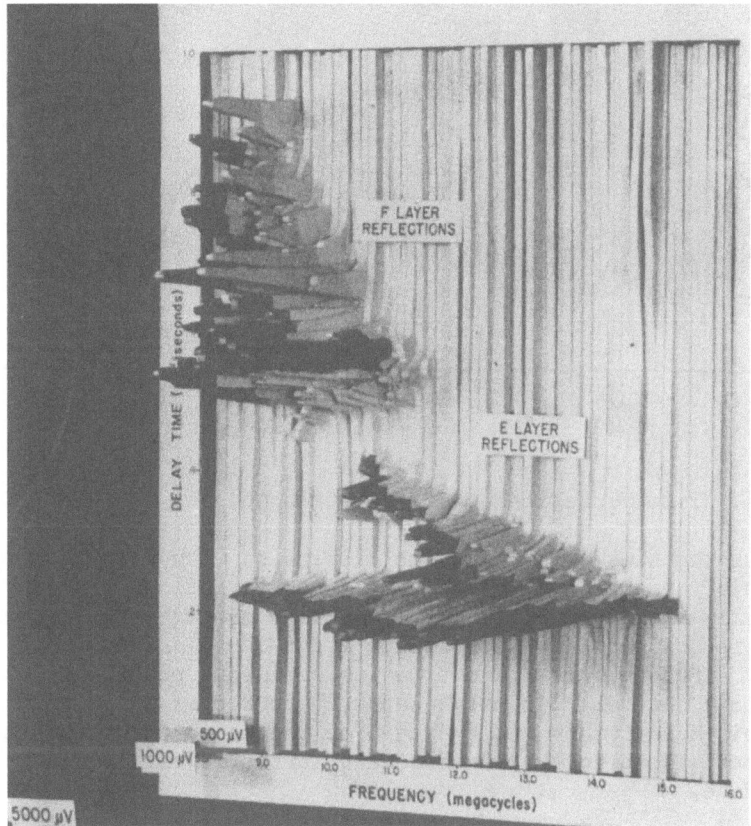

Figure 5. Dynamic ionogram.

served frequency (MOF) and the lowest observed frequency (LOF) at
noon on the 1st and 15th of month for the Ottawa-Resolute Bay path.
The extreme fluctuations shown here are not revealed by the average values that are usually presented. The variability of the HF
spectrum available is further illustrated in Figure 3. This figure shows the variation of the MOF and LOF for January 15, 1960,
for the same path.

    Not only does the extent of the band vary greatly but the de-
tailed characteristics of propagation varies greatly within this
band. Figure 4 illustrates the variation of the average signal
strength as a function of frequency. This figure does not show
the full extent of instantaneous variation with frequency because
each value plotted is an average over several minutes.

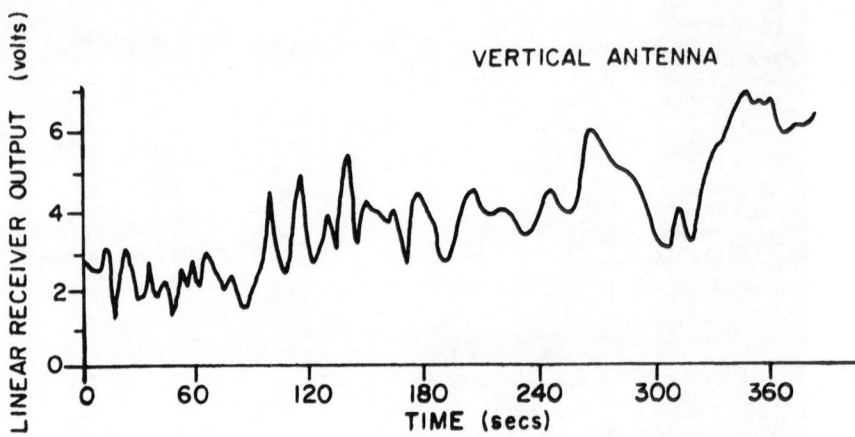

Figure 6. Time variation of signal strength.

Both the time delay and time dispersion vary with frequency
(7). This variation as well as amplitude variation is shown
graphically in Figure 5. This figure is a montage of a series of
impulse responses of the ionosphere at frequencies separated by
100 kHz. Even this is not a true instantaneous picture of the
ionospheric channel response because it took over a minute to complete this measurement and by the time the measurement was made at
the highest frequency the measurement at the lowest frequency was
out of date.

The relevant variation with time of signal strength on a
single frequency is shown in Figure 6.

Another important characteristic of the ionospheric response
is not shown by these figures and that is the frequency dispersion. An example of frequency dispersion as measured by the Stanford Research Institute is shown in Figure 7 (8).

To obtain a more complete illustration of the characteristics
of the HF channel, a series of time-frequency-amplitude responses
would have to be shown where the time delay, time dispersion, and
amplitude of all modes of propagation would simultaneously vary
with time.

To further complicate the picture, the received high frequency
wave is not propagated from a point source but results from reflection from a variable volume. This results in a variation of
the signals in the three space dimensions. A theoretical assess-

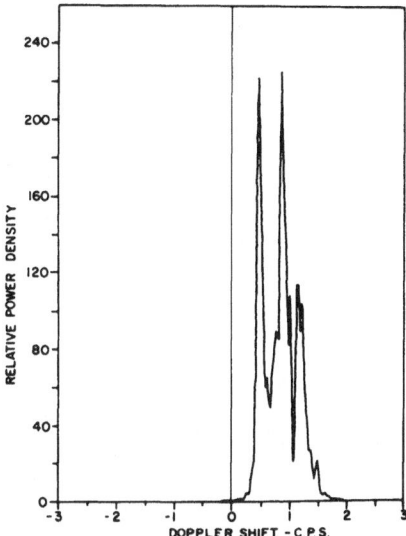

Figure 7. Frequency dispersion.

ment of the total complexity in the time-frequency-space domain is
nearing solution (9).

3. PERFORMANCE THROUGH AN HF CHANNEL

Once a frequency has been selected for use as a radio communication channel, the performance of any given modulation on this
channel depends on the attenuation and time-frequency dispersion
and how these vary with time (10). To give a better appreciation
of how these affect modulation systems, a simplified look will be
taken at the performance of CW, DSB, DSB suppressed carrier, SSB,
FSK, and delta phase modulation in the presence of time or frequency dispersion. An exact analysis will not be attempted but
rather a simplified physical picture will be presented.

It is necessary to point out before proceeding that the fourier transform of the pulse response gives the frequency correlation of the channel and also that the transform of the frequency
dispersion gives the time correlation of the channel. That is, a
time dispersion causes variation of response with frequency and a
frequency dispersion causes variations with time.

Figure 8 shows the effects of dispersion on a CW on-off signal. With two modes of differing time delays, the first period of
the received signal is equal to the first mode, the second is

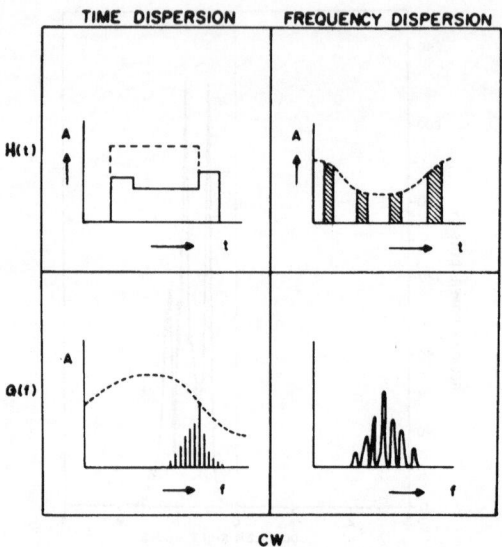

Figure 8. Effects of time & frequency dispersion on CW system.

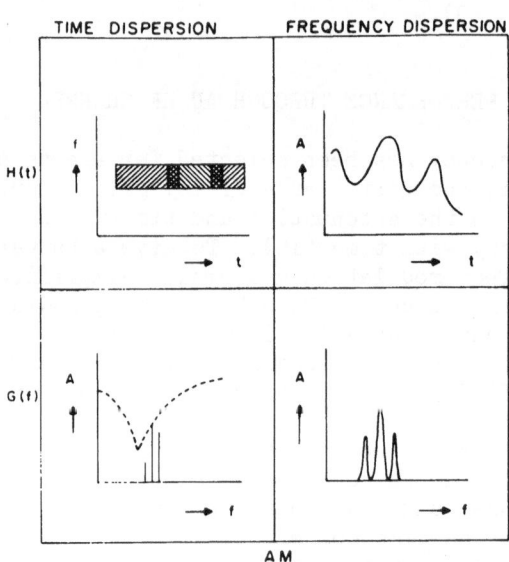

Figure 9. Effects of dispersion on AM system.

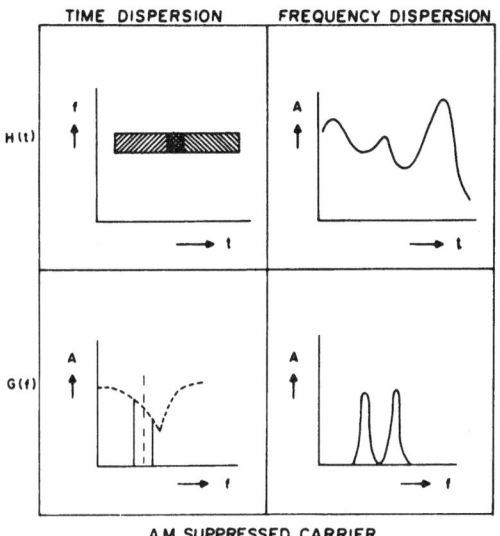

Figure 10. Effects of dispersion on suppressed carrier system.

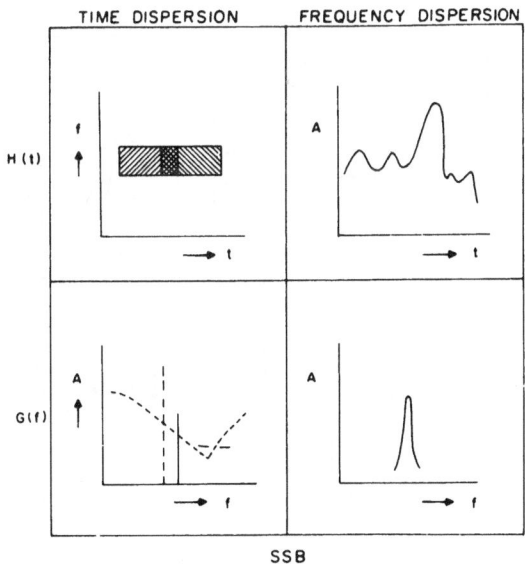

Figure 11. Effects of dispersion on SSB system.

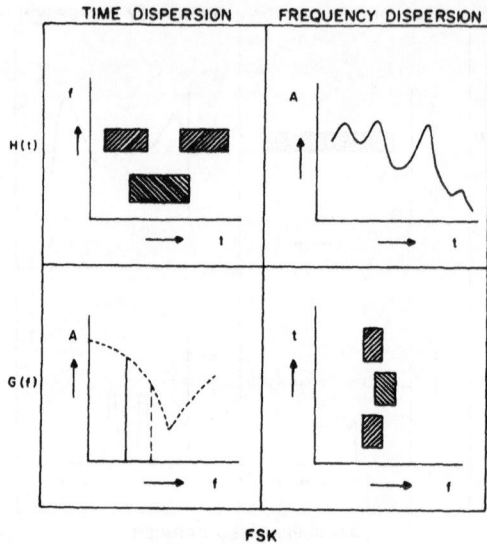

Figure 12. Effects of dispersion on FSK system.

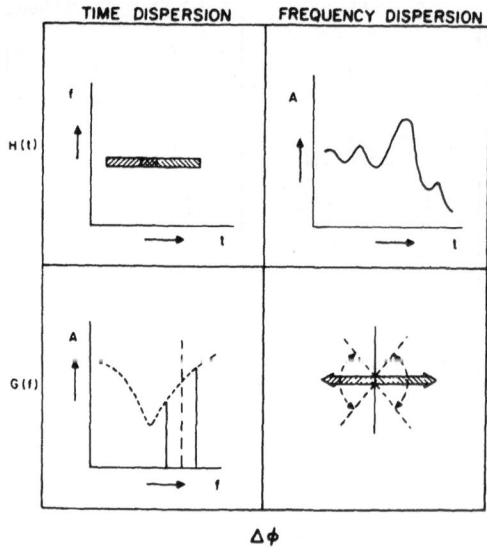

Figure 13. Effects of dispersion on differential phase system.

equal to the vector sum of the two modes, and the last portion is
equal to the second mode of propagation. Since normal on-off
transmissions are for hand keyed morse code the delays are short
with respect to the length of the baud and the major part of the
signal is the middle portion. This effect can be more easily seen
from the frequency response due to time dispersion. This shows a
variation with frequency and this variation will fluctuate as the
relative time delays vary. The frequency dispersion produces variations of signal strength with time which is also shown in
Figure 12.

Figure 9 shows the effects of DSB transmissions. In addition
to the time fading caused by frequency dispersion, intersymbol interference and distortion of opposite sidebands is caused by time
dispersion.

Figure 10 shows similar effects for suppressed carrier as for
normal DSB except that since the carrier is missing distortion can
be reduced.

In Figure 11, SSB is shown to be superior to both DSB systems
because only one side band is transmitted. Intersymbol interference and variations with time and frequency remain.

As shown in Figure 12, FSK or frequency shift modulation is
affected by time and frequency fading and intersymbol interference. The effects of fading however can be reduced by the use of a
symmetrical decision process in the detection of energy in two
frequency channels.

Finally in Figure 13, the effects of dispersion are shown on
differential phase modulation. In addition to intersymbol interference and fading a new effect is produced by frequency dispersion. In differential phase modulation, the phases of the signal
at two different times are compared and since the relative phase
of a signal varies with time separation errors can result.

4. CURES

The representation used to show the effects of time and frequency dispersion immediately suggest many ways of combating the
resulting deterioration of performance.

The methods used fall into two categories, direct and indirect.

The most obvious of the direct methods is the selection of a
good frequency. If there is no dispersion in the channel, the on-

ly problem left is ensuring an adequate signal-to-noise ratio.
The selection of a best channel can at the same time ensure the
best signal-to-noise ratio.

To combat time dispersion a large number of direct antidotes
are available. Some, such as using transmissions short compared
to the baud length, using long bauds, sampling the signal in the
middle of the baud, or integrating energy over only a portion of
the baud, are simple. Others such as QFM, (11), RAKE, (12), inverse ionospheres, equalization, or ADAPTACOM (13) are more complicated.

To combat frequency dispersion directly, simple cures are
available such as large frequency shifts in FSK and short bauds in
differential phase modulation.

In the part, most methods of combating the ills of HF propagation were indirect. These depended on the statistical independence or inverse dependence of the signals. These included diversity techniques such as space, polarization (14), frequency and
time (15, 16, 17) and error correction and detection (18) and ARQ
(19). ARQ in the simple form of asking for a repeat has been used
for the improvement of communications since the earliest times.

5. SUMMARY

The HF mode of propagation is a time-variable, frequency-
variable dispersive non-stationary process which has great potential for economical communications at all ranges but only if the
channel characteristics are allowed for in the design of HF communication system.

6. REFERENCES

1. C.A. Parry, Chairman; D.P. Worthington, G. Brunette, R.C.
   Kirby, R. Kulinyi, O.G. Villard, W.R. Vincent and W.L.
   Hatton, Panelists on: "The Present Status and Future Utilization of Advanced Technology in High Frequency Systems Engineering". 1966 IEEE International Convention, March 1966.

2. The Editor, "The Characterization of transmission channels",
   IEE Trans. on Communication Systems, vol. CS-11, September
   1963, pp 259-260.

3. W.L. Hatton, "Oblique-sounding and HF radio communication",
   IRE Trans. on Communication Systems, Vol. CS-9, September
   1961, pp 275-279.

4. P.A. Bello, "Characterization of randomly time-variant linear channels", IEEE Trans. on Communication Systems, vol. CS-11, December 1963, pp 360-393.

5. D.L. Lucas and J.D. Harpei, Jr., "A numerical representation of CCIR Report 322 High frequency (3-30 Mc/s) atmosphere radio noise data". NBS Technical Note 318, August 1965.

6. D.H. Jelly, "The effects of polar-cap absorption on HF oblique-incidence circuits", J. Geophys. Res. vol. 68, March 1963, pp 1705-1714.

7. D.J. Doyle, E.D. DuCharme and G.W. Jull, "The structure of HF signals revealed by an oblique incidence sweep frequency sounder", IRE Trans. on Antennas and Propagation, vol. AP-8, July 1960, pp 449-451.

8. R.F. Daly, "A power spectrum program for estimating the doppler profile of a radio channel", Stanford Research Institute Research Memorandum 15, Ocrober 1964.

9. G.O. Young and A.A. Ksienski, "Space time correlation for information-carrying signals", IEEE Trans. on Antennas and Propagation, vol. AP-15, January 1967, pp 163-171.

10. W.R. Vincent, R.F. Daly and B.M. Sifford, "Modeling Communication Systems", These Proceedings.

11. A.J. Strassman and A.C. Chapman, "A long range digital communication system", IRE Trans. on Communication Systems, vol. CS-9, December 1961, pp 383-389.

12. R. Price and P.E. Green, Jr., "A communication technique for multipath channels, Proc IRE, vol. 46, March 1958, pp 555-570.

13. B. Goldberg, "300 kHz - 30 MHz MF/HF", IEEE Trans. on Communication Technology, vol. COM-14, December 1966, pp 767-784.

14. J. Ames and P. Kennedy, "A new look at polarization diversity reception at high frequencies", Conference Record IEEE Annual Communications Convention, June 1965, pp 665-670.

15. K.W. Otten, "Reliable long-distance data-transmission system with frequency and time diversity", 1965 IEEE Digest International Symposium on Global Communications, June 3, 1964, p 57 & p 100.

16. B.D. Fritchman and J.F. Leonard, "Test results of a time-dispersed forward error control system", IEEE Trans. on Communication Technology, vol. COM-13, June 1965, pp 233-234.

17. Walter Lyons, "Error correction via time diversity", IEEE
     Trans. on Communication Technology, vol. COM-13, June 1965,
     pp 234-237.

18. A. Kohlenberg and A.S. Berner, "An experimental comparison of
     coding versus frequency diversity for HF Telegraphy transmis-
     sion", IEEE Trans. on Communication Technology, vol. COM-14,
     August 1966, pp 532-533.

19. H.C.A. van Duuren, "Error probability and transmission speed
     on circuits using error detection and automatic repetition of
     signals", IRE Trans. on Communication Systems, vol. CS-9,
     March 1961, pp 38-50.

7. BIBLIOGRAPHY

D.A. Ashford and V.V. Vilips, "Data transmission on HF radio using
parallel duobinary FM", IEEE Digest, 1964 International Symposium
on Global Communications, p 58, June 3, 1964.

Bruce B. Barrow, "Diversity combination of fading signals with unequal mean strengths", IEEE Trans. on Communications Systems, vol.
CS-11, pp 73-78, March 1963.

P.A. Bello and D.B. Melin, "The effect of frequency-selective fading on the binary error probabilities of incoherent and differen-
tially coherent matched filter receivers", IEEE Trans. on Communication Systems, vol. CS-11, pp 170-186, June 1963.

P.A. Bello, "Binary error probabilities over selectively fading
channels containing specular components", IEEE Trans. on Communication Technology, vol. COM-14, pp 400-406, August 1966.

P.A. Bello, "Error probabilities due to atmospheric noise and flat
fading in HF ionospheric communications", IEEE Trans. on Communication Technology, Vol. COM-13, pp 266-279, September 1965.

P.A. Bello, "Some techniques for the instantaneous real-time measurement of multipath and doppler spread", IEEE Trans. on Communication Technology, vol. COM-13, pp 285-292, September 1965.

P.A. Bello, "Selective fading limitations of the Kathryn modem and
some system design considerations", IEEE Trans. on Communication
Technology, vol. COM-13, pp 320-333, September 1965.

P.J. Brice and G.O. Evans, "The performance of long-distance high-
frequency radio links terminating in the United Kingdom", Record
IEE Convention on HF Communication, pp 7-19, March 1963.

A.C. Croisdale, "Automatic error correcting systems used on h.f. radio links", IEE Convention on HF Communication, pp 187-210, March 1963.

R.D. Chipp and F. Cosgrove, "Economic analysis of communication systems", IRE Trans. on Communication Systems, vol. CS-10, pp 416-421, December 1962.

H. Fiege-Kollmann, "The optimum bit length for HF-transmission", Proc. of Eight IEEE National Communications Symposium, pp 108-115, October 1962.

N.C. Gerson, "High latitude hf communications", IEEE Trans. on Communication Systems, vol. CS-12, pp 107-109, March 1964.

B. Goldberg, "HF radio data transmission", IRE Trans. on Communication Systems, vol. CS-9, pp 21-28, March 1961.

C.J. Hughes and D.W. Morris, "The phase characteristics of hf radio waves received after propagation", Record IEE Convention on HF Communication, pp 20-46, 25 March, 1963.

G. Jacobs and E.T. Martin, "The dwindling high-frequency spectrum", IRE Trans. on Communication Systems, vol. CS-9, pp 399-408, December 1961.

F.M. Lightfoot and P.S. Kogut, "A compact hf digital modem with interference and multipath discrimination characteristics", IEEE Trans. on Communication Technology, vol. COM-13, pp 104-108, March 1965.

W.C. Lindsey, "Coding for specular and scatter channels", IEEE Trans. on Communication Technology, vol. COM-13, pp 237-238, June 1965.

Walter Lyon, "Optimizing high-frequency telegraph transmission", IEEE Trans. on Communication Systems, vol. CS-12, pp 104-107, March 1964.

P.M. Ridout and L.K. Wheeler, "The choice of multi-channel telegraph systems for use on hf radio links", Record IEE Convention on HF Communication, pp 153-167, March 1963.

H.K. Robin, D. Boyley, T.L. Murray and J.D. Ralphs, "A multitone signalling system employing quenched resonators for use on radio-teleprinter circuits", Record IEE Convention on HF Communication, pp 168-186, March 1963.

R.K. Salaman, "A new ionospheric multipath reduction factor (mrf)",
IEEE Trans. on Communication Systems, vol. CS-10, pp 220-221,
June 1962.

H.N. Shaver, B.C. Tupper and J.B. Lomax, "Evaluation of a Gaussian
HF Channel Model", IEEE Trans. on Communication Technology, vol.
COM-15, pp 79-88, February 1967.

R.W. Steele, "Performance of vocoder equipments during transmission errors", IEEE Digest, 1964 International Symposium on Global
Communication, p 89, June 4, 1964.

HF SPATIAL AND TEMPORAL PROPAGATION CHARACTERISTICS AND SOUNDING

ASSISTED COMMUNICATIONS

G. W. Jull

Defence Research Telecommunications Establishment

Ottawa, Canada

Abstract: This paper reviews and extends knowledge of the temporal and spatial characteristics of HF signals propagated via the
ionosphere. The characteristics of HF signals are considered with
reference to the problem of predicting signal strength on a group
of assigned communications frequencies, using a channel sounding
system. The particular application of sounding discussed is to
provide predictions applicable (i) for many minutes (ii) for a
communications path separated from the channel sounding path and
(iii) for a communications path in a direction opposite to the
channel sounding path. Studies are reported which showed that
movements of small and intermediate sizes of irregularities resulted in non-stationary HF signal statistics. They also resulted
in differences in averaged signal strength at receiving stations
separated by 2 to 32 km. For 22 to 32 km separations, the differences were 4 to 5 db (for 68% of sample periods), when signal levels were averaged for eight minutes.

Other studies are reported which showed that ionospheric non-
reciprocity resulted in unequal levels of signals travelling in
opposite directions between linear antennas. The path non-reciprocity which arises when polarization fading was present was
appreciably reduced by averaging signal levels for eight minutes.

It is concluded that inaccuracies in quantitative signal
strength prediction can arise due to the effects of ionospheric
irregularities and path non-reciprocity. These inaccuracies can
be appreciably reduced by sampling and averaging signal strength
measurements for at least eight minutes. Additional communications power margins must also be provided to allow for the remain-

ing uncertainties in signal strength during the time of message
transmission.

1. INTRODUCTION

Over the past five years, a number of studies (1, 2) have
been conducted to establish the usefulness of multi-frequency
sounding systems for selecting optimum, short-term operating frequencies for HF communication systems. These studies have confirmed that knowledge of the maximum and lowest frequencies, propagation modes, and multipath structure is indeed valuable for
selecting communications frequencies. This is particularly the
case during moderate and severe ionospheric disturbances. Because
these parameters of a propagation path are slowly-varying functions of the ionospheric structure, they are valid over wide areas,
and by continuously up-dated observations of these parameters, it
is often possible to make useful extrapolations for many minutes
into the future. This has led to the concept of common user
sounding, particularly by the U.S. Defence Communication Agency
and Stanford Research Institute (3). The common user concept presupposes that averaged propagation characteristics are correlated
over wide areas, and for usefully long periods of time, and therefore one sounding system can furnish averaged propagation information which is applicable for widely separated communications systems.

A quite different type of prediction is required when path
sounding is required to predict quantitative signal strength information on particular assignments. Signal strength information,
combined with interference measurements, in principle makes it
possible to predict communications quality of particular channels.
In this case much more must be known, _a priori_, and through up-
dated measurements, about the temporal and spatial fine structure
of signals. This paper outlines some of the propagation and related operational problems which are presented by this requirement.
It extends work previously reported on the effect of irregularities and ionospheric non-reciprocity on prediction of HF channel
characteristics (4).

It will be immediately apparent that some form of statistical
measurements and predictions are required, since the length of
typical messages are longer than the correlation time for important short-term ionospheric changes. Secondly, since HF signal
statistics are non-stationary, the additional power margins required to allow for uncertainties due to non-stationary statistics
must be known. Finally, the uncertainties in prediction which
arise due to sounding on a path which is not the same as the communications path must also be known. The paths may be different
either because the sounding and communications systems are not

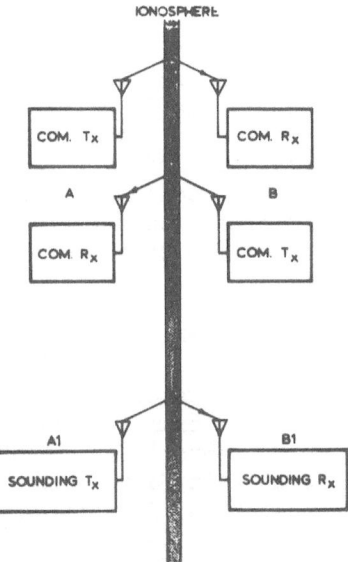

Figure 1. Representation of a duplex communications system
assisted by a sounding system.

colocated, or, for situations in which a path is sounded in one
direction, and communication takes place in the opposite direction.

This paper reviews and extends present knowledge of HF signal
characteristics which can give rise to inaccuracies in prediction
of signal strength on a channel. This study is important in de-
ciding whether these inaccuracies are small enough to justify
quantitative measurements for some applications. One possible
system for which these considerations apply is illustrated in
Figure 1. A duplex communications system operating between terminals A and B is assisted by measurements obtained with a one-way
sounding system transmitting from A1 to B1. The sounding system
must provide appropriate signal strength information which is correlated between the separated paths used for sounding and communications. This application has recently been studied at DRTE, and
will be discussed by Stevens (5). The communications system was
an RCA air-ground-air system, with sounding only from ground to
air. The present paper discusses the propagation problems important in this application. The studies to be reported have been
carried out on mid-latitude paths during relatively undisturbed
propagation conditions. They are preliminary to studies of propagation over auroral zone paths during disturbances.

2. REVIEW OF TEMPORAL AND SPATIAL CHARACTERISTICS OF HF SIGNALS

HF signals vary with time and distance due to diffraction and reflection by continuously-changing irregularities in ionospheric structure. The time and distance scales of these variations are presented in Figure 2. These variations arise from (i) small scale irregularities, (ii) intermediate and large scale irregularities and (iii) large scale averaged electron density variations. These scale sizes are subdivided somewhat arbitrarily, with overlaps in the effects for different ionospheric processes, and they pertain to undisturbed propagation conditions.

2.1 Small Scale Irregularities

The characteristics of small scale irregularities have been studied for many years (see, for example, Ratcliffe (6)). They have an average scale size of approximately 300 m. The movements of these irregularities result in fast fading of HF signals, which is uncorrelated on the ground for distances greater than 600 m, and time intervals greater than 1-10 sec. This fading is usually assumed to arise from time-varying phase interference between component waves making up the signal. The envelope distribution function for the signal has been observed (6), to be approximated by a Rayleigh distribution (when all component waves have the same amplitude), or by a Nakagami-Rice distribution (when one component wave has a power greater than that of all other waves). On some occasions, there can be two or more components of comparable strength in the signal. In these cases periodic or quasi-periodic fading results (7).

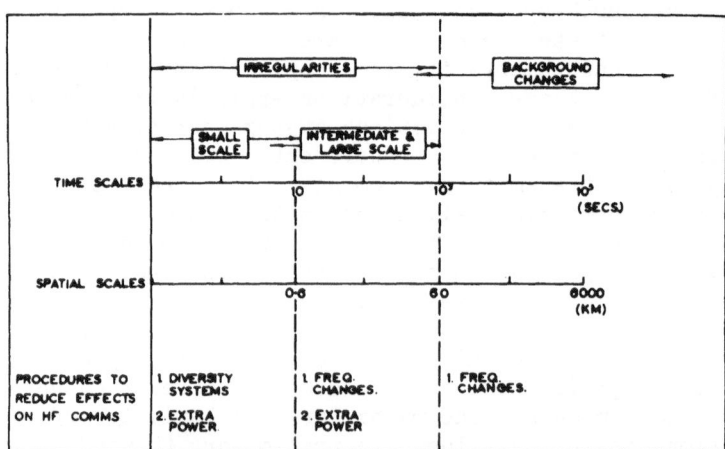

Figure 2. Time and spatial scales of ionospheric variations, and
           procedures to reduce their effects on the quality of
           HF communications.

The effects of small scale irregularities on communications systems are reduced by space (or polarization) diversity antenna systems, or by time (or frequency) diversity transmission systems. Their effects on quantitative channel measurement are to introduce sampling errors which in principle, can be reduced by averaging over many correlation periods.

2.2 Intermediate and Large Scale Irregularities

Small scale irregularities are often imposed on a wrinkled electron density background, due to intermediate, or large-scale irregularities. These take the form of randomly shaped blobs, travelling disturbances or layer tilts. They can be many tens of km across, have velocities of 50-350 m/sec, and periods of 10-100 sec (8). Their movement results in changes of the rms levels, as well as the phases of component waves making up the signal. For example, well-marked increases in signal level are observed when a concave part of the reflecting surface moves across the signal path. This scale of irregularities can be important for two reasons. First, the fading of signals will be correlated on space diversity antennas, since these are normally separated by less than the correlation distance. Secondly the short-term signal levels on one path used for sounding may not be the same as the signal levels on the path for communications, because these paths are separated by more than the correlation distance. Therefore the scale of these irregularities are such as to present a possible limitation on the accuracy of signal strength prediction. The results of studies of these irregularities are reported in Section 3.

2.3 Large Scale, Averaged Electron Density Variations

The average background electron density which is solar-controlled has a diurnal cycle, although important changes in absorption and maximum frequencies for HF signals occur in times of one hour or less. These time scales of variation in HF signal characteristics are successfully measured, and can in part be predicted, by using multifrequency pulse sounding systems, adopted for assisting communications, to select appropriate operating frequency bands.

3. THE DIFFERENCES IN SIGNAL LEVELS ON SEPARATED PATHS

There have been very few studies concerned with the effects of intermediate and large scale irregularities on HF signals. Meadows and Moorat (9) discussed their significance in studies to measure absorption of 1E (one-hop E Layer) signals. Balser and

Smith (10) found that a high fraction (~30%) of 10 min amplitude
distributions of pulsed signals did not conform to Makagami-Rice
or Rayleigh distributions. They concluded that the statistics are
best exemplified as non-stationary. Their observations probably
showed effects of these irregularities.

Studies have recently been conducted to compare the levels of
pulsed signals received at separated sites near Ottawa, Ontario
from a transmitter at Halifax, Nova Scotia (a 960 km 1F path) and
on other occasions, from a transmitter in Yorkshire, England (a
5400 km 2F path). The signal levels received at each site were
averaged for periods of 4, 8 and 12 min. If only small scale irregularities were present, 8-12 min averaged levels should be
equal, within sampling errors, since these time scales are many
times greater than the correlation times for these irregularities.
On the other hand, if larger irregularities were present, the correlation times would be much longer and 8-12 min average signal
levels could be different on the separated paths.

3.1 The Variation of Signal Levels on Separated Paths

Observed differences in 8 min averaged signal levels on separated 1F paths are illustrated in Figure 3. The reflection
points of these paths were separated by 3, 6 and 16 km. It is
highly probable that the observed differences arise from the pre-

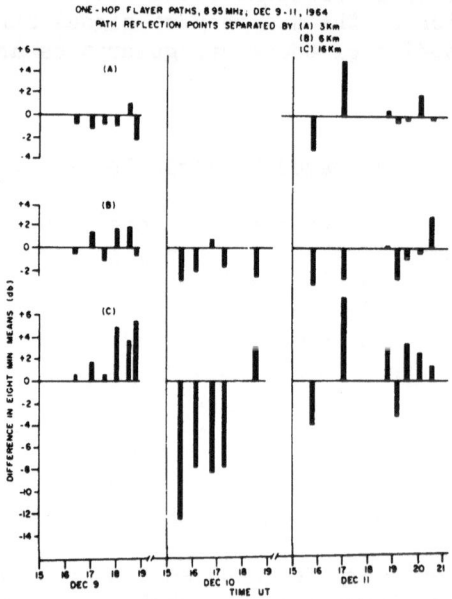

Figure 3. Differences in eight-minute averaged signal levels on
           1F paths.

Date	Averaged Standard Deviations, σ of mean, m		
	4 min	8 min	12 min
December 9	0.42 m	0.46 m	0.53 m
December 10	0.49 m	0.58 m	0.62 m

Table 1.   The increase in standard deviation of 1F signals with
increase in time.   Ottawa - Halifax 1F path, 8.95 MHz,
Dec. 9-10, 1964.

sence of intermediate sized irregularities during the recording
periods. For the paths separated by 16 km, the levels differed by
-12 to +3 db (Dec. 10) and -4 to +7 db (Dec. 11). For smaller
separations of paths, the differences were less, showing that the
larger variability for greater path separations is due to intermediate sizes of irregularities, and not sampling errors.

These irregularities were responsible for nonstationary signal statistics in periods of 4 to 12 min. Table 1 presents the
standard deviation, σ, of the mean, m, for the recording days represented in Figure 3 (σ was the averaged standard deviation for
intervals of 4, 8 and 12 min on Dec. 9 and 10). The increase in σ
with increasing length of interval results from a change in mean
levels with time: that is to say, the rms levels of the component
waves in the signal were changing slowly, but appreciably, during
time intervals of less than 12 min.

3.2 The Distribution of Differences in Signal Levels

The results of a statistical study of the differences in signal levels are summarized below:

The Effect of Averaging Time. The statistical spread of signal level differences on separated paths are presented in
Figure 4. The spread is specified in terms of the db ranges
which contain 68 and 95 per cent of the samples. Comparison
of curves (A), (B) and (C) shows that the large spread in 4
min averaged levels was considerably reduced by averaging for
8 to 12 min.

The Dependence on Path Orientation. Tests were carried out
to determine whether or not the spread in signal levels were
dependant on the orientation of the receiving stations relative to the transmission direction. No dependence was found,

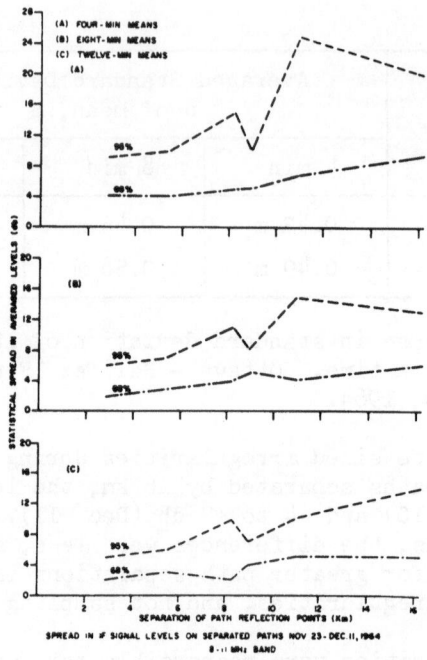

Figure 4.   Spread in 1F averaged signal levels on separated paths.
(A) four-minute averages   (B) eight-minute averages
(C) twelve-minute averages.

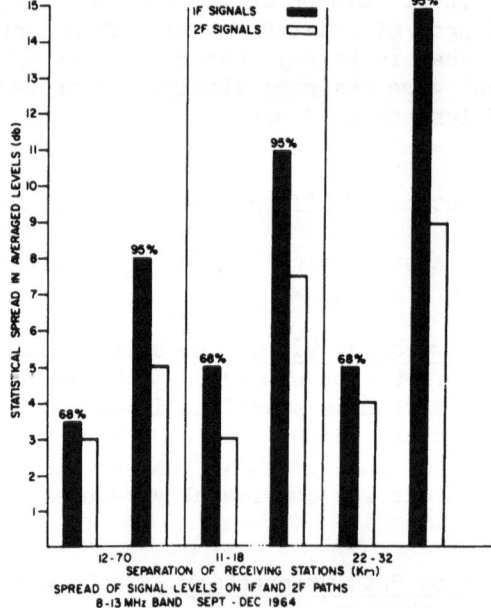

Figure 5.   Comparison of the statistical spread of signal levels
on 1F and 2F paths.

either for the 1F Halifax-Ottawa path, or the 2F Yorkshire-Ottawa path.

Comparison between the 1F and 2F Paths. The statistical spread in signal levels on the 1F and 2F paths are compared in Figure 5. It is seen that the 68 percentile spreads for the 2F signals increased from 3 db (1.2 to 7 km site separation) to 4 db (22 to 32 km site separation). Approximately the same spreads were found for 1F signals. On the other hand, the 95 percentile valves for the 2F signals were significantly lower than found for 1F signals.

4. THE DIFFERENCES IN LEVELS OF SIGNALS TRAVELLING IN OPPOSITE DIRECTIONS

4.1 Predictions of Ray Theory

Over the past few years much progress has been made in understanding the reciprocal, and non reciprocal characteristics of HF ionospheric ray paths. A recent ray theory of reciprocity (11) predicts the following characteristics of ionospheric paths:

(1) A path between two antennas connected by an ionospheric path can be non reciprocal, but this is not due to differences in (i) the ray paths in opposite directions, (ii) the attenuation in opposite directions or (iii) the phase paths in opposite directions. On the basis of ray theory, these are the same for signals travelling in the two directions.

(2) The reciprocity, or nonreciprocity of a path arises only because of the interaction of the downcoming, or upgoing, magnetoionic waves with the antennas at each side of the path.

(3) A reciprocal ray path is formed between the terminals of linearly polarized antennas if the polarizations of the antennas are both in, or perpendicular to, the magnetic planes at each side of the path (a magnetic plane is defined by the Earth's magnetic field direction and the wave normal direction). Otherwise the path can be nonreciprocal.

A nonreciprocal one-hop path is illustrated in Figure 6. Figure 6A shows ordinary and extraordinary ray paths between antennas at a and b which are horizontally polarized along the x axis. Figure 6B shows the angles between the antennas and the magnetic planes at the points A and B (where the rays enter and leave the ionosphere). These angles, designated ψ_A and ψ_B at A and B, define the antenna polarizations, relative to the magnetic planes. When $\psi_A = \psi_B = \frac{n\pi}{2}$ (where n is an integer) the path is

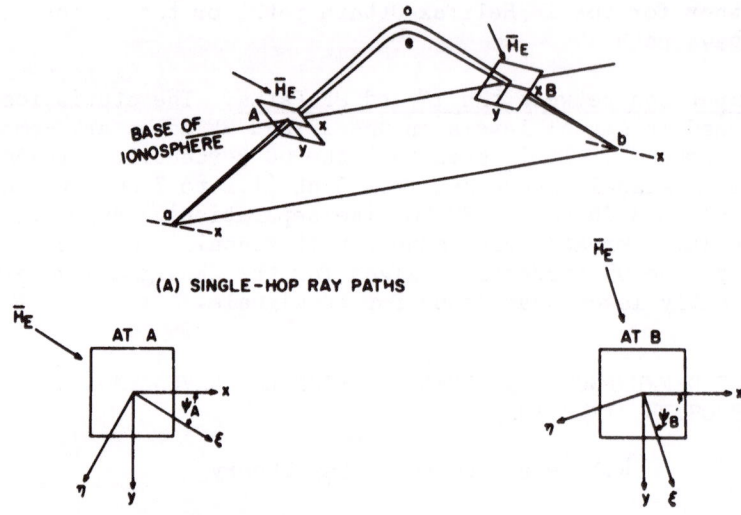

(A) SINGLE-HOP RAY PATHS

(B) ORIENTATION OF MAGNETIC PLANES

Figure 6. The orientation of magnetic planes at each side of a
           single-hop ionospheric path.
           (A) Ordinary, o, and extraordinary, e, ray paths be-
           tween antennas which are horizontally polarized along
           the x direction.
           (B) Angles, ψ, in the plane of the wavefronts, between
           the coordinate axis, ξ, (in the magnetic plane) and the
           x axis, at the points A and B, where the rays enter or
           leave the ionosphere.

reciprocal, independant of wave polarization. Most HF antennas
are designed to radiate and receive either horizontally, or ver-
tically, polarized waves. For the majority of azimuthal propaga-
tion directions, $\psi_A \neq \psi_B \neq \frac{n\pi}{2}$, and therefore most paths are nonreciprocal.

     Characteristics of nonreciprocal one-hop paths are summarized
in Table 2 for the case of circularily polarized magnetoionic waves. (This is approximately the case for many HF paths). The antenna currents, i_o and i_e, due to ordinary, o, and extraordinary,
e, waves are compared for signals travelling in the directions
(ab) and (ba). Two classes of antenna polarization are distinguished; linearity polarized antennas (for which ψ_A and ψ_B are
real), and elliptically polarized antennas (which ψ_A and ψ_B are
complex or imaginary).

     <u>Linearily Polarized Antennas</u>. For this class of antenna po-
larizations, the magnitude of currents are equal for the two di-

ANTENNA POLARIZATIONS	MAGNITUDES, $|i|$			PHASES, ϕ	
	o	e	o + e	o	e
LINEAR ψ_A, ψ_B real	$|i_o|_{,ab} = |i_o|_{,ba}$	$|i_e|_{,ab} = |i_e|_{,ba}$	$|i_o+i_e|_{,ba} \neq |i_o+i_e|_{,ba}$	$\phi_{o,ab} > < \phi_{o,ba}$	$\phi_{e,ab} < > \phi_{e,ba}$
ELLIPTICAL ψ_A, ψ_B complex (or imaginary)	$|i_o|_{,ab} \neq |i_o|_{,ba}$	$|i_e|_{,ab} \neq |i_e|_{,ba}$	"	"	"

Table 2. Received antenna currents on nonreciprocal paths.

rections, if the signal contains only one type of magneto-ionic wave. However, the phases are unequal. It is significant that if the phase path is <u>longer</u> in one direction for one type of wave (say the o wave), it is <u>shorter</u> in that direction for the e wave. If the signal contains only one type of wave, the reciprocity in phase is undetected, since only magnitudes, $|i|$, and not phases, ϕ, of currents can be compared.

If the signal contains both o and e waves with comparable strength, differential changes in o and e ionospheric phase paths result in polarization fading. In this case, the difference in two-way phase paths is readily observed as a time displacement of polarization fading cycles, (with the result that $|i_o + i_e|$, ab $\neq$ $|i_o + i_e|$, ba).

<u>Elliptically Polarized Antennas</u>. For this class of antenna polarizations, the magnitude of the currents, as well as the phases, can be unequal in opposite directions, when only one type of wave is present in the signal. As a limiting case, it is well-known that a circularly polarized antenna can transmit, but not receive, one type of magneto-ionic wave. Path nonreciprocity therefore results when either type of wave is present in the signal (12).

4.2 Observed Nonreciprocal Path Characteristics

The present experimental evidence supports ray theory predictions of nonreciprocal path characteristics. Evidence to be discussed was obtained during two-way pulse transmission tests between horizontally polarized antennas on two nonreciprocal paths. The first was a path between Ottawa, Ontario and Halifax, Nova Scotia (a 960 km path), the second was a path between Atlanta, Georgia and Ipswich, Massachusetts (13) (a 1500 km path). Calculations show that these paths are expected to be about equally nonreciprocal.

An example of observed nonreciprocity on the Ottawa-Halifax path is presented in Figure 7. During this recording period the 1F signals exhibited periodic polarization fading, thus revealing the nonreciprocity in phase paths. The figure presents the cross-correlation function, $r_{ab}(\tau)$ between signals received at the two ends of the path, also the autocorrelation functions, $r_{aa}(\tau)$ and $r_{bb}(\tau)$. The time τ_p between maxima in $r_{aa}(\tau)$ or $r_{bb}(\tau)$ provides a measure of the averaged polarization fading period. The maximum of $r_{ab}(\tau)$ is not at $\tau = 0$ but at a delay corresponding to 0.5 τ_p. Calculations based on ray theory predict a displacement of ± 0.33 τ_p for this path, in fair first order agreement with observations. Whenever short-term periodic polarization fading occurred similar results were found for this path and the Atlanta-Ipswich path.

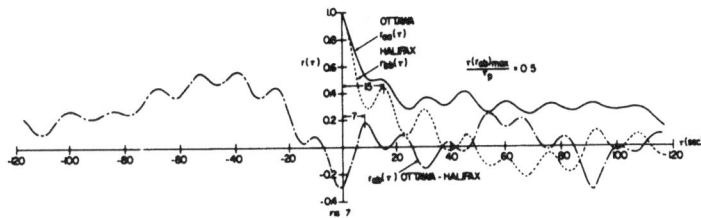

Figure 7.   Correlation of two-way fading of single-hop F layer
            signals, Ottawa-Halifax path, during a period in which
            the signals exhibited polarization fading. 11.4 MHz,
            18: 15 UT, March 28, 1962.

On the other hand, when there was only one type of wave in
the signal, the two-way fading was correlated. This was also the
case when both types of wave were present in the signal, whenever
the polarization fading periods were long compared with the length
of recording (10 min).

Data from the two paths were analyzed to determine whether
averaging polarization fading effects over 8 min intervals would
reduce the differences in signal strength in opposite directions.
Some results of this analysis for the Atlanta-Ipswich path are
presented in Table 3. This data was obtained during recording
periods for which 1E, 1F, 1E-1F (identified as propagation by one
E and one F layer reflection) and 2F signals were simultaneously
present. All signals (except two out of the three 1E layer signals) showed evidence of polarization fading, with low crosscorrelation coefficients $r_{ab}(o)$, and the maximum of $r_{ab}(\tau)$ displaced
from $\tau = 0$. These displacements in polarization fading cycles are
in fair agreement with predicted displacements of $\pm 0.35\ \tau_p$ for
this path. The 8 min averaged signal levels in opposite directions were compared, and the differences in averaged levels are
presented in Table 3. These differences were much lower for 1E-1F
and 2F signals than for 1F signals. This can be accounted for by
shorter polarization fading periods for the 1E-1F and 2F signals
(10-20 sec), compared with those for 1F signals (up to 108 sec).

The results of these tests are in general agreement with
those reported by Laver and Stanesby (14), who observed short-term
differences in two-way averaged path less of a few db on a UK-US
path. On occasion differences up to 10 db were observed, and
these can be attributed to insufficient averaging of polarization
fading effects. However, they did find occasions on which the
signals received at the US end of the path were 5 db or more higher than at the UK end, for hours at a time. These systematic dif-

Date (1960)	Time (UT)	Freq. (MHz)	Prop. Mode	Fading obs. Random-R Polar-P	Corr. time τ_r(sec)	Cross. Coss. Coeff. $r_{ab}(o)$	Polar Fading Period τ_p(sec)	Disp. Polar Fading (cycles)	Diff. in Levels (corrected)+ (db)
5/3	19:35	8.3	1E	R	40	0.90	-	-	+5
			1F	P+R	13	0.25	108	+0.5	-2
			1E-1F	R+P(?)	8	0.53	12(?)	+0.2	+2
			2F	R+P	6	0.37	24	+0.2	0
26/2	19:25	8.4	1E	R	60	0.88	-	-	+3
			1F	P+R	20	0.41	36	-0.4	-3
			1E-1F	P+R	6	0.32	18	+0.2	-3
			2F	P+R	4	0.50	12	+0.3(?)	-2
5/3	19:50	8.3	1E	R+P(?)	2	0.27	68(?)	+0.2	+4
			1F	R+P(?)	19	0.15	50	-0.4	+7
			1E-1F	R+P	5	0.27	10	+0.4	+1
			2F	R+P(?)	4	0.48	20	-	0

+ corrected by allowing for calculated mean difference in received power (+2 db).

(?) uncertain identification.

Table 3.   The comparison of short-term and averaged characteristics of signals travelling in opposite directions. 8 MHz band, Atlanta-Ipswich (Eight minute average levels: Ipswich relative to Atlanta) Balser and Smith (13).

ferences could be explained if (i) the rhombic antennas used were elliptically polarized for transmission paths that deviated from the great circle direction, and (ii) the transmission paths were off the great circle due to large scale layer tilts (a common occurrence on the transatlantic path).

5. CONCLUSIONS

The present evidence shows that HF signal statistics can be nonstationary over periods of 8-12 min due to the effects of ionospheric irregularities of various scale sizes. These irregularities are sufficiently small to result in significant signal level differences on paths separated by 32 km or less, but still sufficiently large that most diversity antenna separations can not reduce their effects on the fading of signals.

Furthermore, most ionospheric paths are nonreciprocal, with the results that there can be large instantaneous differences in the level of signals travelling in opposite directions.

These ionospheric propagation characteristics can result in inaccuracies in channel sounding prediction of signal characteristics (i) for many minutes (ii) for a communications path separated from the channel sounding path and (iii) for a communications path in a direction opposite to the channel sounding path. The inaccuracies are appreciably reduced by sampling and averaging signal strength measurements for at least 8 min. Additional communications power margins must also be provided to allow for variation in averaged signal strength. These margins will be lower than would be required in the absence of quantitative signal strength prediction. Therefore channel sounding is potentially valuable, particularly for under-powered communications systems.

Acknowledgement: The author acknowledges the large contribution to these studies by his colleagues at DRTE, particularly Mr J.P. Murray and Mr G.F. Poaps who carried out a major share of the experimental work and data analysis. He also acknowledges the contribution of Dr M. Balser and Mr W.B. Smith of the MIT Lincoln Laboratory Lexington, Massachusetts, who provided DRTE with unpublished data from the Atlanta-Ipswich path. Also the contribution of Mr W. Joy and his colleagues at A.S.W.E., Portsmouth, Enland, who constructed equipment and provided for the pulse transmissions from Yorkshire, England.

6. REFERENCES

1. G.W. Jull, D.J. Doyle, G.W. Irvine and J.P. Murray, "Frequency sounding techniques for JF communications over auroral zone paths", Proc IRE, Vol 50, July 1962, pp 1676-1682.

2. R.D. Egan and D.S. Pratt, "Oblique sounding and HF communications", Record of Convention on HF Communications, Inst. of Elect. Eng. March 25-27, Savoy Place, London, U.K. 1963.

3. W.R. Vincent, unpublished communication, 1963.

4. G.W. Jull, G.E. Poaps and J.P. Murray, "Predictions of HF channel characteristics with frequency sounding information", Convention Proceedings of Globecom VII, IEEE Communications Convention, University of Colorado, U.S.A., June 7-9, 1965, pp 653-656.

5. E.E. Stevens, "The CHEC System", Presentation at NATO Advanced Study Institute on "Ionospheric Radio Communications in the Arctic", Finse, Norway. April 12-20, 1967.

6. J.A. Ratcliffe, "Some aspects of diffraction theory and their application to the ionosphere", Reports on Progress in Physics, Vol. 19, 1956, pp 188-267.

7. J. Ames, "Spatial properties of the amplitude fading of continuous HF radio waves", Jr. Res. NBS, Vol. 68D, 1964, pp 1309-1318.

8. E.N. Bramley, "Direction finding studies of large-scale ionospheric irregularities", Proc. Roy. Soc. A, Vol. 220, 1953, pp 39-61.

9. R.W. Meadows and A.J. Moorat, "The effect of fading on the accuracy of measurement of ionospheric absorption", Proc. IEE, Vol. 1050, 1958, pp 27-32.

10. M.Balser and W.B. Smith, "Some statistical properties of pulsed oblique HF ionospheric transmissions", Jr. Res. NBS, Vol. 66D, 1962, pp 721-730.

11. K.G. Budden and G.W. Jull, "Reciprocity and nonreciprocity with magnetoionic rays", Can. J. Phys. Vol. 42, 1964, pp 113-139.

12. G.W. Jull, "Short-term and averaged characteristics of non-reciprocal HF ionospheric paths", Presentation at the Eleventh Symposium of the AGARD Electromagnetic Wave Propagation Committee on "Ionospheric Oblique Radio Wave Propagation Near the Lowest Usable Frequency", Leicester, U.K. July 25-29, 1966. Also Trans. IEEE, AP-15, No. 2, 1967, pp 268-277.

13. M. Balser and W.B. Smith, Presentation at U.R.S.I.-I.R.E. Meeting 1960 (unpublished) also private communication, 1963.

14. K.J.M. Laver and H. Stanesby, "An experimental test of reciprocal transmission over two long-distance radio circuits", Proc. IEE, Vol. 103B, 1956, pp 227-231.

THE PROPAGATION OF HIGH FREQUENCY WAVES ON THE WINNIPEG-RESOLUTE
BAY OBLIQUE SOUNDER CIRCUIT

L. E. Petrie and E. S. Warren

Defence Research Telecommunications Establishment

Ottawa, Canada

Abstract: The observed frequency limits for propagation of high
frequency waves across the auroral zone between terminals at
Winnipeg and Resolute Bay are presented. The accuracy of techniques for predicting the monthly median values of these limits,
the effect of Spread F upon the F layer junction frequency, and
the nature of the mechanism determining the lowest observed frequencies of the propagation paths are discussed.

1. INTRODUCTION

Data from vertical incidence ionosondes have been extensively
used to estimate the behaviour of ionospheric parameters at many
geographic locations. From these estimates, predictions of the
high frequencies suitable for point-to-point communications are
made. To obtain detailed information that could be used to assess
the adequacy of the prediction techniques for high frequency radio
propagation at high latitudes, an oblique sounder circuit was installed transverse to the auroral zone. The northern terminal was
located at Resolute Bay; the southern terminal was located at
Winnipeg from August 1957 until 1960 and at Ottawa from August
1960 until September 1963. The oblique sounding equipment (1) was
programmed to record ionograms at 20 and 40 minutes past each
hour. A representative sample of maximum observed frequencies
(MOF's) and minimum observed frequencies (LOF's) for these circuits has been published (2). This paper deals with the variability and predictability of high frequency wave propagation on the
2760 kilometer path between Winnipeg and Resolute Bay.

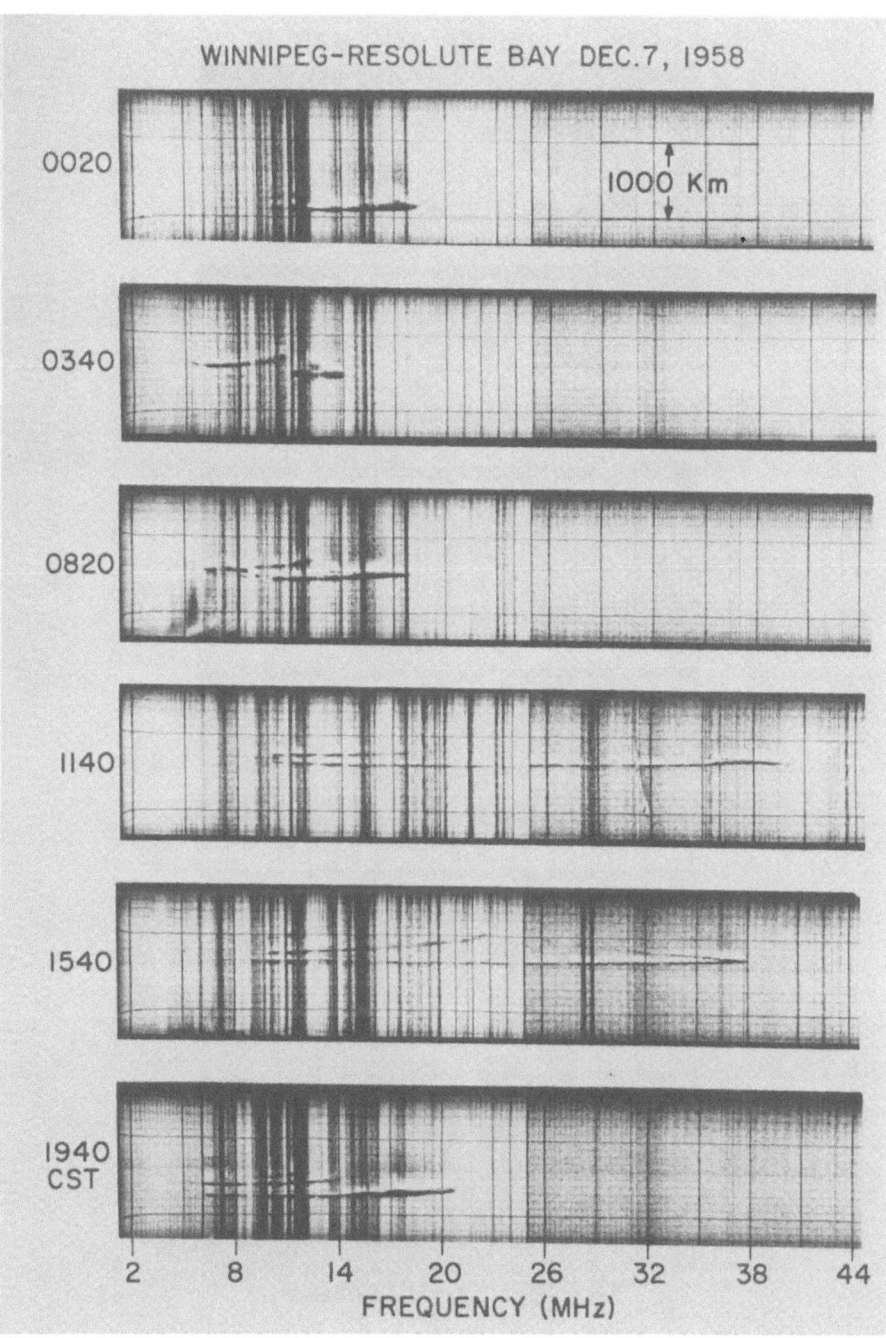

Figure 1. Propagation modes in the winter.

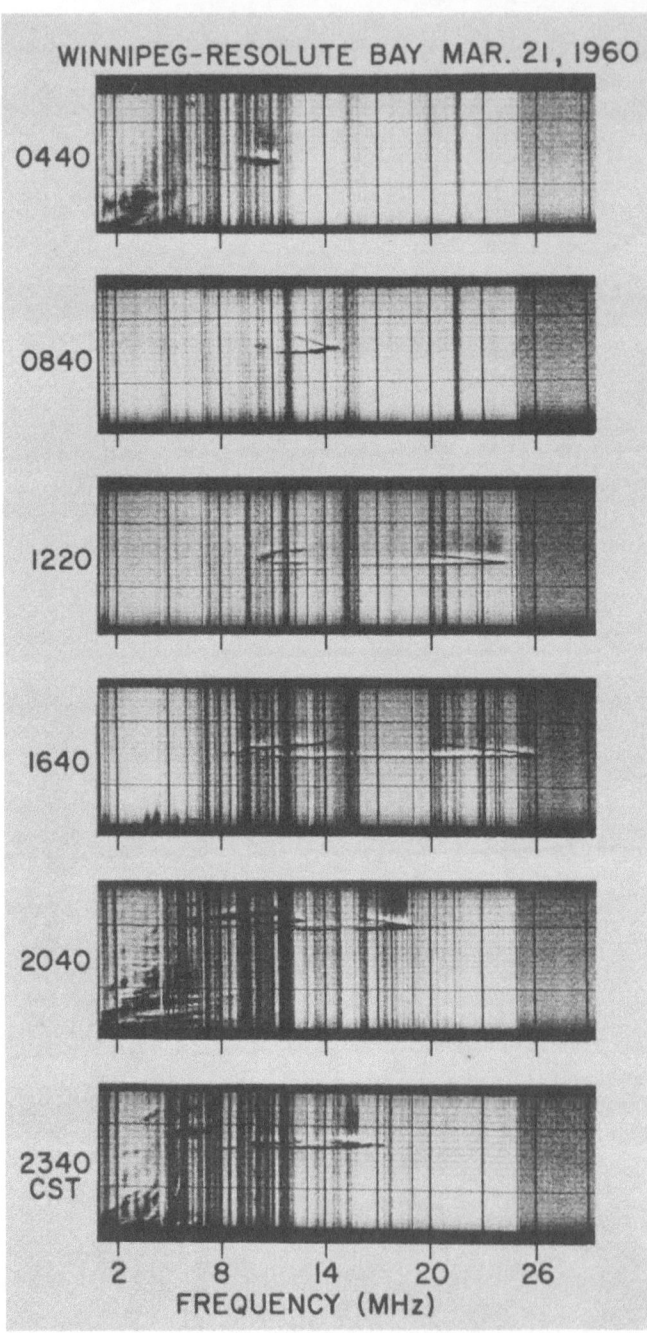

Figure 2.   Propagation modes in the spring equinox.

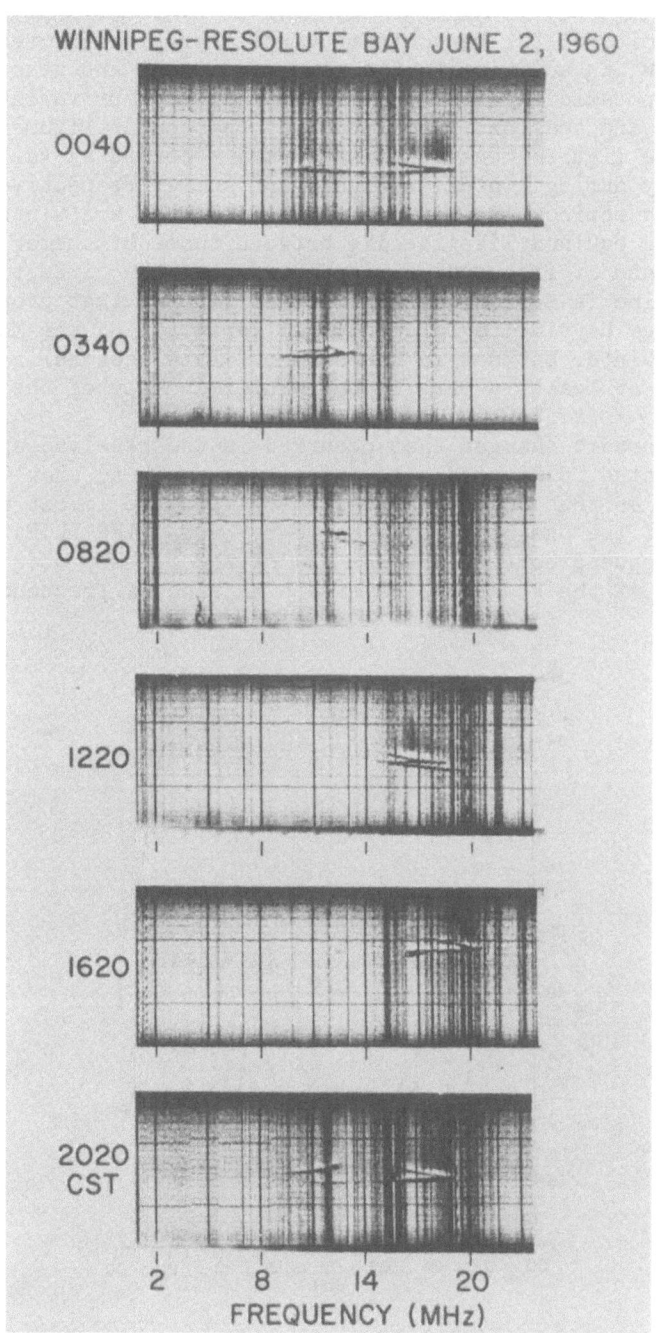

Figure 3. Propagation modes in the summer.

2. VARIABILITY

The limits of the band of frequencies propagated over this
circuit show, in addition to the regular diurnal and seasonal variations, sporadic day-to-day and even hour-to-hour variations.
The diurnal and seasonal variations may be seen in Figures 1, 2
and 3; these figures consist of ionograms recorded at various
times of day during winter, equinox, and summer respectively. The
MOF's during equinox lie between those of summer and winter. The
LOF's during equinox likewise lie between those of summer and winter. The band of frequencies propagated over the circuit in the
winter daytime is approximately twice as wide as that propagated
in the summer daytime; a large diurnal variation of the MOF occurred in the winter but not in the summer. This seasonal variation
is produced at least in part by the winter anomaly of the F region.

The sporadic changes that occurred in the received spectrum
are illustrated in Figures 4 to 7 inclusive. These show the band
of frequencies that could be received between the lowest observed
frequency of the one-hop F2 low-angle-ray mode (1F2 LLOF) and the
junction frequencies (1F1 JF, 1F2 JF) of the one-hop mode propagated by each of the F1 and F2 layers. The band of frequencies be-

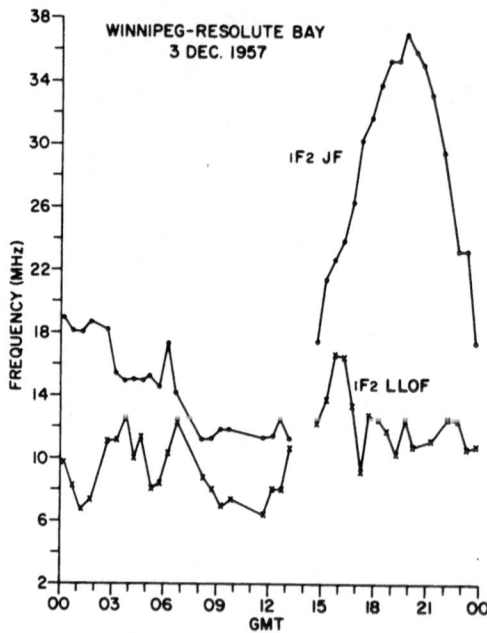

Figure 4. Propagation conditions during an undisturbed day in
           winter.

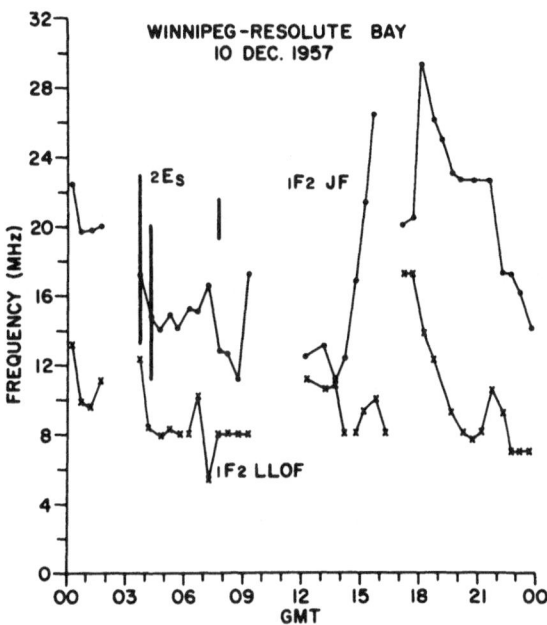

Figure 5. Propagation conditions during a disturbed day in
           winter.

tween this LLOF and the junction frequencies represent rather
well the spectrum of usable frequencies for this circuit as a consequence of the following considerations. The lowest observed
frequency of the F2 high angle ray (1F2 HLOF) was generally greater than the 1F2 LLOF. When absorption was high, the multihop
modes were not observed because they were attenuated below the am-
bient noise level. At times of day when absorption was small, the
LOF's of the different F layer modes did not differ greatly. Thus
the 1F2 LLOF approximates the lowest frequency and the junction
frequencies approximate the highest frequencies that were observed
to propagate on this circuit. Much higher power or advanced modulation techniques can be expected to extend the band of usable
frequencies.

     During December 1957, the daily planetary magnetic index,
Ap, had values ranging between 4 and 53. Figure 4 shows the propagation conditions during the relatively quiet day (Ap = 13) of
December 3, 1957. Except for an interval near 1400 GMT, the band
of propagated frequencies is continuous throughout the 24 hour
period. No E-Layer-propagated signals were observed. Figure 5
shows the frequencies propagated during the moderately disturbed

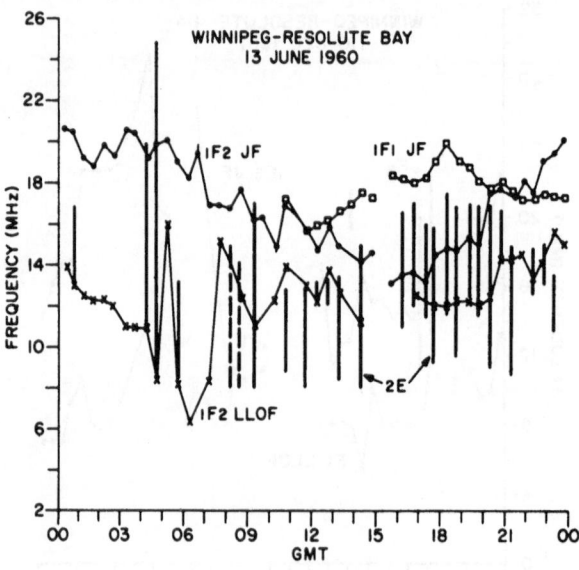

Figure 6. Propagation conditions during an undisturbed day in
summer.

day (Ap = 22) of December 10, 1957. During this 24 hour period
there were three periods of time when, because of ionospheric conditions, high frequency wave propagation was not observed on any
frequency. Two-hop E-layer signals were also observed; they are
represented in Figure 5 by solid vertical lines. This and other
diagrams indicate a tendency for a sporadic increase in the 1F2
LLOF to be accompanied by a decrease in the 1F2JF. For example,
in Figure 5, at 1700 hours following a propagation failure, the
LLOF was higher and the 1F2JF lower than is typical for this time
of day.

During June 1960, the Ap values ranged from 5 to 65. Figure
6 shows the limits of the band of frequencies received during the
quiet day (Ap = 6) of June 13, 1960. High frequency signals were
received throughout the day except during the sounding that occurred at 1520 GMT; it is unlikely that this gap was produced by
ionospheric conditions. A two-hop E mode is usually observed
during the daylight hours. The dotted vertical lines represent
an N-mode. Propagation during the somewhat disturbed day (Ap =
28) of June 1, 1960 is represented in Figure 7. The major difference between propagation during this disturbed day and during
quiet days is that radio blackout occurred during a large fraction

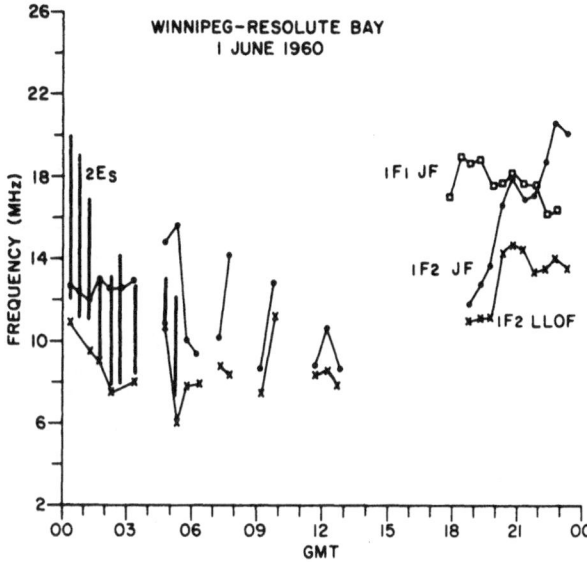

Figure 7. Propagation conditions during a disturbed day in
           summer.

of the disturbed day. The daytime two-hop E mode did not appear
during the disturbed day, probably because of enhanced absorption.
The junction frequencies were lower and the nighttime two-hop E
mode occurred more often on 1 June than during the undisturbed
night of June 13.

     For a large fraction of the time the band of frequencies pro-
pagating between Winnipeg and Resolute Bay was narrow and suffered
sporadic fluctuations; these factors render difficult the efficient use of the high frequency band for communications on this
circuit.

3. LOWEST OBSERVED FREQUENCIES

     Muldrew and Maliphant (3) have shown that no upper limit
exists for the range of path lengths of rays propagated in ionospheres for which the distribution in height of the electron number density has continuous derivatives and a slope, dN/dh, that at
some height has zero value. The curves that for each frequency
relate propagation distance to elevation angle of the path at the
transmitter are continuous if the electron density distribution

has the properties specified by Muldrew and Maliphant. It follows
that the lowest frequency propagated to any given distance by reflection from an ionospheric layer is the vertical incidence penetration frequency of the next lower layer if the electron density
distribution of the lower layer has the specified properties. The
work of Titheridge (4) indicates that the slope attains zero values in the upper E region. It could therefore be expected that
the 1F1 LLOF would be equal to the vertical incidence penetration
frequency of the E layer. However, the values of the 1F1 LLOF observed on the Winnipeg Resolute Bay oblique ionograms are several
times greater than the vertical incidence penetration frequency of
the E layer.

   This circumstance cannot always be explained by the failure
of the slope to approach zero value. Rocket measurements commonly
reveal thin horizontally stratified structures in the E region; at
the maxima of the stratifications dN/dh has zero values. Extreme
defocussing in such stratifications can explain the observed values of the 1F1 LLOF although absorption and scatter cannot be excluded as factors that affect the magnitude of the 1F1 LLOF.

   Such defocussing can be discussed conveniently in terms of a
vertical electron density distribution with a discontinuous slope.
A discontinuous slope can limit the propagation distance. This
may be illustrated by the ray path geometry for reflection of a
ray at the bottom of the F layer in an ionosphere in which the E
layer electron density does not change with height. For these
conditions, the ray paths both within and below the E region are
straight lines; and such straight line ray paths can propagate
energy only to limited distances. The lowest frequency propagated
by the upper layer to a finite distance is then not usually the
vertical incidence E layer penetration frequency.

   The slightest rounding of the shape of the density distribution at the bottom of this constant density E region would result
in the lowest frequency propagated by the F layer becoming equal
to the vertical incidence penetration frequency of the E layer.
This can be a considerably lower frequency than the LLOF calculated for the discontinuous constant distribution discussed above.
The rays corresponding to the frequencies that propagate for the
rounded distribution but not for the discontinuous distribution
are of low signal strength since the density distribution has been
modified only to a minor extent by the rounding. The band of frequencies for which the signal will be large will be that appropriate to the corresponding distribution with discontinuous slope.
Experimental values of the 1F1 LLOF can therefore be expected to
correspond to those computed with distributions of discontinuous
slope since very weak signals will be obscured in the noise environment. This discussion was not intended to promote a constant

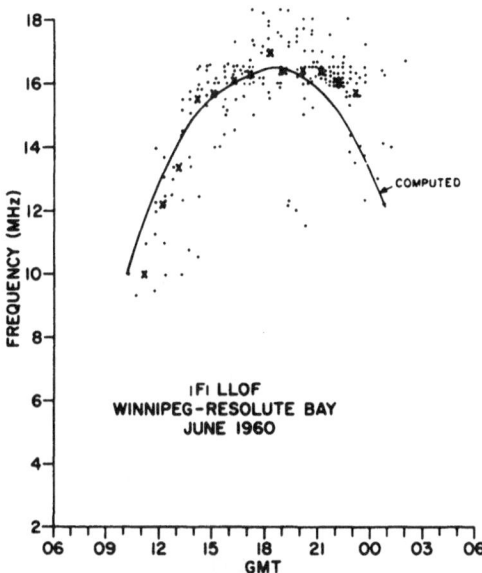

Figure 8.   Comparison of the observed 1-hop F2 layer low angle ray
            LOF's with the estimated monthly median value.

density model of the E region, but rather to indicate the propagation effects produced by vertical electron density gradients that
approximate discontinuous functions.

     It is possible to compute LLOF's that approximate the ob-
served median values (shown by X's in Figures 8 to 14) on the assumption of a straight line ray path from the ground to the minimum apparent reflection height in the F region at the center of
the path together with an obscuring height of 160 kilometers in
the E region. This is an empirical method; the E layer obscuring
height was chosen to provide best fit to the experimental data.
The straight line ray path approximation is a discontinuous slope
calculation of the F1 layer LLOF appropriate to an E layer electron density distribution in which E layer refraction may be to-
tally neglected. The estimated LOF's shown by the solid line of
Figure 8 were computed by multiplying foE by sec ϕ_o (corrected)
(5) for the obscuring height in the E region. The observed values
of h'F1 at Churchill, which is near the center of the path, were
used in this calculation. The values of foE for the appropriate

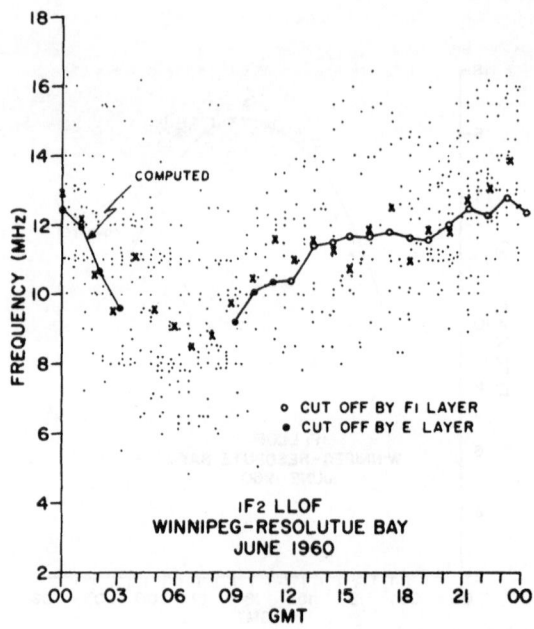

Figure 9. Comparison of the observed 1-hop F1 layer low angle ray
           LOF's with the estimated monthly median value.

geographical location and sunspot number were obtained from DRTE
Report 1-1-3 (6). The computed values are high for the period
1200 to 1600 GMT and low for 1900 to 2200 GMT; this indicates that
a higher obscuring height in the morning and a lower obscuring
height in the afternoon is required for a more precise fit to the
experimental data.

An obscuring height of 160 km was also used for computing the
1F2 LLOF by the same technique for the hours 0100 to 0300 GMT and
0900 to 1100 GMT. The calculated and observed values are shown in
Figure 9. The r.m.s. percentage deviation of the observed values
of the 1F1 LLOF and 1F2 LLOF from the estimated median value computed on the basis of an E region obscuring height of 160 km is
6.6 percent.

For the period 1200 to 2300 GMT the 1F2 LLOF is determined by
the characteristics of the F1 layer. An examination of Churchill
ionograms reveals that the apparent height of reflection of the
strongest signal at foF1 was usually within 200 km of the minimum
apparent height of the F2 layer. The LLOF's for the hours 1200 to

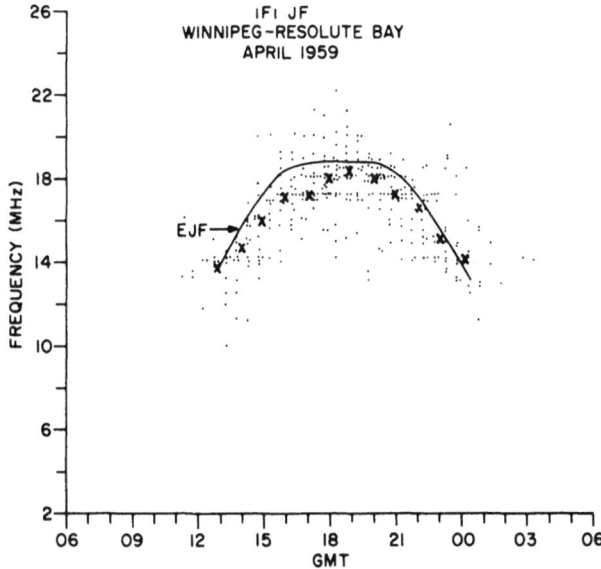

Figure 10. Comparison of the 1F1 junction frequency for April with
            the estimated monthly median value.

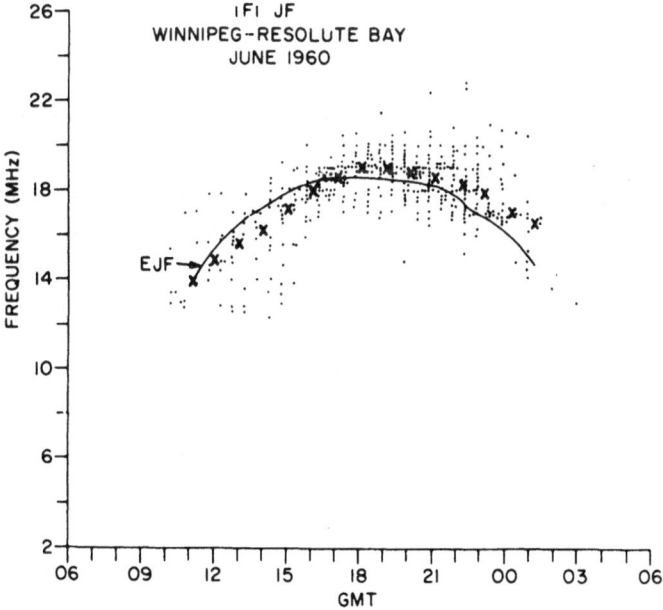

Figure 11. Comparison of the 1F1 junction frequency for June with
            the estimated monthly median value.

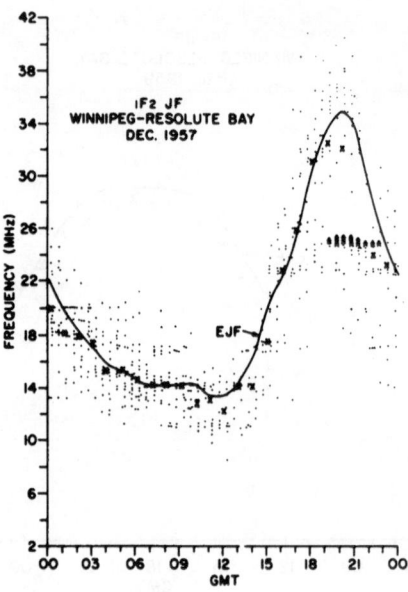

Figure 12. Comparison of the 1F2 junction frequency for December
with the estimated monthly median value.

2300 GMT were computed from Churchill vertical incidence data by
multiplying the monthly median values of foF1 by the value of
sec ϕ_0 (corrected) appropriate to the median value of h'F2. The
calculated values are shown by a solid line between 1200 and 2300
GMT in Figure 9. For these hours the median observed 1F2 LLOF's
are randomly distributed about the computed values with an r.m.s.
percentage deviation of 6.3 percent.

4. MAXIMUM TRANSMISSION FREQUENCIES

Estimates of the 1F1 JF on the Winnipeg-Resolute Bay circuit
were made using an F1 layer prediction technique (7) and were
found to differ systematically from the median observed values.
The estimated values for April 1959 are high throughout the day
with an r.m.s. percentage deviation of 5.3 percent (Figure 10).

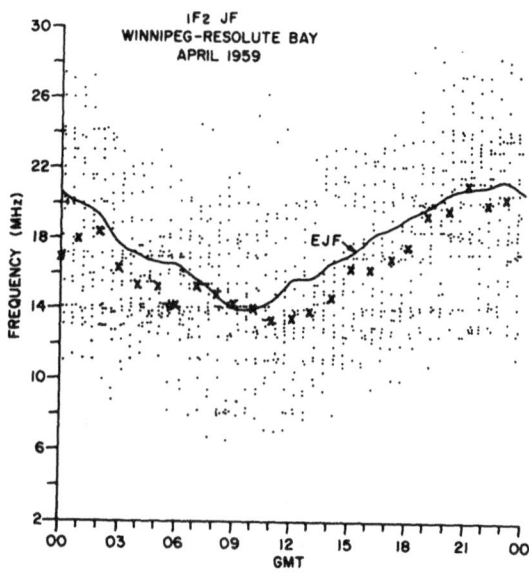

Figure 13. Comparison of the 1F2 junction frequency for April
            with the estimated monthly median value.

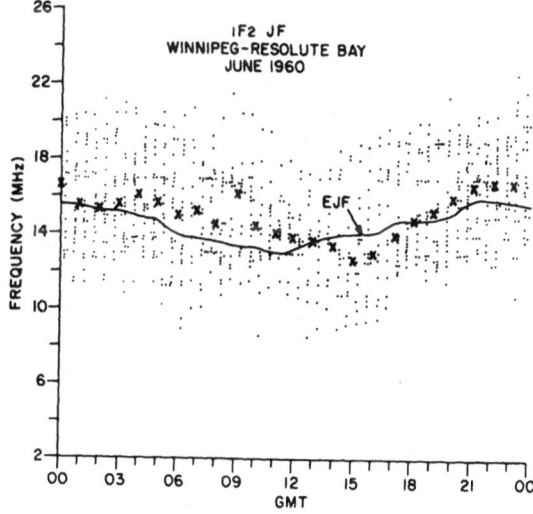

Figure 14. Comparison of the 1F2 junction frequency for June
            with the estimated monthly median value.

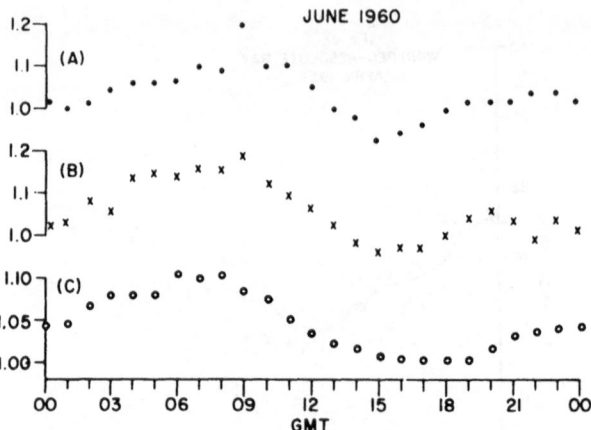

Figure 15.   The similarity of the diurnal behaviour of Spread F
             to that of the inaccuracy of the F2 layer EJF.

        (A)   Ratio of the median JF to estimated monthly
              median JF

        (B)   Median of the ratios of JF to EJF computed from
              Churchill ionograms

        (C)   Median of the ratios of foF2 plus frequency
              spread to foF2 for Churchill.

The estimated values for June 1960 are displaced toward the morning hours from the median observed values with an average r.m.s.
deviation of 3.5 percent (Figure 11). The sources of these discrepancies have not been identified.

     The values of the 1F2 JF observed during December, April, and
June are compared in Figures 12, 13 and 14 respectively with
monthly median values computed using the method of DRTE Report
1-1-3. The computation is conventional, involving the use of
median values of fxF2 together with factors for sunspot number and
for distance dependence of the JF. The r.m.s. percentage deviation of the observed values from the computed values are 5.0, 7.1
and 5.8 respectively. The variability of the junction frequencies
shown by the dispersal of the observed values on these diagrams is
large, particularly during equinox and summer.

     Figure 15 (a) shows the ratio of the median observed JF for
June 1960 to the monthly median value predicted using DRTE Report
1-1-3. To discover why the ratio departs from unity to Winnipeg-
Resolute Bay estimated junction frequencies (EJF's) were computed

using Churchill vertical incidence ionograms and a transmission
slider. The ratios of individual observed values of the JF to the
corresponding EJF's were computed. Curve B of Figure 15 shows the
median of these junction frequency ratios. That is, curve A is
the ratio of medians; curve B is the median of ratios. Because
curves A and B have the same form, it is believed that the discrepancy of the predicted and observed values is due in part to an
inappropriate application of the transmission slider technique,
the common element of these calculations. To aid in identifying
the source of error, the frequency range of Spread F near the foF2
was scaled for an apparent height 300 kilometers greater than h'F.
The ionograms used were those recorded hourly with the standardized medium receiver gain. The median values of the ratio of the
upper frequency limit of the Spread F to foF2 are plotted as curve
C in Figure 15. The similarity of the shape of curve C to that of
B indicates that the practice of positioning the transmission curve at the lower frequency edge of the Spread F leads to the observed discrepancy between the JF and the EJF of curve B. The
recommendation to use the lower frequency edge is one that was
made with reservations (8).

The monthly median foF2 is not completely determined by the
sunspot number; random fluctuations in foF2 of approximately 10
percent of its magnitude occur about the best linear curve relating foF2 to the sunspot number. The effects of these fluctuations appear in predictions. The differences in magnitude between
curves A and B are due in part to such errors in estimating foF2.
At Churchill the values of foF2 from predictions were 8 percent
higher than observed median values for June 1960.

5. CONCLUSIONS

The following conclusions can be drawn concerning radio propagation on the Winnipeg-Resolute Bay circuit:

1. The frequency limits of the band of high frequencies that
   propagate between the terminals of the circuit show large
   sporadic day-to-day and hour-to-hour fluctuations, typically
   several megahertz in magnitude.

2. The accuracy of the MOF prediction was checked by comparing
   predicted values with oblique incidence sounding data for the
   months December 1957, April 1959 and June 1960. The r.m.s.
   deviation of the observed values from the predicted values
   was 4.3 percent for the F1 layer MOF and 5.9 percent for the
   F2 layer MOF. Because these discrepancies are less than the
   presently unpredictable fluctuations in monthly median values

of foF2, it is probable that errors for these months partially compensated each other.

3.   The practice of positioning transmission curves on ionograms
      at the lower rather than at the upper frequency edge of the
      Spread F is the cause of some of the error in predicting
      junction frequencies.

4.   A calculation of the 1F1 LLOF based upon straight line ray
      paths from the ground to the minimum apparent height of the
      F layer together with an E region obscuring height of 160
      kilometers was found to yield values agreeing within 6.6
      percent with the median of the observed values. The 1F2 LLOF
      produced by the obscuring effect of the F1 layer can be esti-
      mated by multiplying foF1 by the value of sec ϕ_0 (corrected)
      appropriate to h'F2 and the circuit length.

Acknowledgement: The authors gratefully acknowledge helpful discussion with Mr. R. Piggott.

6. REFERENCES

1. R. Southern, An Automatic Stepped Frequency Ionospheric
    Sounder for Oblique Incidence Measurements in the 1 to 49 Mc
    Band, DRTE Report 1058, May 1961.

2. Canadian Oblique Ionospheric Data, Defence Research Telecom-
    munications Establishment, Nos. 1-6.

3. D.B. Muldrew and R.G. Maliphant, Long-Distance One-Hop Iono-
    spheric Radio-Wave Propagation, J. Geophys. Res., 67, May
    1962, p 1805.

4. J.E. Titheridge, The Use of the Extraordinary Ray in the Ana-
    lysis of the Ionospheric Records, J. Atmos. Terr. Phys. 17,
    1959-60, p 96.

5. K. Davies, Ionospheric Radio Propagation, National Bureau of
    Standards Monograph 80, April 1965, p 171.

6. Prediction of Optimum Traffic Frequencies for Northern Lati-
    tudes, Defence Research Telecommunications Establishment
    Report 1-1-3, November 1954.

7. L.E. Petrie and E.E. Stevens, An Fl-layer MUF Prediction System for Northern Latitudes, IEEE Trans. Ant. and Prop. 13, July 1965, p 542.

8. W.R. Piggott and K. Rawer, URSI Handbook of Ionogram Interpretation and Reduction, Elsevier Publishing Co. 1961.

ROYAL AIR FORCE COASTAL COMMAND MARITIME AIR/GROUND COMMUNICATIONS

TO AND FROM AURORAL REGIONS

P. J. Wright

Headquarters Coastal Command Royal Air Force

Northwood, Middlesex, England

Abstract: Procedures established by RAF Coastal Command in the United Kingdom for operating air-ground-air communication are presented together with a description the experience obtained. Possible means of improvement are commented upon.

1. COMMUNICATION PROCEDURES

RAF Coastal Command in the United Kingdom operates long-range aircraft over sea areas extending from the Bay of Biscay to the limits of permanent ice in the Norwegian Sea. The aircraft generally operate below 5000 feet. Ideally, we require solid air-ground-air communications all the time throughout this area, at these low altitudes.

We find little difficulty in communicating with aircraft to the west and south of the British Isles but we sometimes have poor communications to aircraft flying north of the British Isles. The reason for this are well known to you, so I will concentrate on how we attempt to make the best use of the facilities we have available.

Firstly we operate a ground-to-air hand-speed morse broadcast from stations in the United Kingdom. For a pre-arranged period in every hour, this broadcast is transmitted simultaneously on frequencies in the 3, 6 and 9 MHz bands. The aircraft operator listens on the frequency appropriate to the aircraft's position and the time of day. Frequency prediction tables and charts have been found to be of little use when flying over arctic regions, because such aids cannot take account of the unpredictable fluctuations of the arctic ionosphere. Aircraft operators therefore rely on their experience in choosing a frequency to monitor, and when flying out from the United Kingdom there is the temptation to remain on the 3 MHz frequency until it is made unusable by fading and interference, thus delaying the change to higher frequencies beyond the time when it should be made.

The aircraft operator also has a separate family of frequencies in the 3, 6 and 9 MHz bands for two-way air-ground-air morse communications with his controlling Headquarters in the United Kingdom. These frequencies are continuously monitored by the controlling Headquarters. The aircraft operator is taught to choose an appropriate frequency by "sampling" the hourly broadcast I have just mentioned. Thus, if the most recent broadcast was received best on 6 MHz, the aircraft operator will select a 6 MHz frequency when attempting to contact his controlling headquarters.

You will have realized that the effectiveness of our air-to-ground communications depends upon two things. First, the aircraft operator must take a complete sample of the broadcast frequencies each hour and must make a correct decision as to which is best. Second, at the time when the aircraft operator wishes to communicate with his control, ionospheric conditions must not be significantly different from those which affected the radio path between the control station and the aircraft's operating area when the operator sampled the broadcast frequencies. If the ionosphere is stable these requirements can be met but the probability of meeting them falls off as the ionosphere becomes less stable.

To sum up, our air-ground-air communications to the north of the British Isles are not as reliable as we should like, even though we maintain a large establishment of men and equipment on the ground to make separate families of frequencies continuously available for the broadcast system and for the air-ground-air system.

2. POSSIBLE MEANS OF IMPROVEMENTS

To improve our communications, especially to the north of the British Isles, we have been considering the use of the LF/MF bands and also the use of HF oblique ionospheric sounding. We can see a promising future for an LF/MF broadcast for ground-to-air communications, similar to the Prestwick LF meteorological broadcast to civil airliners over the Atlantic. For air-to-ground communications, however, we do not yet know whether the performance of LF/MF would justify the installation in our aircraft of the heavy transmitters and very long trailing aerials required for efficient performance on these frequencies.

As regards HF, we are very interested in the work that has been done by DRTE Canada in the field of HF oblique ionospheric sounding and consider that his technique may significantly ease our frequency selection problems and permit an appreciable improvement in communications. In particular we have studied the CHEC system and we are impressed by the facility it may offer in allowing us to select for air-ground-air communications the best assigned frequency available from minute to minute. In addition, new equipment is becoming available which can change frequency quickly, and the CHEC system should enable us to exploit this facility fully and perhaps to make more effective use of the large number of frequencies at present available without having to increase our establishment of men or equipment on the ground.

3. CONCLUDING REMARKS

By way of summary, to overcome the difficulties of air-ground-air HF communications to the north of the British Isles we are obliged to run separate ground-to-air broadcast and air-ground-air communications nets, each working simultaneously on different families of frequencies in the 3, 6 and 9 MHz bands. Further, having at present no really satisfactory method of indicating the optium frequency to the aircraft operator, we have to rely on a method of selecting an operating frequency which is based on extrapolation from an ionospheric situation which may no longer apply. This system is expensive in men and equipment. We are considering the use of LF/MF for ground-to-air broadcast communications, but for air-to-ground communications we do not at present know whether the advantages of LF/MF outweigh the penalties. Our main hope for improving HF air-to-ground communications lies in the use of transmitters and receivers which can change frequency quickly, allied with the use of a technique for oblique ionospheric sounding.

DEVELOPMENTS OF HF PREDICTIONS FOR THE ARCTIC

L. E. Petrie

Defence Research Telecommunications Establishment

Ottawa, Canada

Abstract: A technique is described for predicting the Fl layer
junction frequencies for latitudes greater than 35°N. In addition
the calculation of more reliable factors for use in predicting the
optimum traffic frequency is discussed and charts of these factors
are presented.

1. INTRODUCTION

A prediction system developed by the Defence Research Telecommunications Establishment (DRTE) was published in 1954 (1). It
was designed for the prediction of traffic frequencies for communication circuits operating at latitudes greater than 35°N. This
paper describes two of the improvements made to the prediction
system in recent years; the prediction of a more reliable Fl layer
junction frequency (Fl EJF) and the prediction of factors for computing the optimum traffic frequency (FOT).

2. PREDICTION OF THE Fl LAYER JUNCTION FREQUENCY

In the past it has been customary to calculate the Fl layer
EJF using a combined E-Fl layer prediction method in which the Fl
layer prediction is treated as an extension of an E layer prediction system. In this method it is assumed that propagation via
the Fl layer occurs whenever the regular E layer is present. This
is not always true; in winter months, the Fl layer disappears
while the E layer is often observed. The Fl layer dependence on
the solar zenith angle differs from that of the E layer.

263

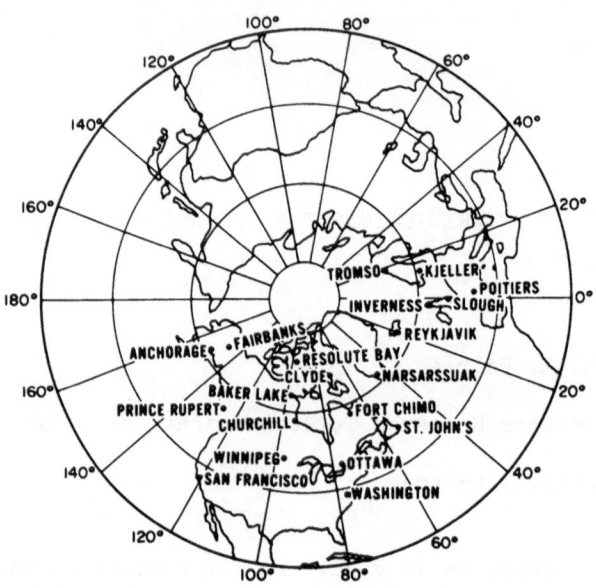

Figure 1.   Source of data for foF1 study.

The E-F1 layer prediction method also contains the assumption
that the product of the penetration frequencies and EJF factors
for the F1 layer is the same as the product of the corresponding
quantities for the E layer. This assumption can lead to systematic errors of about 10 percent between the predicted F1 layer EJF
and the observed 1F JF on the Winnipeg-Resolute Bay and Ottawa-
Saskatoon circuits.

At DRTE an F1 prediction system was developed whereby the F1
layer prediction is considered separately and not as an extension
of an E layer prediction system. F1 layer penetration frequencies
(foF1) measured between 1949 and 1961 at the stations shown in
Figure 1 were used in the investigation. For a given month, time
of day and geographic location, a linear relationship is found between foF1 and the twelve month running average Zurich sunspot
number, R_{12}. Although a slight longitudinal variation in the foF1
at a given latitude is observed, the variation is too small to
warrant inclusion in the prediction system. Also, the disappear-
ance of the F1 layer near sunrise and sunset at high sunspot numbers has not been taken into account. Contours of constant values
of foF1 for $R_{12} = 0$ and $R_{12} = 100$ were plotted as a function of
local time and latitude for each month. Charts for the months of
June and August are shown in Figures 2 and 3. From these charts
the foF1 can be calculated for any location greater than 35°N and

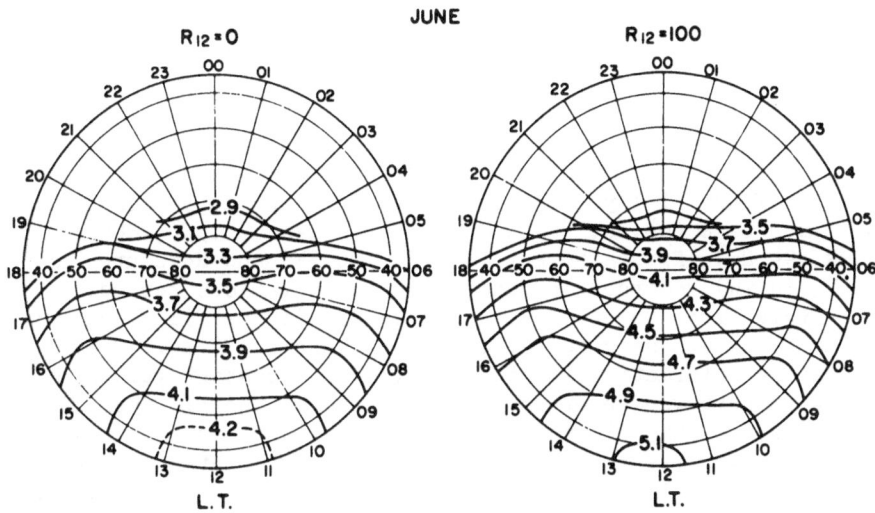

Figure 2. Countour charts of foF1 for June.

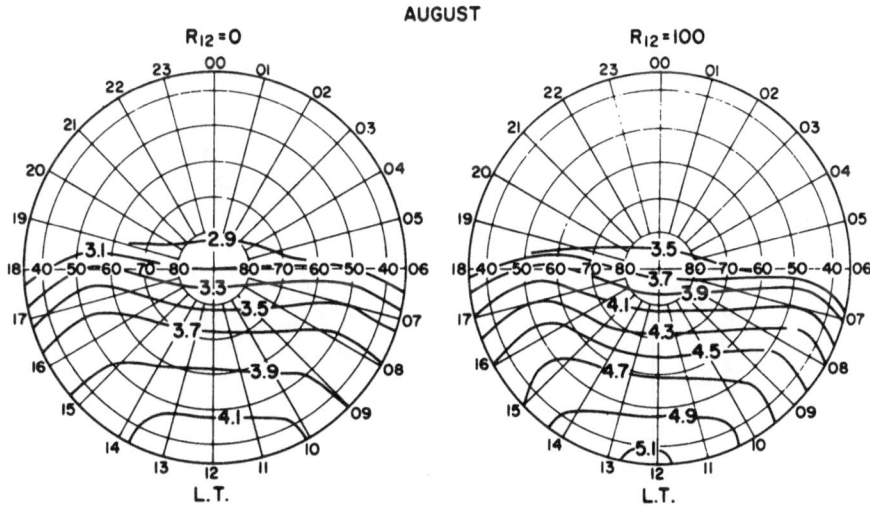

Figure 3. Contour charts of foF1 for August.

for any sunspot number.

The junction frequency for an ionospheric layer is usually
estimated by multiplying the vertical incidence penetration frequency by an EJF factor. This factor is a function of the transmitter-receiver distance, and height and shape of the ionospheric
layer. The Fl layer EJF factors obtained using the standard
transmission curves (4) were found to be consistently too low in
value. Ray tracing (5) was used to find means of improving the
accuracy of the Fl layer EJF factors. Representative ionograms
from Baker Lake (64.3N, 96.0W), Ottawa (45.4N, 75.9W), Churchill
(58.8N, 94.2W), Prince Rupert (54.3N, 130.3W) and Resolute Bay
(74.7N, 94.9W) recorded in June, 1200 LT at various periods of the
solar cycle were converted to electron density profiles and ray
tracing calculations were performed on these profiles. From the
ray tracing information smoothed curves for constant values of
sunspot numbers were plotted as a function of the Fl layer EJF
factor and the skip distance. These curves are shown in Figure 4.
The ray tracing results indicate that the earth intercepts the
one-hop Fl-layer low-angle-ray mode of propagation at distances
greater than about 3500 km. The prediction of the Fl layer EJF
for distances greater than 3500 km is not included in this prediction system. However, one-hop Fl-layer high-angle ray propagation
has been observed at greater distances (6).

Since the Fl layer EJF factors shown in Figure 4 are only
appropriate for June at 1200 LT, an adjustment in these factors is
necessary at other times to take into account the diurnal and seasonal variation in the height and shape of the Fl layer. The

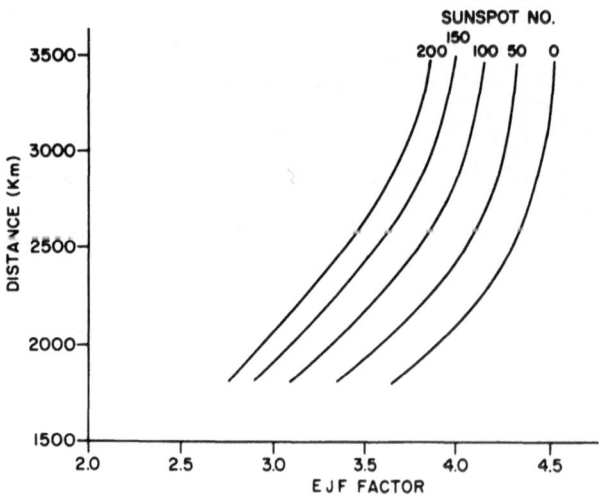

Figure 4. EJF factors for June 1200 LT.

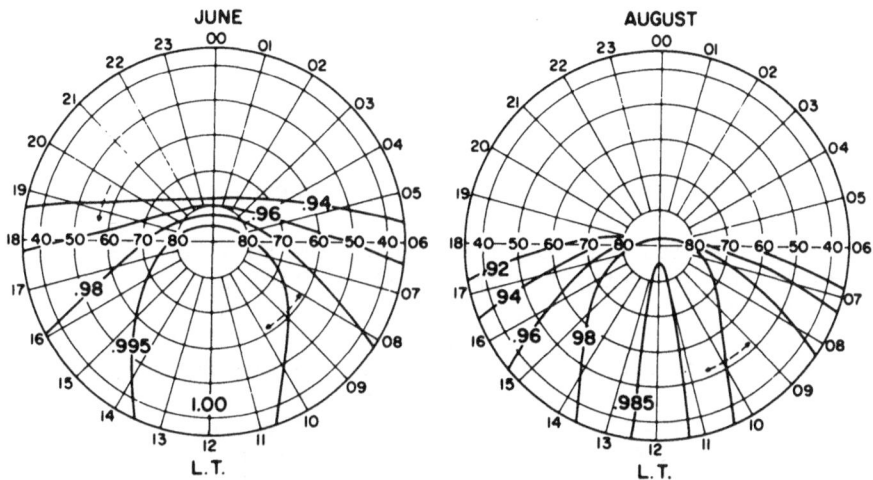

Figure 5. Adjustment factors $\Gamma_{T,M}$.

M(3000)Fl is in error by about 10 percent even for transmitter-receiver distances of 3000 km. However, the M(3000)Fl does exhibit diurnal and seasonal variations that are similar to the Fl layer EJF factors obtained from representative ray tracing calculations. It was therefore assumed that the variations of the M(3000)Fl could be used to adjust the Fl layer EJF factors for June, 1200 LT for all other times. The resulting adjustment coefficients, which are independent of sunspot number are shown in Figure 5 for the months of June and August.

The one-hop Fl layer EJF is calculated using the information in Figures 2 to 5. Thus

$$1Fl \ EJF \ = \ foFl \times Z \times \Gamma_{T,M}$$

where

$$foFl \ = \ (foFl_{100} - foFl_{0}) \ \frac{R_{12}}{100} + foFl_{0}$$

and

$foFl_{100}$ is the foFl at $R_{12} = 100$ from Figures 2 or 3
$foFl_{0}$ is the foFl at $R_{12} = 0$ from Figures 2 or 3
Z is the EJF factor from Figure 4
$\Gamma_{T,M}$ is the adjustment coefficient from Figure 5.

The accuracy of the Fl layer EJF has been confirmed by oblique sounding observations. Comparisons between the predicted Fl layer EJF and the monthly median JF have been presented in the literature (3, 7). For example, the average percentage deviation and the rms percentage deviation of the EJF from the monthly median JF on the Winnipeg-Resolute Bay circuit from July 1959 to July 1960 is about +1 percent and 5 percent respectively.

3. F2-LAYER OPTIMUM TRAFFIC FREQUENCY

In prediction systems, it is customary to allow for the variations of the JF around the predicted junction frequencies for a given time of day and month by calculating an Optimum Traffic Frequency (FOT). The FOT is the frequency that is less than the junction frequencies 90 percent of the time and greater than the junction frequencies 10 percent of the time. For the F2 layer the predicted FOT is usually calculated by multiplying the monthly median EJF by a factor of 0.85. At high latitudes the value of 0.85 leads to significant errors in the predicted FOT.

Since variations in the minimum height and the shape of the F2 layer are rather small except near sunrise and sunset, improved FOT factors can be computed on the assumption that the variations of the foF2 are solely responsible for the variations of the JF. Ionospheric data recorded between 1942 and 1960 for the stations shown in Figure 6 have been used in an investigation of the variation of the foF2. For each month, hour of the day and geographic location, the ratio of the 10 percentile foF2 to the 50 percentile (median) foF2 was determined from the density distribution of foF2.

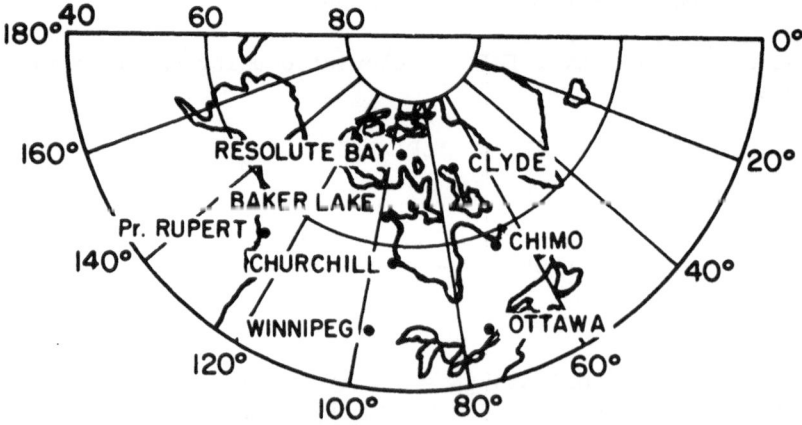

Figure 6.   Source of data for FOT study.

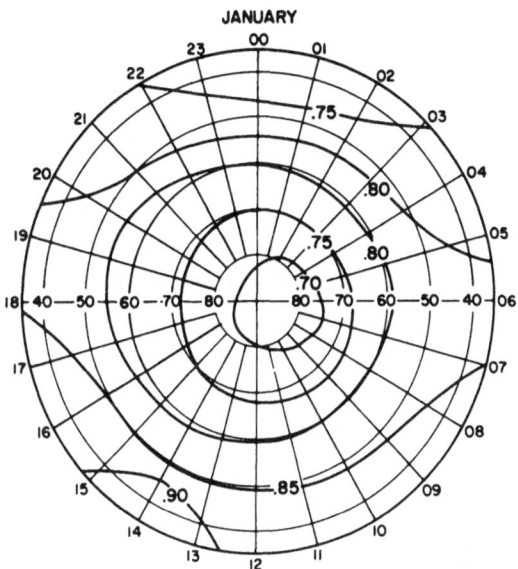

Figure 7. Contour charts of FOT factors for January.

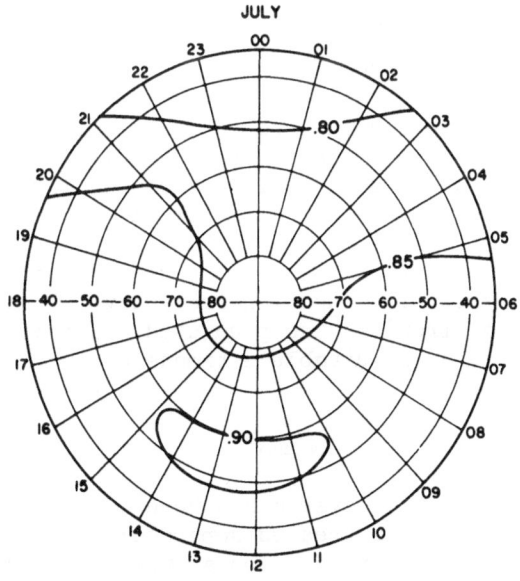

Figure 8. Contour charts of FOT factors for July.

For a given local time, month and various sunspot numbers the rms
percentage deviation of these ratios from their average value is
generally less than 4 percent. Thus the ratio or FOT factor is
nearly independent of sunspot number and can be treated as a constant for all sunspot numbers. For each month, contours of the
FOT factor were plotted as a function of local time and latitude.
Contour charts for January and July are shown in Figures 7 and 8.
The FOT factors vary in magnitude between .75 and .90 for January
and between .80 and .90 for June. Therefore, on the assumption
that the variations of the foF2 are solely responsible for the
variations of the JF, the value of .85 used in many prediction
systems is an appropriate average value but produces at times a
systematic error in the estimated FOT.

   The day-to-day variations in the F2 layer JF are due to variations of foF2, of the minimum height and of the shape of the
F2 layer. Variations of the minimum height and the shape of the
F2 layer are reflected as variations in the F2 layer EJF factor.
The EJF factor has been determined by applying the standard transmission curve for a distance of 1000 km to representative ionograms recorded at Ottawa and Resolute Bay from 1959 to 1963. Variations of the EJF factors from the monthly median value would
result in a 3 percent rms deviation of the JF while variations of
the foF2 from the monthly median value would result in an 18 percent deviation of the JF. Thus, if the assumption is made that
the variations of the foF2 are solely responsible for variation of
the JF, negligible error will be introduced in the calculation of
the FOT for distances less than 1000 km and small errors will be
introduced for longer distances.

   The FOT factors have been checked using oblique sounder observations on the Winnipeg-Resolute Bay (2760 km) and Ottawa-
Resolute Bay circuits (3400 km). The predicted FOT factors shown
in Figures 7 and 8 exhibit about a +2 percent average deviation
and a 6 percent rms deviation from the observed FOT factors, that
is, from the observed ratios of the 10 percentile junction frequency to the 50 percentile junction frequency. Thus, even for
these long distance circuits, the predicted FOT factors shown in
Figures 7 and 8 provide a suitable estimate of the FOT.

                        4. SUMMARY

1.   An F1 layer prediction system has been developed in which the
     F1 layer prediction is treated separately and not as an ex-
     tension of an E layer prediction system. Contour charts of
     the F1 layer critical frequency and EJF factors are provided
     for use in the prediction of the F1 layer EJF.

2.   Improved factors for predicting the FOT at high latitudes
     have been deduced using vertical incidence sounding data.
     These factors have been tested using oblique sounder data
     from the Winnipeg-Resolute Bay and Ottawa-Resolute Bay cir-
     cuits and have been found to exhibit a +2 percent average
     deviation and a 6 percent rms deviation from the observed FOT
     factors.

5.   REFERENCES

1.   Prediction of Optimum Traffic Frequencies for Northern Lati-
     tudes, DRTE RPL Report 1-1-3, November 1954.

2.   K. Davies, Ionospheric Radio Propagation, National Bureau of
     Standards Monograph 80, Aptil 1965.

3.   L.E. Petrie and E.E. Stevens, An Fl Layer MUF Prediction Sys-
     tem for Northern Latitudes, IEEE Trans. Ant. and Prop. AP-13,
     July 1965, p 542.

4.   N. Smith, The Relation of Radio Sky-wave Transmission to Iono-
     sphere Measurements, Proc. IRE 27, May 1939, p 332.

5.   D.B. Muldrew, An Ionospheric Ray Tracing Technique and its
     Application to a Problem in Long Distance Radio Propagation,
     IRE Trans. Ant. and Prop. AP-7, October 1959, p 393.

6.   L.H. Tveten, Long Distance One Hop Fl Propagation Through the
     Auroral Zone, J. Geophys. Res. 66, June 1961, p 1683.

7.   L.E. Petrie and E.S. Warren, The Propagation of High Frequency
     Waves on the Winnipeg-Resolute Bay Oblique Sounder Circuit,
     NATO Advanced Study Institute on Ionospheric Radio Communica-
     tions in the Arctic, Finse, Norway, April 1967.

COMPARISON OF PROPAGATION PREDICTIONS AND MEASURED PERFORMANCE FOR TWO PATHS OVER THE NORWEGIAN SEA

K. W. Blake and N. H. Knudtzon

SHAPE Technical Centre, Haag, Netherlands

Abstract: This presentation deals with the outcome of an experiment designed to compare frequency predictions with propagation on two HF paths across the Norwegian Sea.

Circuit performance is analysed in terms of monthly reliabilities defined in the paper.

1. LOCATION OF CIRCUITS AND TEST PARAMETERS

This paper describes trials which have been carried out on two paths over the Norwegian Sea in order to compare frequency predictions with actual propagation performance in the HF band. As shown in Figure 1, the one path follows closely the central region of the auroral zone, whereas the other has its mid-point outside the auroral zone; both are about 1700 km long. Transmissions were made regularly from existing stations at Oslo and Andøya in Norway on 3.8, 5.4, 7.6, 11.6, 15.8 and 20.3 MHz from July 1964 to July 1965, which included a sunspot minimum period. The signal strengths were recorded by six fixed-frequency receivers specially installed for this purpose at Keflavik on Iceland.

For various reasons it was preferred to use existing transmitting stations rather than to install special oblique sounders. The transmitter power was 10 kW and the transmitting antennas were of the vertical logarithmic-periodic type. The receiving antenna was a broad-band monopole.

A CW signal was transmitted on each of the six frequencies for ten minutes in every hour on two consecutive days of the week

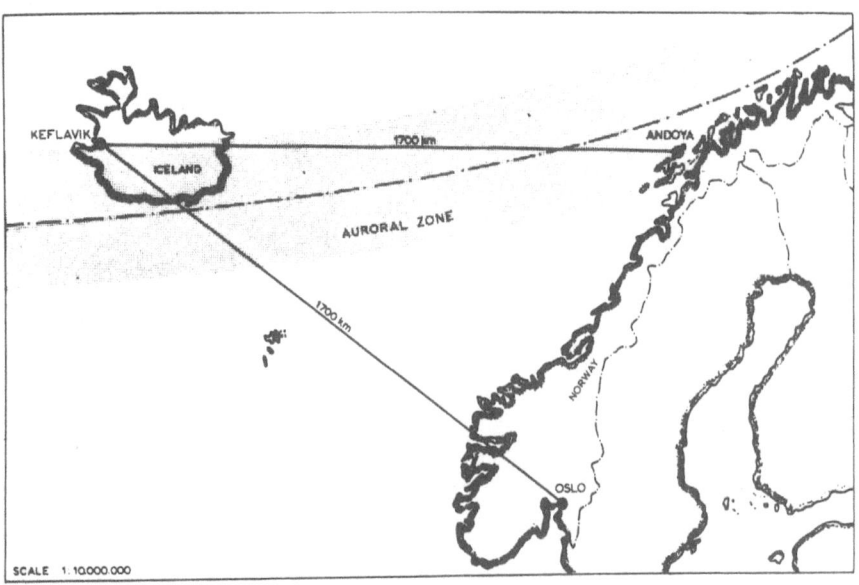

Figure 1.   The two test paths.

from each station.   The signal strength was recorded every 14 seconds during these ten-minute periods, as dots on a paper chart,
different colors being used for the different frequencies.   The
receiving system was fully automatic; to maintain synchronism both
the recorder and the receiver sequencing was operated from a bat-
tery-driven frequency divider chain.

    From these recordings were read for each of the 14,000 ten-
minute periods for each path:

- maximum signal level
- median signal level
- minimum signal level
- noise level

in decibels relative to 1 microvolt/meter, to the nearest 5 db.
These data, together with corresponding hour, frequency, station
identification and interference assessment were transferred to
tapes for computer analysis.   Of the various results thus obtained,
only those concerning the monthly reliability will be discussed in
this paper.

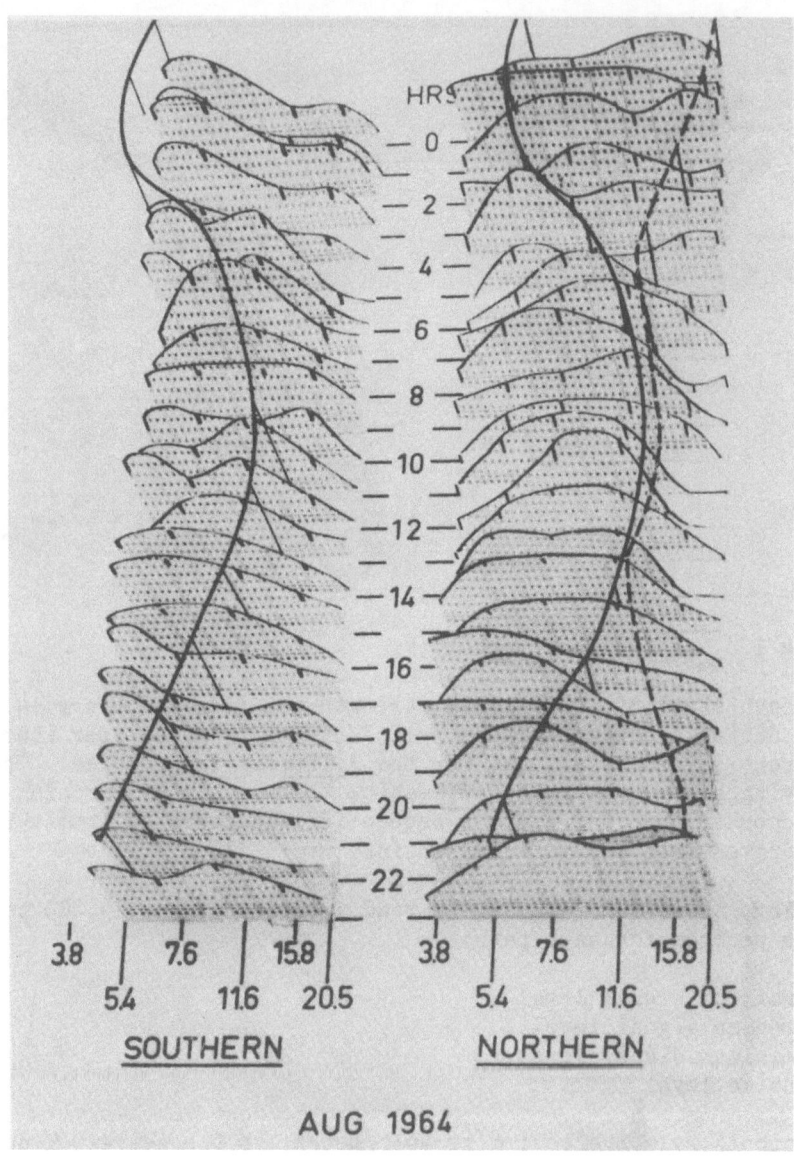

Figure 2. Three-dimensional plot reliability/frequency/hour
for the month of August 1964.

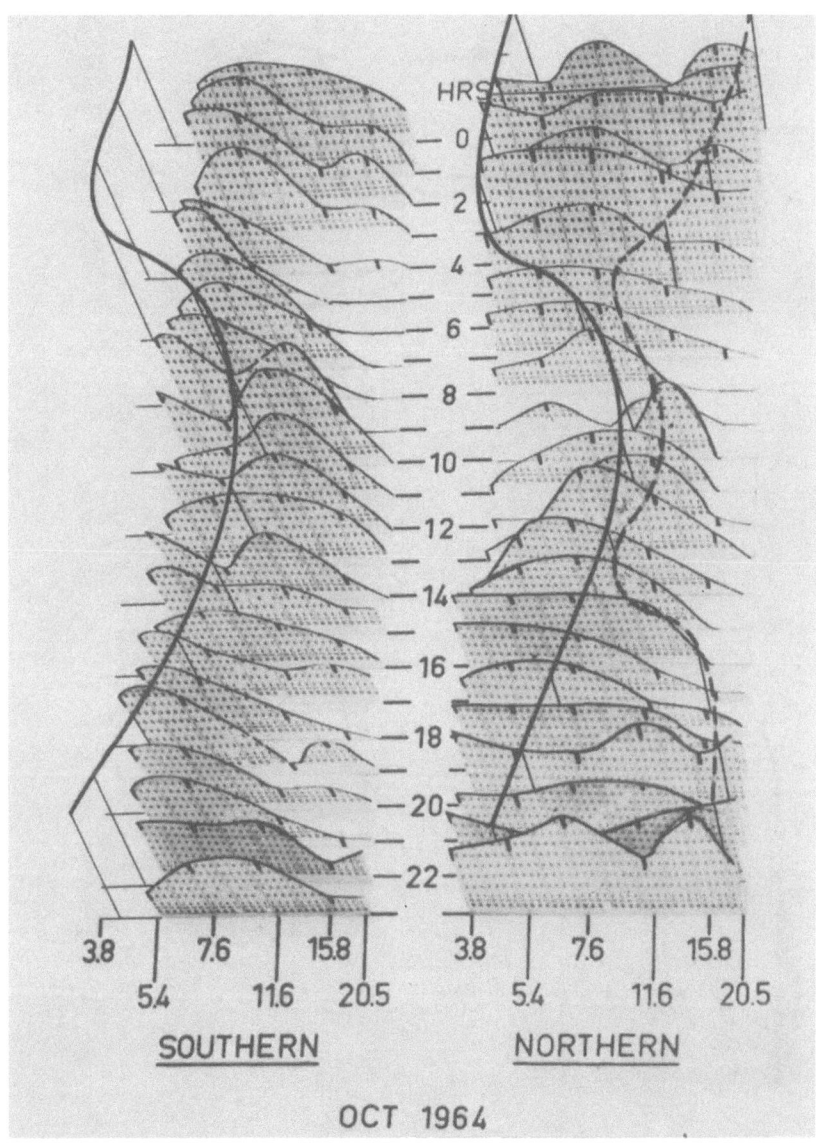

Figure 3.   Three-dimensional plot reliability/frequency/hour
for the month of October 1964.

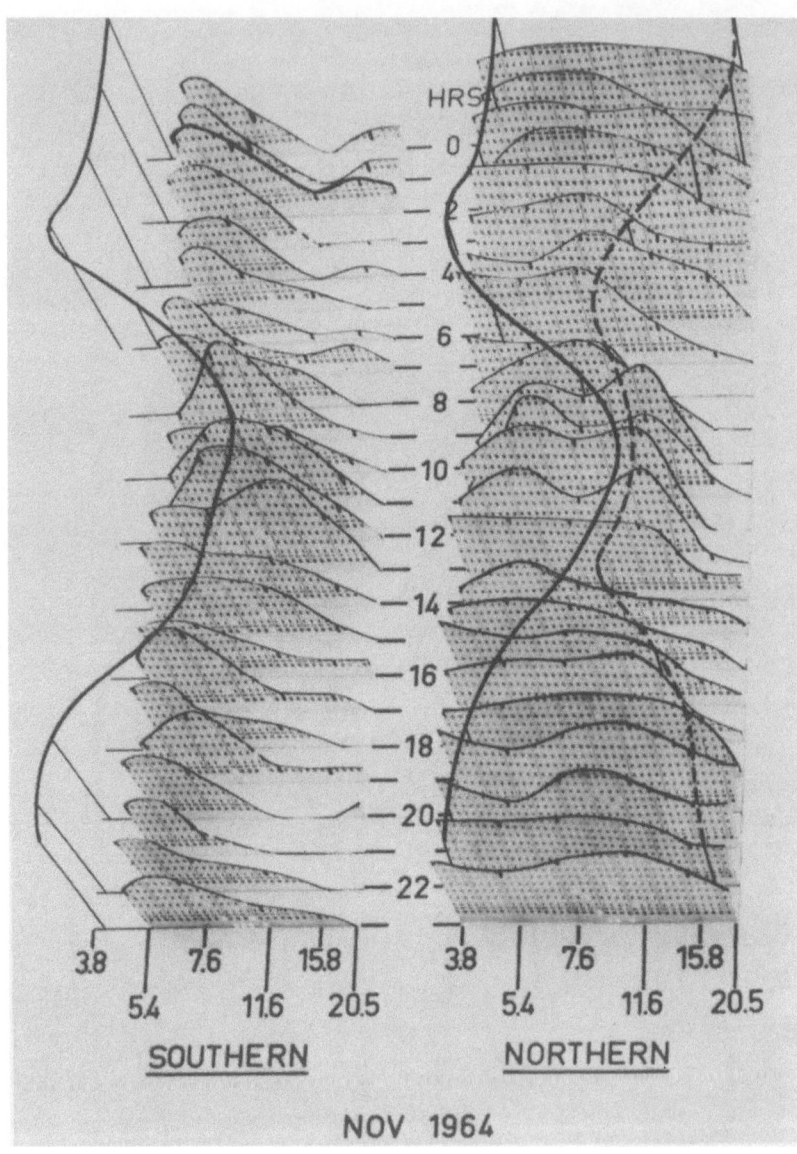

Figure 4.   Three-dimensional plot reliability/frequency/hour
            for the month of November 1964.

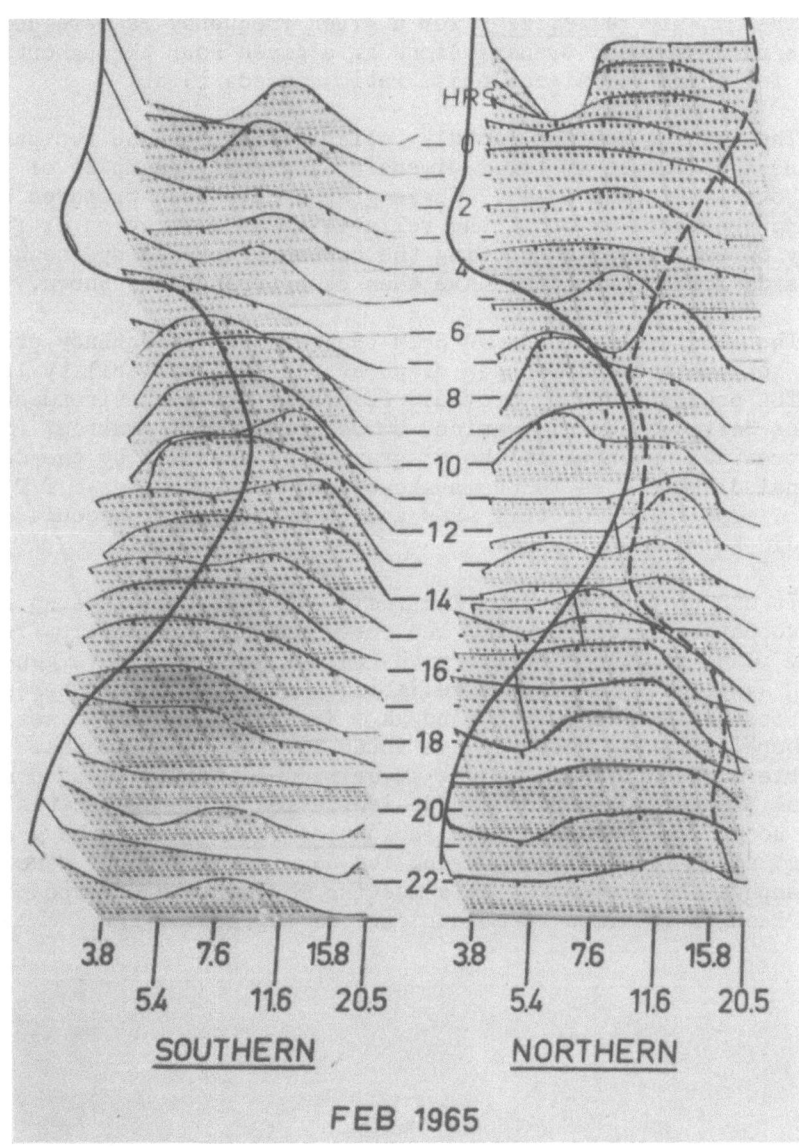

Figure 5.   Three-dimensional plot reliability/frequency/hour
            for the month of February 1965.

2. RELIABILITY PATTERNS

The monthly <u>reliability</u> for a given frequency is here defined
as the percentage of transmissions at a given hour throughout the
month for which the signal/noise ratio exceeds 15 db.

The results for the monthly reliabilities for the two paths
have been displayed in three-dimensional graphs, examples of these
are shown in Figures 2 - 5. These graphs have been prepared by
letting the computer punch the reliability as a function of frequency on each card and cutting the contours defined by the holes.
The cards for the 24 hours have then been arranged as shown.

The ITSA (full line) and SPIM (dotted line) frequency predic-
tions for the month have been drawn at the 100% reliability level.
The ITSA predictions are provided regularly by the Environmental
Science Services Administration, US Department of Commerce. The
SPIM predictions shown in the diagrams were provided by the Centre
National d'Etudes des Télécommunications, Ministère des P.T.T.,
France, especially for this area and to take special account of
sporadic-E.

It can be observed from Figures 2 - 5 that the patterns for
the two paths are different. The curves in the reliability/fre-
quency plane for the southern path (Oslo - Keflavik) have generally a ⋂ shape, i.e. a maximum reliability at about noon, whereas
those for the northern path (Andøya - Keflavik) in many cases have
a <u>W</u> shape. Similar patterns are found for the other months. A
possible explanation may be the superimposed influences of normal
diurnal E-layer ionization and de-ionization which culminates at
about noon, and the sporadic-E incidences which tend to be greatest at night and smallest during the day in the auroral zone.
The geophysists present at this meeting may be able to expound on
these results obtained by communications engineers.

RESULTS FROM STEPPED FREQUENCY OBLIQUE SOUNDINGS AT HIGH LATITUDES

K. Folkestad

Norwegian Defence Research Establishment

Kjeller, Norway

Abstract: This paper presents results from comparing measured maximum propagated frequencies with predictions on long-distance polar circuits. Samples are given of circuit behaviour during disturbed conditions. Detection of backscatter E_s echoes is briefly discussed.

1. INTRODUCTION

A Granger stepped frequency sounder was operated on a regular basis from October 1963 to July 1966 at Andøya in Northern Norway. For parts of the operating period the sounder was synchronized with similar equipment at College, New Jersey, Boston and Pullman. This presentation deals with a brief description of some practical aspects of the outcome of this sounding experiment.

2. EXPERIMENTAL DATA

The geographical positions of the sounders involved are displayed in Figure 1. For the two paths, Andøya-College and Andøya-New Jersey, the great-circle distances are 4800 km and 5900 km respectively. It is seen that the Andøya-College circuit traverses the auroral zone at approximately right angles, whereas the Andøya-New Jersey path, which essentially comes out as an elongation of the Andøya-Boston link, is orientated more along the tangential direction to the auroral belt. The latter path may appropriately be termed an "auroral link" in the following treatment; the Andøya-College path will be referred to as the "transpolar circuit".

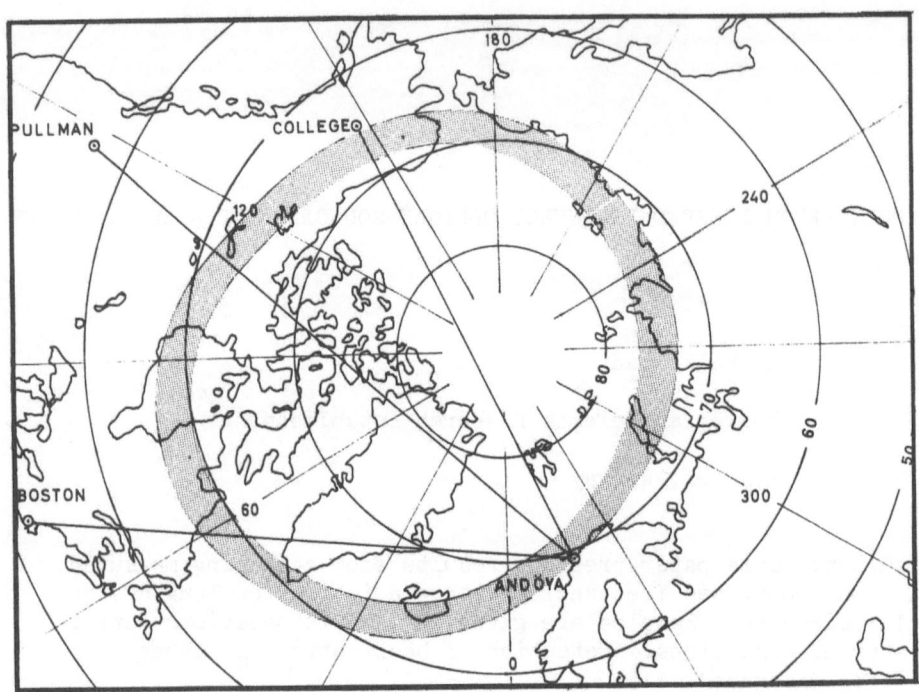

Figure 1. Geographical position of sounding stations.

The frequency range covered extended from 4 to 64 Mc/s. A conventional pulsed signal technique was employed with a pulse repetition rate ranging from 10 to 50 pps.

3.   RESULTS

3.1 Comparison of Measurements with Predictions

In the running data reduction main emphasis was put on the behaviour of the highest propagated frequencies. Plots exhibiting the diurnal and seasonal trends in the distributions of the MOFs for the winter, spring and summer of 1964 are presented in Figures 2, a to d. The spotted areas represent the distributions of all the read MOFs. Median frequencies are given by the broken curves. The thick horizontal lines indicate periods with predicted screening by the E-layer. Drawn also are the MUFs predicted by a standard program based on punched card data from the Institute for Telecommunication Sciences and Aeronomy, Boulder. Vertical lines at the bottom of each display indicate the numbers of detectable signals as percentage of the total number of readings.

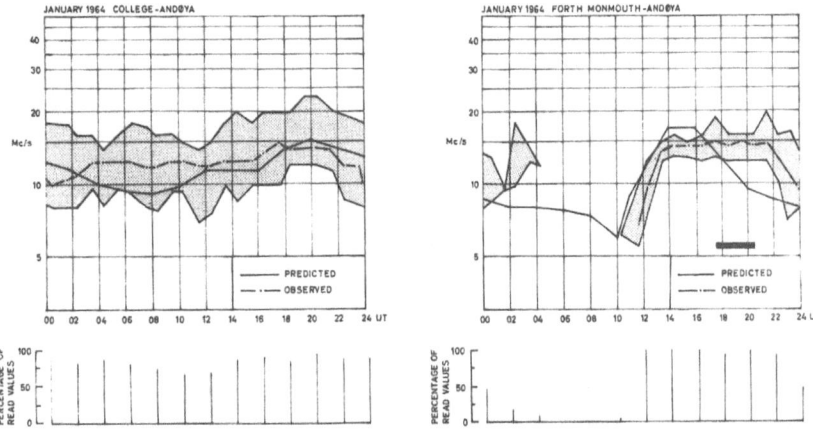

a.

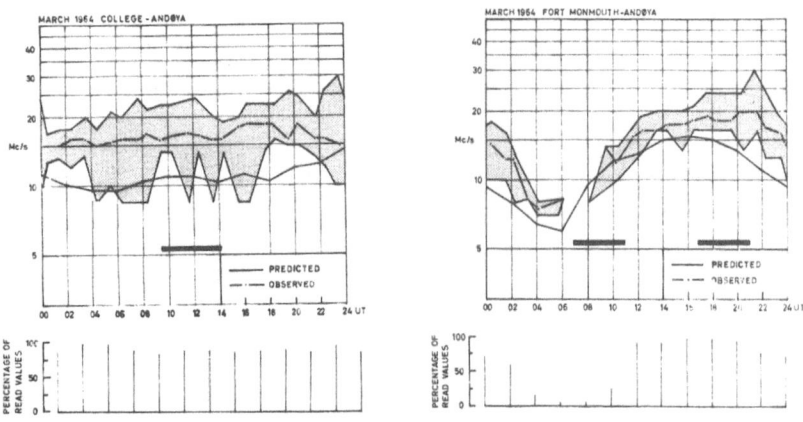

b.

Figure 2, a and b.   Comparison of measured and predicted
                     frequencies.

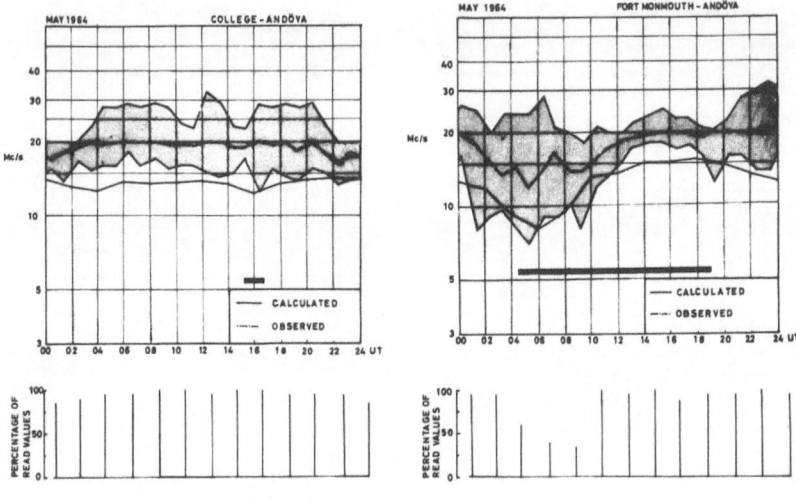

c.

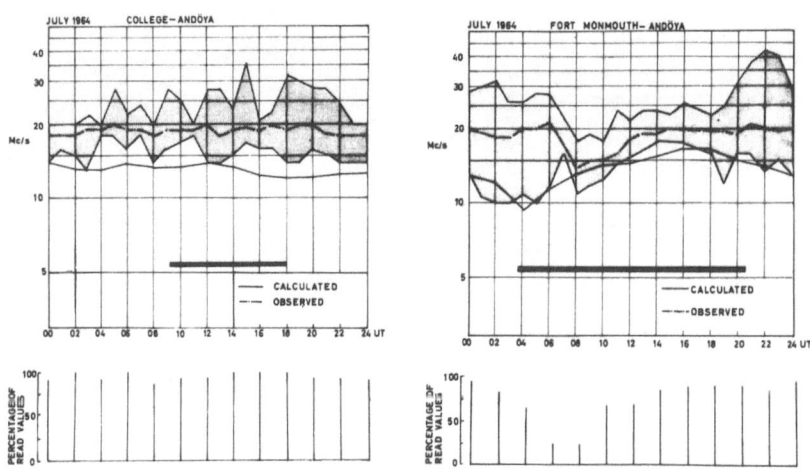

d.

Figure 2, c and d.   Comparison of measured and predicted
                     frequencies.

Inspection of the plots readily reveals the following characteristics:

1)   For both circuits the observed median values of the maximum
     observed frequencies were, for the greater part of the time,
     substantially above the predicted MUFs.

2)   The most pronounced discrepancies between observed and pre-
     dicted maximum frequencies are demonstrated in the data from
     the spring- and summer-months. On the average, for these
     months, the predicted maximum usable frequencies were about
     5 Mc/s below the corresponding observed medians.

3)   During the early morning hours, particularly in the winter
     months, the transmissions on the transpolar circuit were
     found to be more reliable than signals propagation along the
     auroral link.

4)   Poor reception and noticeable scattering of the read frequen-
     cies for the morning hours is exhibited in the late spring
     and summer plots for the auroral path. This is an indication
     that the propagational mechanisms involved are of a rather
     unstable and unpredictable nature.

4. BEHAVIOUR DURING DISTURBED CONDITIONS

From a practical communication point of view the abilities of
the circuits to propagate information at times of disturbances are
of particular importance.

The plots displayed in Figures 3a and 3b have been prepared
to gain some insight into the behaviour of the circuits considered
during periods with perturbations in the lower ionosphere. As a
measure of the ionospheric activity the riometer data from Andøya,
Tromsø and Longyearbyen were used. The latter station is well
within the auroral belt, about 900 km to the north of Andøya.
Simultaneous observations of absorption at all three stations
should warrant that the disturbed region is rather extended. In
examining the propagation spectra of the circuits it should be
kept in mind, that absorption events involved primarily occur at
the times of the day when the transmission conditions for the auroral path normally are unfavourable.

It might be concluded nonetheless that, for the type of disturbance under consideration, the auroral link is more severely
affected than is the transpolar circuit.

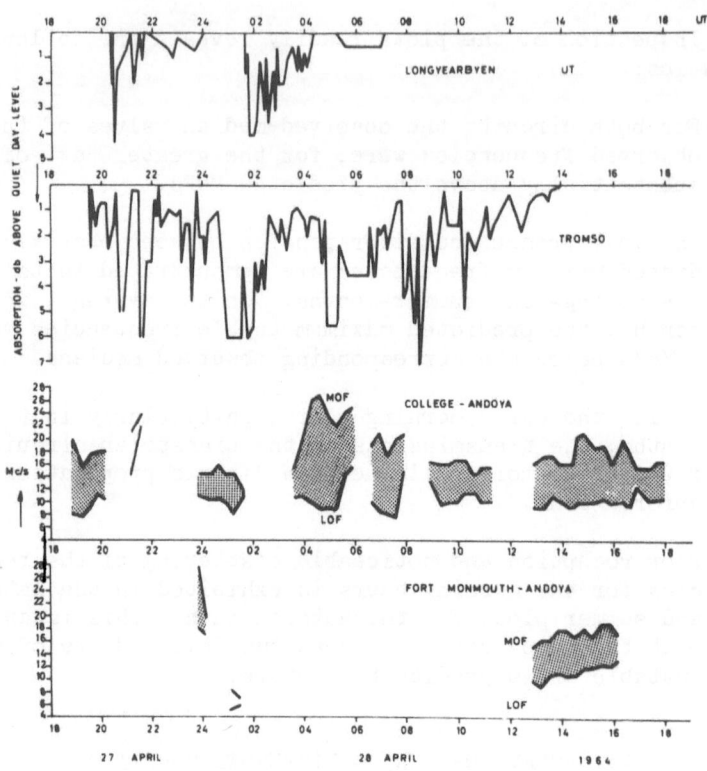

Figure 3a.   Circuit behaviour during disturbed conditions.

5.   CONSIDERATION OF PROPAGATIONAL AGENCIES

Generally, a detailed study of the types and morphologies of the ionizing agencies controlling the MOFs of auroral and polar radio circuits is impossible. Supplementing information, first of all in the form of vertical ionograms, is scanty indeed for the regions involved. In addition, with conventional modulation and detection methods, the oblique ionograms are frequently of such a blurred nature so as to preclude reliable mode identification.

There is experimental evidence, however, that widespread sporadic-E formations play a very important role in sustaining transmissions on high-latitude paths (1). At the Andøya site Es-echoes were regularly observed in the backscatter modes during the winter nights. A situation where both 1 and 2 hop Es-components are detected is depicted in Figure 4. It might be inferred from the observed time-delays that the Es layer in this case had a horizon-

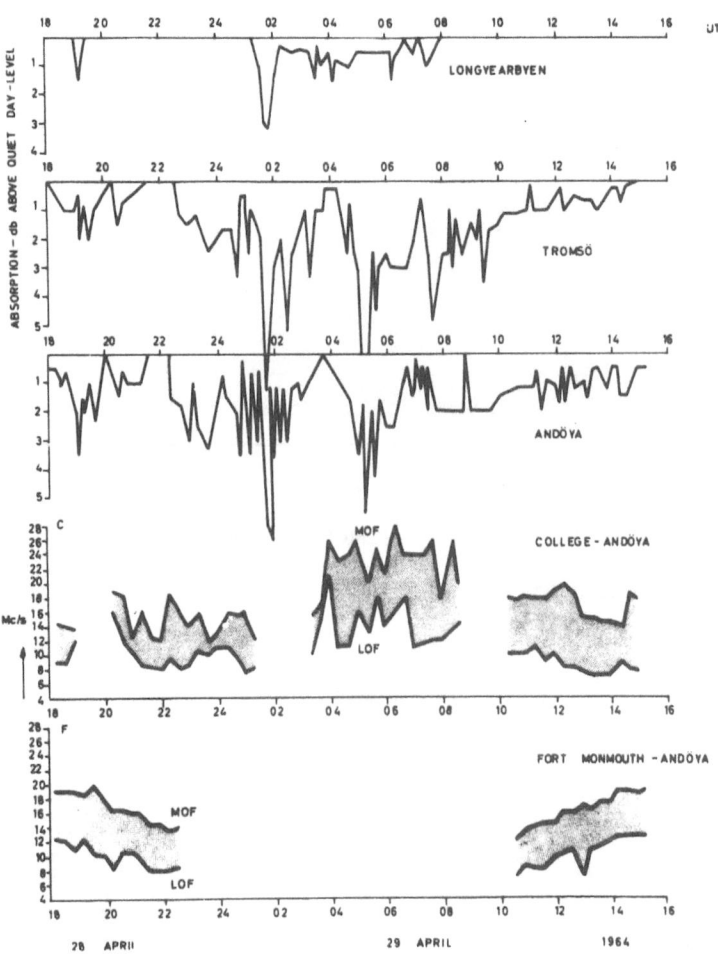

Figure 3b.   Circuit behaviour during disturbed conditions.

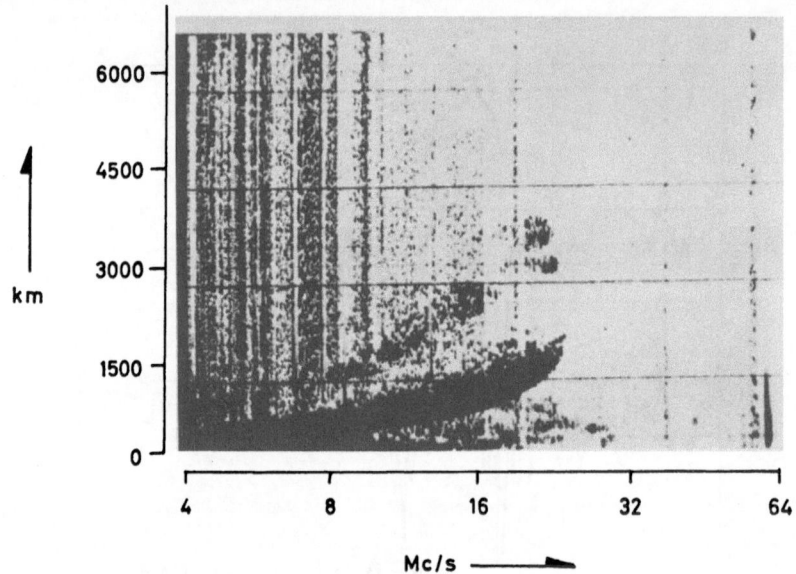

Figure 4. Ground-scatter Es echoes.

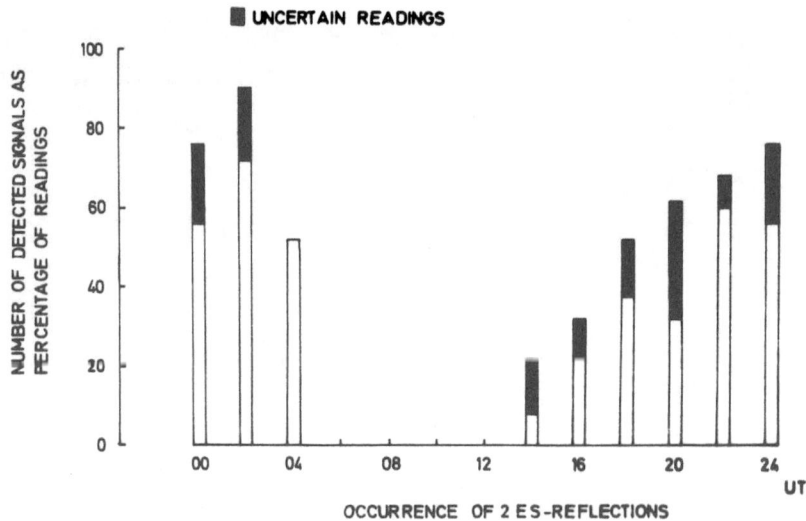

Figure 5. Diurnal occurrence of ground-scatter Es echoes.

tal extent of at least 1500 km.

Results from an analysis of the diurnal occurrence of ground scatter Es-echoes for the months February-March 1964 are given in Figure 5. Maximum occurrence is at night hours with peak around 0200.

It may be worth noting that sporadic E-formations hitherto have been given but scarce attendance in standard frequency predictions.

6. RESULTS OF OTHER WORKERS

Analyses of the transmissions in 1964 from Andøya to College have recently been made by Bartholomew at Stanford Research Institute (2). His findings seem to be consistent with our results. He observed MOFs substantially above the predicted MUFs; he noted furthermore, that the variations in the median frequencies were more dependent upon season that upon the diurnal pattern.

Other features revealed by Bartholomew's investigations may be summarized as follows:

1) Median MOFs for disturbed days, as measured by Kp-indices, did not differ significantly from those observed for quiet days

2) Median LOFs were generally higher on disturbed days than on quiet days

3) The median LOFs were found to be just slightly above the predicted lowest usable frequencies.

7. CONCLUDING REMARKS

From the features disclosed in the analysis described it is inferred that in operating long-range HF radio-links at high latitudes working frequencies above the MUFs, determined by standard computing programs in their present stage of development, might be used with advantage.

Furthermore the experimental findings stress the need for having sporadic E-formations incorporated in the prediction procedures.

Finally, in establishing direct transmissions on the basis of continuous operation between a high-latitude European station and the American Continent a transpolar circuit seems to be preferable to a link along the auroral belt.

Acknowledgement: This work was supported in part by the Air Force Cambridge Research Laboratories OAR, through its European Office under Contracts AF 61(052)-810 and AF 61(052)-835.

8. REFERENCES

1. H.F. Bates and R.D. Hunsucker, HF/VHF Auroral and Polar Zone
    Forward Soundings, Scientific Report No 3, UAG-R150, Univer-
    sity of Alaska, College, 1964.

2. R.R. Bartholomew, Results of a High-Latitude HF Backscatter
    Study, Scientific Report 2, SRI Project 5538, Stanford Re-
    search Institute, 1966.

THE SIGNIFICANCE OF SPORADIC E PROPAGATION IN DETERMINING THE MUF

E. E. Stevens

Defence Research Telecommunications Establishment

Ottawa, Canada

Abstract: This paper deals with two aspects of sporadic E propagation. One concerns the application of sporadic E tabulated data to propagation studies while the other refers to the significance of sporadic E propagation for paths in excess of 2000 kms.

1. PROPAGATION STUDIES

When conducting propagation studies for a particular path or within a specific area, it is necessary to arrive at a fairly definite conclusion concerning the upper frequency limit of sky-wave propagated signals. This must not only include the conventional modes but must take into account the effect of sporadic E propagation. This is of particular importance when it is necessary to provide a prediction of the probability of interference from sky-wave propagated signals on a given frequency assignment or vice versa. The problem of intercept may or may not be significant, depending upon the application.

There are two sources of information that may be used in conducting such a study. One is the frequency prediction systems that include probability predictions for sporadic E propagation. The other is the actual tabulated data of sporadic E characteristics observed at the various ionospheric stations throughout the world.

If it is possible to select a station that is representative of the area concerned, direct application of tabulated data may be the best approach. However, we might ask: How valid are tabulations of sporadic E characteristics in conducting such a study?

For example, would the product of foEs by a factor of 5 yield a
result that would be representative of sporadic E propagation for
a 2000 km path, centered on the observing station, or would fbEs
be a more accurate parameter to use?

Due to the lack of uniformity in the various types of equipment at ionospheric stations, the conditions reflected in the data
from one station may vary considerably with that from another,
even though actual conditions may be very similar. Consequently,
there is the possibility that this factor may influence the results of propagation studies, despite standardization of scaling
rules for the tabulation of sporadic E characteristics.

2. SPORADIC E PROPAGATION OVER PATHS IN EXCESS OF 2000 KMS

The other area discussed concerns the significance of sporadic E propagation over high latitude transmission paths in excess
of 2000 kms. A brief outline is given of results obtained from
the analysis of oblique incidence ionograms for the transmission
paths, Ottawa to Resolute Bay 3300 kms, and Ottawa to Hague 5500
kms for the years 1960 and 1961. In this investigation an effort
has been made to establish the percentage time multi-hop sporadic
E is the dominant mode, or where sporadic E propagation contribu-
tes to a frequency increase beyond the regular F layer mode, by a
combination of Es and F layer reflections.

Figures 1 A and B show the results of an analysis of oblique
incidence ionospheric sounding data for the Ottawa-Resolute path,
for the last half of 1960 and for the complete year of 1961, with
respect to the percentage time sporadic E propagation was a significant factor in determining the upper frequency limit of the MOF.
The percentage time data were available for this study is also indicated.

Figures 2 A and B are similar plots for the Ottawa-Hague path,
for the years 1960 and 1961. As would be expected, there is less
evidence of the effect of sporadic E propagation over this longer
path.

The overall results are summarized in tables 1 and 2.

This study indicates that, for the areas and paths concerned,
the effect of sporadic E propagation is not a significant factor
in determining the overall MUF, except during the summer months;
and, as would be expected, the effect of sporadic E is more pronounced in the shorter, trans-auroral path.

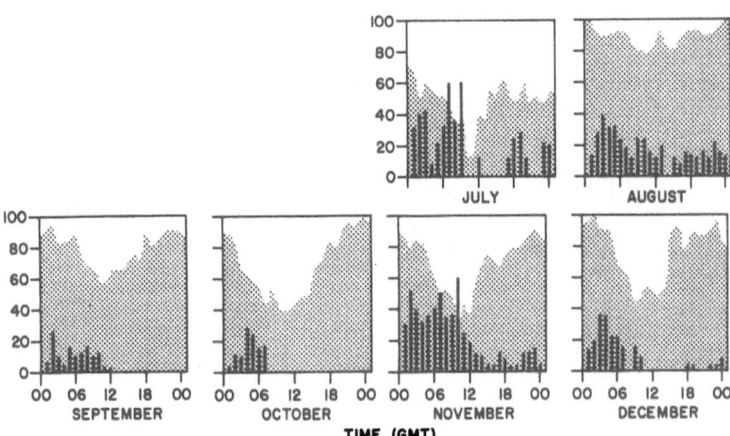

Figure 1A. Percentage of occurrences where Es controls or enhances the MOF. Percentage of time data available indicated by spotted area.

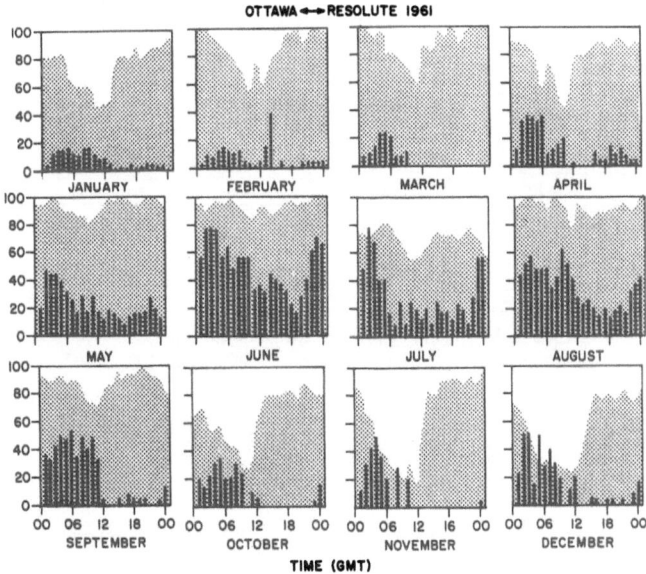

Figure 1B. Percentage of occurrences where Es controls or enhances the MOF. Percentage of time data available indicated by spotted area.

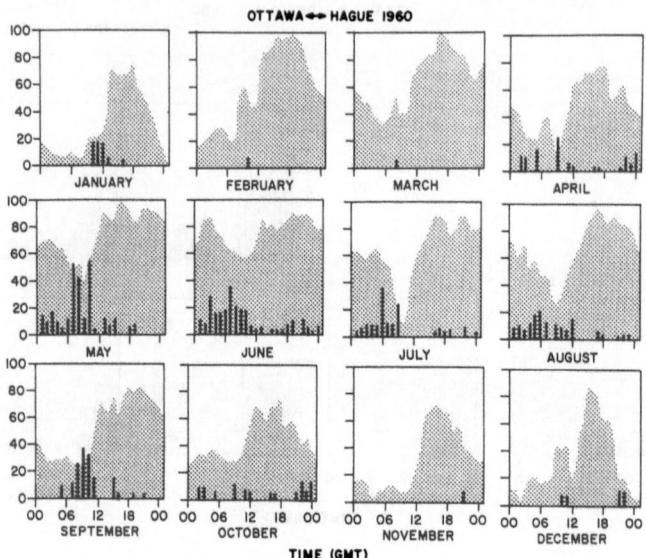

Figure 2A. Percentage of occurrences where Es controls or enhances MOF. Percentage of time data available indicated by spotted area.

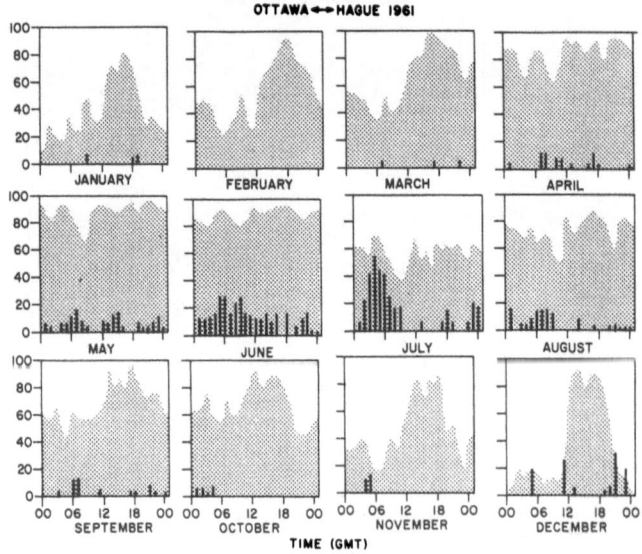

Figure 2B. Percentage of occurrences where Es controls or enhances MOF. Percentage of time data available indicated by spotted area.

Table 1. Percentage time sporadic E significant   Ottawa–Resolute.

Year	Monthly Average												Yearly Average
	Jan	Feb	Mar	Apr	May	June	July	Aug	Sept	Oct	Nov	Dec	
1960	#	#	#	#	#	#	19.5	19.0	6.0	5.0	23.0	8.5	13.0
1961	7.0		5.0	14.0	26.0	47.0	28.0	35.0	31.0	10.0	9.0	16.0	19.0

Data not available

Table 2. Percentage time sporadic E significant   Ottawa–Hague

Year	Monthly Average												Yearly Average
	Jan	Feb	Mar	Apr	May	June	July	Aug	Sept	Oct	Nov	Dec	
1960	2.5	-	-	4.0	11.0	12.0	6.0	6.0	6.0	4.0	-	1.5	4.4
1961	1.0	-	.5	3.0	6.0	13.0	14.0	4.0	2.0	1.0	1.0	4.5	4.0

MORPHOLOGY OF FADING OF RADIO WAVES TRAVERSING THE AURORAL IONOSPHERE

Richard S. Allen

Air Force Cambridge Research Laboratories

Bedford, Massachusetts, USA.

Abstract: The morphology of scintillation fading is developed for HF and VHF radio waves which have traversed the ionosphere at high latitudes. Forms are indicated for the variation of the depth of fading with sub-ionospheric latitude, geomagnetic conditions, seasonal and diurnal time and solar cycle. Techniques are suggested for normalizing data obtained by different observatories with varying radio frequency or geometric parameters. Interdependence of several parameters indicates that a common cause controls the details of their behaviour.

1. INTRODUCTION

The conclusions to be made are that scintillations caused by ionospheric irregularities in the poler auroral region are quite important to high latitude propagation, that they are not well documented and that a determined effort is being devoted to their study.

One of the important aspects of propagation in the polar region during the next decade will be transmission through the ionosphere. Signals which have traversed localized irregularities of electron density show amplitude and phase deviations, commonly called scintillation. Many satellites presently transmit data to polar ground stations. Manned Orbiting Laboratories are being planned which will require assured voice communication to a net of ground receivers, some within the polar regions. Perhaps most concerned at present are those communication and surveillance systems which will rely on VHF transmission from an aircraft, to a

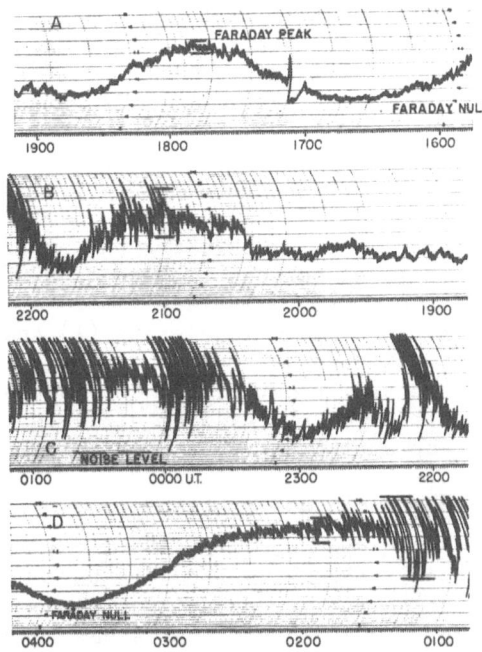

Figure 1. Scintillation of 136 MHz beacon transmitted from the
           synchronous satellite Earlybird, June 12, 13, 1965.
           Observations of the Kiruna Geophysical Observatory.
           Long slow fading is the result of Faraday rotation
           within the ionosphere.

satellite such as Early Bird, and then to a master ground station.

A particular example of this would be a 136 MHz navigation
and surveillance system planned for air traffic control. Pulse
and voice signals would be relayed from ground control to the aircraft via a special air traffic control and navigation satellite
located over the mid Atlantic. For an SST flying the polar route
between Seattle and Copenhagen, for example, or a standard jet
flying a minimum time transatlantic route, part of the propagation
path often would be through regions containing strong ionospheric
irregularities.

Since the transceiver and antenna in both vehicles will be
limited by weight considerations, signal to noise ratios near 10
db are very probable for the period through the 1970's and 1980's.
Samples of scintillation records from auroral regions (Figure 1)
showing very deep fading are not uncommon. The statistical behaviour of 3, 6 and 10 db scintillation fading at high latitudes

is, therefore, a very vital current study.

This report will attempt to show what is presently known of high latitude scintillation, establish some methods of comparing prior published data, and will attempt to indicate which new observational procedures seem promising to the author.

2. THEORETICAL BACKGROUND

Scintillation is a popular term for the amplitude and phase deviations of a radio wave that has traversed an irregularity in the ionosphere. If the characteristic size (L) of the irregularity is much larger than the radio wavelength (λ) and the gradient of ionization is not too great, then the wave undergoes only phase deviations (ϕ_0) within the irregularity. As the wave propagates onward, the superposition of energy from different portions of the irregularity builds up amplitude variations.

It is convenient to discuss distance from the irregularities in terms of the Rayleigh distance, $Z_R = \frac{L^2}{\lambda}$, where L is the representative size of the irregularity and λ is the observing radio wavelength. The ionospheric irregularities responsible for scintillation are generally much larger than a wavelength, in fact, the usual range reported for either correlation or peak to peak shadow distance is from a few tenths to several kilometers. Whether this is an important characteristic of the irregularity medium, such as a characteristic wavelength, or is a result of observational selection does not appear to have been explained. At any rate, the Rayleigh distance is on the order of a few hundred kilometers in the frequency range of interest here, HF through VHF. Thus, for irregularities at heights near the F region peak either the observer or the satellite, or both, can be close to the transition region between near (Fresnel) and far (Fraunhofer) zones and this must be considered when trying to apply theoretical analysis.

Sometimes observational evidence indicates that single irregularities, or at most, a few irregularities scattered along the ray path a distance much less than their Rayleigh distance are responsible for the phase and amplitude measurements made at the ground. In that case, some two dimensional analysis, such as that of Mercier (1962), may be consistent with the observations. Either the numerical results presented graphically or the approximation of Briggs and Parkin (1963) may be used for quantitative analysis.

The salient features of the thin screen, may then be summarized for near field observations, $Z/Z_R < 1$.

1. Phase and amplitude deviations are correlated.

2. The magnitude of amplitude increases at the expense of the magnitude of phase as viewing distance (Z/Z_R) from the irregularity is increased.

3. The scale size of features of the diffraction pattern at the receiver is less than the scale size of features in the irregularity distribution. This is nearly $L/\phi_0^{\frac{1}{2}}$ near the irregularities.

For far field observations $(Z > L^2/\lambda)$, these properties are modified so that:

1. Phase and amplitude deviations become uncorrelated.

2. The energy of the diffracted component tends to be equally shared between amplitude and phase deviations.

3. The diffraction pattern has a scale size the same as the irregularity size for small phase deviations $(\phi_0 < 1$ radian) but a scale size L/ϕ_0 for large phase deviations.

During disturbed periods at sub-auroral latitude and during most periods at high latitudes, there is good evidence to suggest much thicker regions over which the irregularities are distributed. In that case, a three dimensional analysis, such as that of Orhaug (1965) is indicated. If the height interval D is much less than the effective Rayleigh distance for the representative irregularity then the results are the same as for the thin screen analysis. As the ratio D/Z_R increases, the near field behaviour departs drastically from the thin screen analysis. Phase and amplitude components become decorrelated and interact in a complex manner. The scale size of the scintillation pattern is an integrated effect of several irregularities viewed simultaneously, but changing independently with time.

Frequency and geometric dependent factors differ significantly for the two cases. Probably empirical factors are the present best method of comparing different sets of data or extending current data for future needs. The western literature may be easily back-referenced from major theoretical papers by Orhaug (scattering by a thick screen, 1965), Singleton (ray optics, 1964), Briggs and Parkin (anisotropy, 1963), and Mercier (two dimensional diffraction screen, 1962).

3. OBSERVATIONAL PROCEDURES

3.1 Source of Data

When observations are limited to those which have traversed the ionosphere, there are two sources of data; studies made with cosmic sources and studies made with beacon satellites. In a sense, they are complementary, since radio stars have an apparent motion which is only as fast as the earth's rotation, while low altitude satellites move so quickly that a sample may be taken from horizon to horizon in a near instantaneous fashion. In another sense they are contrary, in that seasonal, diurnal, and latitudinal parameters are nearly hopelessly dependent in radio star studies. While latitude effects may be abstracted from satellite observations, the slow change of local satellite time still leaves seasonal and diurnal factors dependent on each other, as shown below. Only with synchronous satellites can latitude, diurnal time and season become independent parameters.

Burst emission from the sun or Jupiter need not be considered for statistical studies since the signals are sporadic and highly irregular.

3.2 Empirical Scintillation Indices

Almost everyone in the field uses a different empirical scintillation index and it is very important to bear this in mind when different data are being compared.

The ultimate, of course, would be a root mean square measurement of the fluctuating component, but few observatories have afforded either the man hours or the expense. Briggs and Parkin (1963) have suggested standardized definitions of the various possible indices and the empirical relations shown in Table 1.

R is the instantaneous amplitude of the wave

S_1	S_2	S_3	S_4
$= \dfrac{1}{\overline{R}}\overline{(R-\overline{R})}$	$\dfrac{1}{\overline{R}}\{\overline{(R-\overline{R})^2}\}^{\frac{1}{2}}$	$= \dfrac{1}{\overline{R^2}}\overline{(R^2-\overline{R^2})}$	$= \dfrac{1}{\overline{R^2}}\{\overline{(R^2-\overline{R^2})^2}\}^{\frac{1}{2}}$
$= 0.42\ S_4$	$= 0.52\ S_4$	$= 0.73\ S_4$	

(After Briggs and Parkin, J. Atmos. Terr. Phys., 25, 1963)

Table 1. Suggested definitions of scintillation indices.

Both radio star and satellite observations reduced at Saga-more Hill, Massachusetts, have been assigned an arbitrary but con-sistent index based on the peak to peak variation of the received power. Briefly the procedures, adequately defined elsewhere (Aarons et al, 1963 and Allen et al, 1965), are as follows:

Rate (R) is determined by counting peaks. A peak is counted only if it is at least one third the amplitude of the previous peak. The amplitude of each peak, therefore, becomes the crite-rion by which the next is considered, without further regard to the past or future. For radio star observations, rate is a combi-nation of irregularity, size, ionospheric motion, and earth rota-tion. For satellites, insofar as the motion of the radio wave path is much faster than ionospheric motion, then rate is some measure of irregularity size.

Scintillation index (S) is also a relative measurement. A sample interval containing at least ten peaks counted by the above rule, and at least five minutes duration is chosen from radio star records. A similar sample at the top of a Faraday peak is chosen for the satellite case. Representative peaks are established by the third peak down from greatest and the third up from the low-est. The power scintillation index is then the percentage ratio of the varying component to the expected steady component.

$$S(\%) \quad = \quad \frac{P_{(\text{third peak up})} - P_{(\text{third peak down})}}{2P_{(\text{average})}} \quad = \quad \frac{\Delta P}{P} \ , \ \%$$

Analysis by planimeter and by point computations has indicated that this arbitrary index is from 1.5 to 2.5 times the rms value of the power variation. For radio star observations, the ratio is nearly the same at several frequencies but varies from day to day. From a statistical view-point, it may be assumed to have a fixed relation to the rms value, probably not far from the value 1.4 empirically determined by Briggs and Parkin.

Several other observatories making regular satellite obser-vations have used a similar index for the depth of scintillation. Some of these, however, are referred to amplitude rather than power.

Finally, the quickest method is to assign subjective indices such as 0, 1, 2 for zero, some and fully scintillating conditions. If class limits are carefully set up, and used, then statistical analysis of an extensive group of these data can be quite fruitful.

3.3 Comparison of Data

The easiest statistic to produce, and the most frustrating
to try to use for comparative analysis, is the mean value.

Almost invariably the data samples are truncated at both ends
of the scale. Small values cannot be read and are defined zero.
Large values of amplitude fluctuation can increase to several
times the mean intensity but can only decrease to the system noise
level. Arbitrary reduction procedures then insure a label of
"full" or "100 percent" on variations which may differ by a decade.

When subjective indices are used, such as 0, 1 and 2 or 1
through 5, etc., the class intervals are very often chosen in a
semi-logarithmic fashion (scale of two). Mean values are then
particularly inappropriate for comparison.

If in the reduced data some of the character of the individual observations is preserved, then often the results may be re-
analyzed to study effects that cannot be considered when only mean
values are available. The cumulative probability distributions
used below are an interesting case of this.

4. MORPHOLOGICAL RESULTS

It is convenient to consider all the observations which have
been made of the scintillation of either radio stars or of satellites as independent samples drawn from a statistical distribution
of the form:

$$S \; = \; S \, (X_1, \, X_2, \, X_3, \, \cdots \cdots X_n)$$

It is hoped that at least some of the variates, X_i, are independent, for instance, the depence on the radio frequency of observation hopefully might be uniquely factored as:

$$S \; = \; S \, (f/fo)^{n(f)} \quad (X_2, \, X_3, \, \cdots \cdots X_n)$$

where the exponent of frequency dependence ($n(f)$) might be dependent on frequency but not on the other parameters.

This is a vain hope but a useful way to start and, besides,
its short-comings do lead to some interesting speculation on some
of the other parameters.

When there is good evidence for dependence of some of the va-
rities, a conditional distribution is useful. As an example, the

depth of scintillation is varied by some parameter itself dependent on short term solar activity. If the degree of disturbance
of the irregularity medium by this unidentified parameter can be
characterized by some other observable whose behaviour it modifies,
such as the K index of magnetic activity, then it is rewarding to
examine conditional distributions, as when the K variate is held
at fixed values of 0 and 1, 2 and 3, 4 and 5, etc.

Finally, in order to comprehend or display the results of
this manipulation, a one, two, or perhaps three dimensional display is needed.

It is usual to form a marginal distribution by averaging or
grouping over an unbiased sample of data for fixed values or ranges of the variable being examined. The diurnal variation of mean
values produced by averaging all data for fixed hours is such a
marginal distribution.

Here lies the greatest hazard in the analysis of both radio
star and satellite data. Diurnal, seasonal, and solar cycle parameters may be quasi periodic. If they are indeed independent variables, then they can be normalized individually by suitable summation over their periods:

$$S(S_5, X_6 \cdots \cdot X_n) = S_0(f/f_0)^n G(X_5, X_6 \cdots \cdot X_n) \{ \sum_0^{11} \sum_0^{12} \sum_0^{24} H(yr, mo, hr) \}$$

If instead of a numerical average, the desired result is a median
or a cumulative distribution, then the above summation symbolism
may be interpreted as a statement that enough values of the variate are present in the master distribution to represent their range without bias.

In fact, this is not true for most satellite and radio star
observations. Given a fixed location in the ionosphere, such as
upper transit for the radio star, Cassiopeia A, then local time
and season are dependent. In the fall, auroral period observations of Cassiopeia A near overhead are restricted to the midnight hours. If several radio sources are used in an attempt to
fill in the gaps, then some normalization must be made for the new
conditional parameters introduced with each source.

Low altitude, high inclination satellites produce similar
circumstances. In a given geographic sample, such as overhead, a
twenty-four hour range may be sampled in about three months by a
satellite in a near circular orbit such as transit 4A or S-66.
Note, however, that different hours occur in each of the months.
There is no such thing as a diurnal or seasonal marginal distribution from such data, only an educated guess. Fortunately, there
is a little more hope of finding compatible observations from several satellites to fill in the gaps. A synchronous satellite

would be particularly useful, as the short period of Early Bird
transmission during 1965 has demonstrated already.

   With this explanation and forewarning, the results of a number of studies are now presented. Naturally, where consistent results can be obtained from several sources, there is a higher probable reliability, otherwise the results must still be regarded
critically.

   Since many of the parameters are quite dependent on each
other, when one is being discussed, the reader must expect that
conditional restriction based on other parameters will be imposed
before a discussion of the nature of the dependence. For instance, latitude variations will be examined for quiet (K index low)
geomagnetic conditions before the K dependence is discussed, since
they are interdependent.

4.1 Latitudinal Variation

   One of the dominating factors in the irregularity distributions producing scintillation is latitude. As early as 1951,
Little noted that the variation of the irregularity medium with
latitude in the sub-auroral region was greater than the geometric
effects expected when viewing Cassiopeia A at low angles. Figure
2 shows the best available data, derived from simultaneous observations previously published by a cooperating group of observatories (Joint Satellite Studies Group, 1965). The original data
have been replotted, using only overhead mean values at each station. The resultant smooth variation is conditional on nearly i-
dentical geometric factors (near vertical propagation), low magnetic activity (K = 0, 1 and 2), and is the mean of observations
spaced from June 1962 to March 1963 by joint observational periods
selected without meaningful bias. During this period 3 db fading
($\frac{\Delta P}{P} \approx 50\%$) was the mean behaviour for the Kiruna station near the
outside of the auroral zone.

   Preliminary results from a current study by the same group
are shown in a different form in Figure 3. All data taken near
overhead for the periods November, December 1965, January 1966,
and November, December 1966, January 1967 are being examined for
40 MHz transmissions from the S-66 satellite. Note that these cumulative distributions are expressed in terms of the probability
of the scintillation index being greater than a given level. At
Kiruna, Sweden, about 50 percent of the time fading greater than
3 db was experienced overhead at 40 MHz, while only about 15 percent of the time at Breisach, Germany. Data from other cooperating stations in the Joint Satellite Studies Group will extend the
coverage from Thule to the magnetic equator.

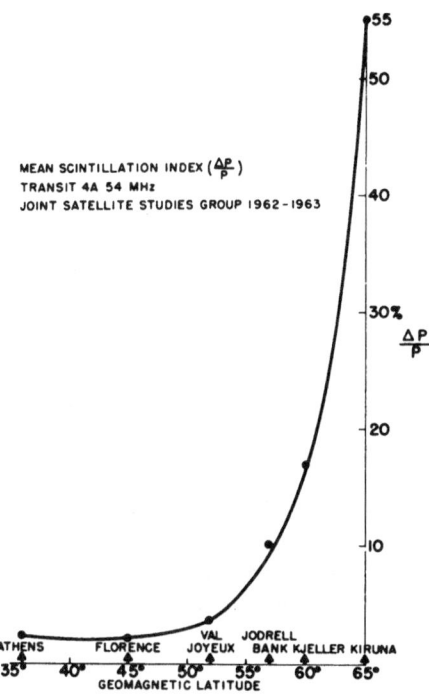

Figure 2. Latitude variation of scintillation index over Western
          Europe in 1962 - 1963 derived from simultaneous over-
          head measurements by stations cooperating in a special
          joint study.

4.2 Magnetic K Index

The variation of high latitude scintillation index with magnetic K index can be represented in many ways. The form in Figure
4 has been chosen because it is so reminiscent of the IGY study
(Gartlein et al, 1960) of the southward extension of aurora. All
observations of Transit 4A taken at Sagamore Hill, Massachusetts,
during regularly scheduled periods each month from January 1962 to
February 1965 have been pooled. Conditional samples were based on
sub-ionospheric (400 km) ranges of ±10° Longitude and 5° Latitude.
Mean values were then produced for all data in subsets determined
by integral values of the local magnetic K index, K_{Fr} from Fredericksburg, Va. Notice particularly that no attempt has been made
to remove geometric effects in these data. The general nature of
the increase of deep fading in the auroral region and the southward extension of the region of occurrence during magnetically
disturbed periods is apparent.

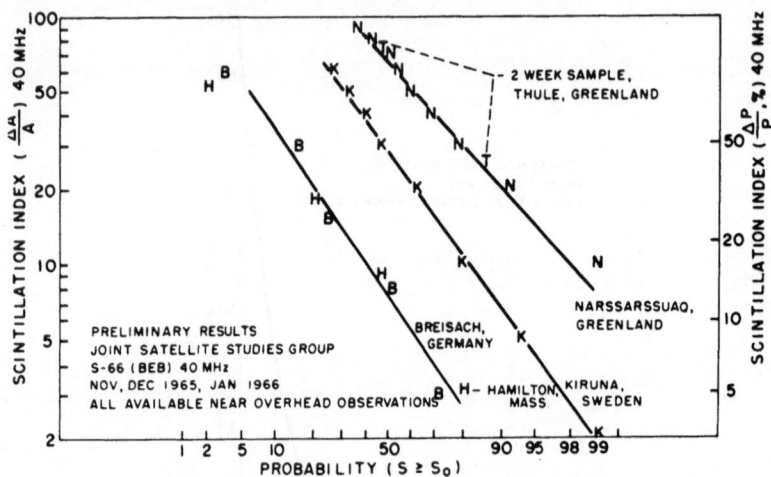

Figure 3.   Initial results of cumulative probability distributions
for nearly simultaneous observation of overhead passes.
Note the decreased occurrence of deep scintillation
southward from the auroral region.

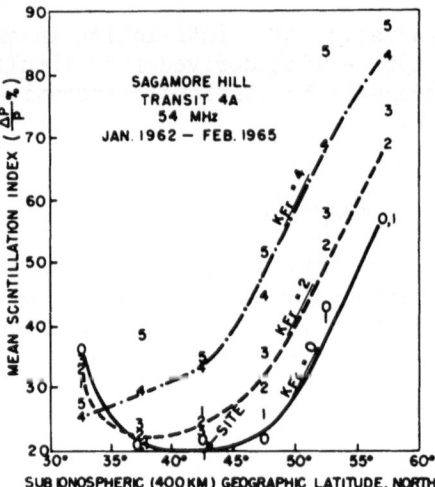

Figure 4.   The latitude and magnetic variation of satellite scintillation at 54°N geomagnetic. Data has not been normalized for geometric effects. K_{Fr} is local
(Fredericksburg, VA.) magnetic index.

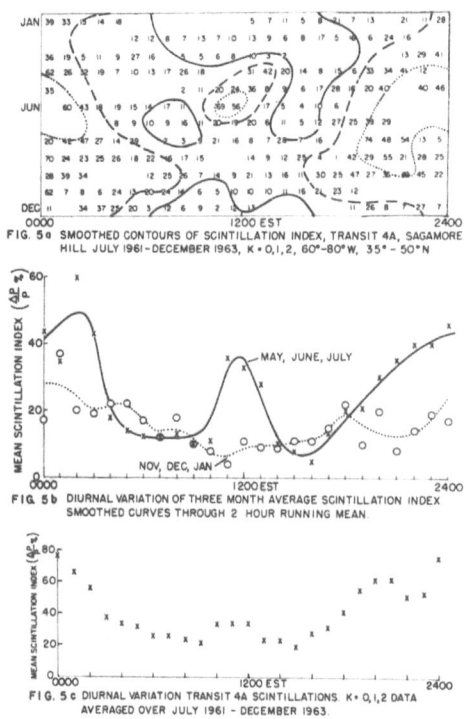

FIG. 5a SMOOTHED CONTOURS OF SCINTILLATION INDEX, TRANSIT 4A, SAGAMORE
HILL JULY 1961-DECEMBER 1963, K = 0,1,2, 60°-80°W, 35° - 50°N

FIG. 5b DIURNAL VARIATION OF THREE MONTH AVERAGE SCINTILLATION INDEX
SMOOTHED CURVES THROUGH 2 HOUR RUNNING MEAN

FIG. 5c DIURNAL VARIATION TRANSIT 4A SCINTILLATIONS. K = 0,1,2 DATA
AVERAGED OVER JULY 1961 - DECEMBER 1963.

Figure 5, a, b and c. Displays of diurnal scintillation effects.

4.3 Seasonal/Diurnal

If a seasonal effect is to be separated from the data, it
must be recurrent over several years and it should be independent
of the other parameters. As mentioned before, statistical techniques such as averaging over a subset of data must be examined
critically for independence of the parameters and for bias in the
sample.

Such an attempt is presented in Figure 5a. This three dimensional display reserves the data points which are the means of
conditional subsets of 54 MHz observations taken in a large seg-
ment of the ionosphere over Sagamore Hill, Massachusetts. First
note again that there are blanks where no data can exist. The
smoothed isopleths are a combination of linear interpolation, value judgment when a single observation is high, or low, or is based on a low number of observations and guessing where blank areas
must be crossed.

Since the occurrence of deep scintillation increases sharply
with increasing latitude, the averaging process has weighted
these data so that it represents the sub-auroral region. With
this in mind, the nighttime maximum near magnetic midnight is a
prominent feature throughout the year and seems to have a localized peak near the winter solstice. The daytime maximum is most
prominent at the summer solstice. It would be interesting to examine a similar collection of data representing the south sub-auroral region, for quiet magnetic conditions.

Although the seasonal and diurnal effects are not independent,
it is instructive to examine the distributions which might be obtained at the margins of the array. The data can be normalized
somewhat by considering the mean for each hour-month subset as an
independent observation. Nothing very valid can be done with the
blanks.

The quasi-marginal distribution of the resultant means for
the diurnal variation, Figure 5b, is consistent with previous displays for radio stars or satellite scintillation. Note that the
daytime maximum, localized to only the summer solstice period, is
almost averaged out and could easily be missed in a reduction of
the data that used larger ranges for the hour-month subsets or
that used indices with wide class intervals.

Perhaps slightly more meaning can be attached to the diurnal
variations conditional on particular seasons as in Figure 5c. The
summer solstice behaviour (May, June, and July) is well substantiated by independent 136 MHz observations made with the Early
Bird satellite during 1965, Figure 5d. These are referenced to a
sub-ionospheric location somewhat southeast of Sagamore Hill
(39°N, 65°W). These data indicate that a careful study of diurnal
behaviour during the year might provide interesting clues to the
features of particle/solar interaction with the upper atmosphere.

The quasi-marginal distributions for the seasonal variations,
Figure 5e, have been made for smaller latitude samples. As mentioned, mean values are particularly inappropriate when a significant number of the individual observations are "100 percent" scintillating. Nevertheless, the seasonal features persist into the
southern edge of the auroral zone.

Since the solstice behaviour has been reported for previous
radio star studies, it is instructive to examine means (Figure 5f)
obtained from Cassiopeia A observations during nearly the same
period. There is general agreement but no significance can be
attached to numerical comparison in light of the seasonal, diurnal,
and latitudinal interdependence within such observations.

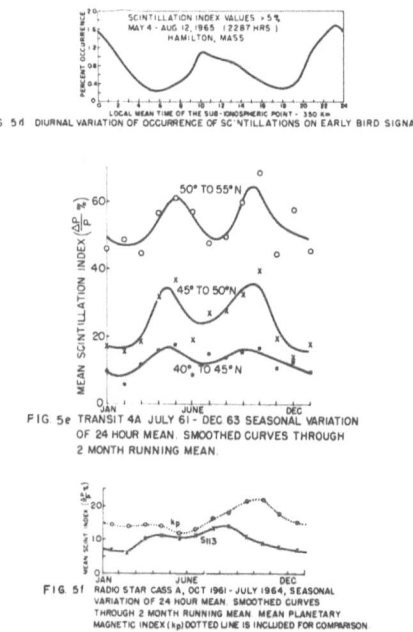

Figure 5, d, e and f.   Diurnal and seasonal scintillation effects.

As a final remark on the seasonal/diurnal problem, not the general agreement between running means for the planetary magnetic index, K_p, and for the radio star or satellite scintillation. This raises the speculation that if there were an index, better than magnetic K index, which could separate the observations into quiet and disturbed periods of solar terrestrial disturbance, then perhaps the solstice behaviour would only be prominent under disturbed conditions.

4.4   Frequency Dependence

When the irregularities can be analyzed in terms of a thin screen producing small phase perturbations, then the frequency dependence can be shown to fall between two limits. For conditions such that the viewing distance is much less than the Rayleigh distance then the depth of scintillation is nearly inversely related to the square of the frequency. Alternatively, in the far field, it is inversely related to the first power of frequency.

For a layer of irregularities of appreciable vertical extent D, the previous conditions apply when the layer is thin, $D/Z_R < 1$.

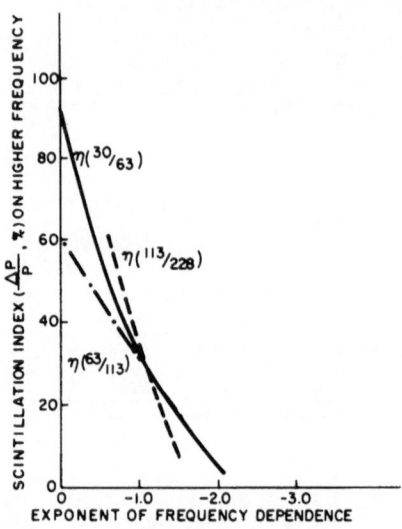

Figure 6. Variation of the mean exponent of frequency dependence
           for radio star scintillation with the depth of scintil-
           lation on the higher frequency. No appreciable change
           was noted between samples of high or low magnetic acti-
           vity when scintillation depth at one frequency was held
           constant.

As the layer parameter, D/Z, increases the inverse frequency dependence changes toward zero exponent. For a thick layer,
D/Z_R >> 1, the near and far field limits, based on Orhaug's weak
scattering analysis, are identical, a depth of scintillation directly related to the square of the frequency.

    The empirical results obtained for observations made at Saga-
more Hill support this picture. The details have been reported
elsewhere (Aarons et al, 1967). Briefly, simultaneous data were
taken at 30, 63, 113 and 228 MHz with the 150 foot antenna for the
auroral region and at 63, 113, 228, 400 and 1200 MHz with the 84
foot antenna for the regions covered by Cassiopeia A. Results for
the ionospheric region near the auroral zone and for the overhead
region were relatively independent of the actual frequency pair
being studied (Figure 6). For very low depths of scintillation,
the dependence was inverse square law. As the depth of scintillation increased the dependence flattened. For high levels, generally somewhere beyond an index of 50 percent, scintillations
sometimes were lower on the lower frequency. This departure from
expected behaviour, often called an "inversion", has been various-
ly explained in terms of decreased irregularity size, multiple

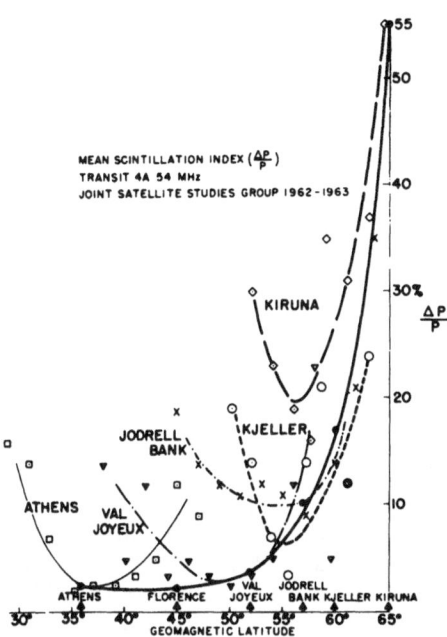

Figure 7. Latitude variation of scintillation index over Western
Europe in 1962 - 1963 derived from simultaneous overhead measurements by stations cooperating in a special
joint study. Off vertical observations show effect of
elevation and propagation angle.

scattering and layer thickness. It was a relatively infrequent
occurrence during the years near solar minimum.

4.5 Geometric Effects

Rather a lot of effort has gone into trying to predict geometric effects from theoretical analysis but, in general, the
restriction of consideration to observations made at a single station has resulted in the conclusion that irregularities vary with
latitude.

When a statistical latitude distribution can be derived from
mutually consistent overhead observations, as in the JSSG study
previously mentioned, then the associated non-overhead observations can be examined for the geometric effects pertinent to the
site latitude. Data from the 1961/1962 JSSG study were restricted
to intensive periods and, therefore, are only useful to illustrate
the main features (Figure 7). At mid-latitudes the depth of scin-

tillation is low and the non-overhead indices rise toward the
horizon more or less as the secant of the zenith angle times the
expected sub-ionospheric depth of scintillation. Apparently, for
geometric purposes, the distribution of irregularities for that
region can be represented by a thin screen placed at some effective height, such as 400 km. At sub-auroral latitudes, the overhead depth of scintillation has increased somewhat, observations
southward are still consistent with a thin screen but observations
northward increase more slowly than might be expected. This is
probably the result of several factors, a thickening of the layer,
a reduction of slant range to the irregularity and perhaps a change in the size of the irregularities.

4.6 Solar Cycle Variation

Briggs (1964) has published a comprehensive study of the
variation of scintillation from 1950 through 1961 on 38 MHz observations of Cassiopeia A at Cambridge, England. Hourly samples
of scintillation index were assigned qualitative indices 0 through
5. The principle results of his study were presented as means of
this semi-logarithmic index. The annual mean for observations
restricted to upper transit was about 0.5 in 1954, near solar minimum and increased to about 1.8 in 1958, near solar maximum. The
correspondence between mean annual index of scintillation and
twelve month running mean sunspot number was striking for both upper transit and lower transit (auroral) regions.

It is not obvious how the mean of indices based on semi-logarithmic class intervals can be compared to mean values based on
a linear scale, as in the present study. However, some information may be retrieved from histograms presented by Briggs, his
Figure 5, for the night hours, 2100 to 0300 during September 1954
to September 1957.

From his sample records, class intervals of the index $\frac{\Delta P}{P}$,
consistent with the present study, may be applied to each of the
Cambridge indices. The histograms then convert to the form given
in Figure 8. From this, an estimate can be made for the range expected during a change from solar minimum to solar maximum. For
instance, it appears that the median scintillation depth at 38 MHz
observed overhead at Cambridge could be expected to increase from
$\frac{\Delta P}{P} \approx 7\%$ to $\frac{\Delta P}{P} \approx 30\%$. Similarly, the occurrence of 3 db fading
might be expected to increase from 3% of the time to about 18% of
the time. Note that no attempt has been made to separate quiet
and disturbed periods in the data presented by Briggs.

5. CHARACTERISTIC PARAMETERS OF THE IRREGULARITIES

For the preceding section, the scintillation index and rate determined from the observations are subjective, as stated. They are integrated effects, in as much as they represent the cumulative effect of the entire ionospheric region traversed by the radio wave.

A different realm is entered when the parameters of individual irregularities must be specified. At mid latitudes isolated irregularities and thin regions may be observed during quiet periods. Features of the scintillation pattern have amplitude and phase changes that suggest only a localized region of the ionosphere is important for particular scintillations. Correlation between frequencies separated by several octaves is often quite high. Then thin screen analysis may be applied to the measurements.

The irregularities responsible for scintillation may be specified by a number of parameters, dependent on the particular model chosen for theoretical analysis. These would include some description of the individual irregularities, such as the variation of electron density with scale size and perhaps the degree of field alignment and an axial ratio. One of the most promising current efforts in this area is being made using both phase and amplitude measurements of satellite signals at the University of Western Ontario (see Turnbull and Forsyth, 1965, for details).

At high latitudes during most periods and at mid latitudes during disturbed periods, the evidence is overwhelmingly in favour of significantly thick layers. Most satellite studies triangulating specific scintillation features have found a rapid scatter of heights which indicate a layer several hundred kilometers thick. Phase and amplitude recording of radio stars have shown complex patterns and correlation of different frequencies has often decreased to very low values for octave separation, both indicating that layer thickness is important in the scintillation process. A recent study based on two frequencies, 54 and 150 MHz, triangulation of satellite signals (Jesperson and Kamas, 1964) illustrates the need for careful discrimination when the thickness of the layer becomes appreciable.

Since extension of existing data to other frequencies and morphological conditions are critically sensitive to the irregularity parameters, it is important that measurements be continued which can be used to determine the characteristics of scintillation of thick layers.

6. DISCUSSION

The basic problem for any morphological study is to insure confidence in the reliability of comparisons made between observations taken under different conditions. Some procedure has to be established first to identify and isolate most of the variables and second to permit quantitative comparison.

The morphological features presented here have been popular for some time. Most have been advocated from observations made at a single station; many times without any qualifying remarks on the interdependence of the parameters. It can be agreed that an individual scintillation event is complex, the result of some direct, localized unknown cause, perhaps particle injection, and that the next scintillation event in the same locality can be entirely different with an entirely different cause, such as a gravity wave, or wedge refraction. Nonetheless, there are statistical features which can be used for quantitative discussion of the phenomenon.

Certainly any magnetic index, Kp included, is inadequate to completely describe the state of solar-terrestrial relationship, yet its use as a conditional variate as in Figure 4 demonstrates that the statistical latitude dependence is very sensitive to the prime causes which perturb the geomagnetic field.

If the behaviour of scintillation depth at fixed latitude for changes in local K index is examined then it is immediately seen that the behaviour varies with latitude. Overhead at 54°N geomagnetic the scintillation depth is insensitive to local K, that is during the years near sunspot minimum. Northward there is an increase, southward a decrease with increasing magnetic activity.

But that description is based on annual means. Is the daytime latitude function during quiet geomagnetic periods observed near the summer solstice different from a similar conditional function obtained for the winter solstice period? Moreover disturbed periods should be examined for the same reason.

Clearly there are a lot of questions.
Future work can easily be directed along three avenues.

First there is a tremendous amount of data, back to at least 1950 for radio stars and to perhaps 1957 or 1958 for satellites. If communication engineers are in need of predictions for scintillation during the next solar maximum, then its too late to take new data, the old will have to do. Some can be used almost directly, for instance Little's observations during solar maximum at college, Alaska (Figure 9). Some will require a little soul searching.

Second, cooperative efforts, such as the Joint Satellite Studies Group in Europe and the Americas can determine gross geographical features. Equally important they can show how the observations of a single station stray from the expected, for instance, how will observations of the same isolated irregularity from two widely separated stations really depend on the propagation angle to the magnetic field?

Third, could be a continued attack on the physical characteristics of the irregularities by single observatories.

The statistical evidence marshalled by Briggs (1964) over a solar cycle, and by other investigators, indicates a common cause for, if not an identity between, the irregularities which result in spread F and those which result in scintillation. With particular regard to high latitudes, studies made with Alouette I have shown frequent occurrence of spread F from at least 1000 km down to the height of the F region peak (Petrie 1964). Granted these, then thin screen analysis should not be used to derive secondary parameters, such as irregularity size or height, unless there is a direct indication that either a single irregularity is being viewed or that contributing irregularities in the layer are very large (L >> $(D\lambda)^{\frac{1}{2}}$). For the same reason it is inappropriate to extend the observations to different frequencies by theoretical approximation or to attempt to use empirical relationships, such as geometric dependence, in regions other than those from which they are derived.

When the irregularity distribution is effectively thin, then there are four major unknowns; height, size, elongation along field lines and electron density distribution. When the irregularity distribution is thick at least a fifth parameter is added. Since some satellites transmit several frequencies, such as 20, 40, 136 and 360 MHz on S-66, and since techniques are well established for measurement of both amplitude and phase simultaneously, the interference is clear.

Acknowledgements: We have not tried to give historical credit but rather have selected current examples which are related to the high latitude region and which seem to have some possibility for engineering application.

Certainly the groups at Manchester and Cambridge, England, established the existence and nature of most of the basic parameters and the basic theory during the early 1950's. It is always distressing to reread one of their reports and be stabbed by something lovingly rediscovered.

Special thanks are due Richard Hoffman for suggestions which

permitted much of the data to be untangled and to Edward Martin
who suggested we stop wasting time and put the data into a form an
engineer could use.

7. REFERENCES

1. J. Aarons, J. Mullen and Sunanda Basu, The statistics of sa-
    tellite scintillations at a subauroral latitude, J. Geophys.
    Res. 69, No. 9, 1785-1794, 1964.

2. J. Aarons, R.S. Allen and T.J. Elkins, Frequency dependence
    of radio star scintillation, J. Geophys. Res., 1967 (in
    press).

3. R.S. Allen, J.A. Klobuchar, W.C. Kidd and H.E. Whitney, Radio
    star scintillation during the July 20, 1963, eclipse, Can. J.
    Phys. 43, 1965, pp 1050-1058.

4. B.H. Briggs and I.A. Parkin, On the variation of radio star
    and satellite scintillation with zenith angle, J. Atmos. Terr.
    Phys. 25, 1963, pp 339-365.

5. B.H. Briggs, Observations of radio star scintillation and
    spread F echoes over a solar cycle, J. Atmos. Terr. Phys. 26,
    1964, pp 1-24.

6. C.G. Little and A. Maxwell, Fluctuations in the intensity of
    radio waves from galactic sources, Phil May 42, 1951, pp 267-
    278.

7. C.G. Little, G.C. Reid, E. Stiltner and R.P. Merritt, An ex-
    perimental investigation of the scintillation of radio stars
    observed at frequencies of 223 and 456 megacycles per second
    from a location close to the auroral zone, J. Geophys. Res.
    67, 1763-1784, 1962.

8. J.L. Jesperson and G. Kamas, Satellite scintillation observa-
    tions at Boulder, Colorado, J. Atmos. Terr. Phys. 26, 1964,
    pp 457-473.

9. Joint Satellite Studies Group, A synoptic study of scintilla-
    tions of ionospheric origin in satellite signals, Planetary
    Space Sci. 13, No. 1, 51-62, 1965.

10. T.A. Orhaug, Scintillation of discrete radio sources,
    Chalmers Tekniska Högskolas Handlingar Nr 229, 1965.

11. D.G. Singleton, Broadband radio star scintillations II inter-
     pretation, Radio Science, Vol. 68D, No. 10, October 1964,
     pp 1995-1108.

12. R.M. Turnbull and P.A. Forsyth, Satellite studies of isolated
     ionospheric irregularities, Can. J. Phys. 43, 1965, pp 800-
     817.

SUMMARY OF DISCUSSION ON HF PAPERS

The nomenclature of Oblique Ionospheric Radio Propagation, in particular that recommended at a meeting at Lindau, Germany, in 1963 was discussed. It was maintained that the nomenclature is excellent for radio propagation experts, but less suitable for the communicators. The communicator is interested in knowing whether he can or can not communicate on a given frequency but he is not interested in the mode of propagation. In Piggott's opinion the present nomenclature should be given years of trial. The terms must be able to be translated into several languages. Whenever the nomenclature is changed, the new terms should be completely different from the old ones so as to avoid confusion. Vincent was not satisfied with the CCIR definition of the LUF and suggested the communicators recommend another definition for this parameter.

A part of the discussion was concerned with the factors determining MUF and LUF of HF circuits. Generally the geometrical configuration as well as absorption for the one hop mode and higher order modes should be taken into account. It was maintained that the forth or even fifth hop mode of propagation might well determine the actual lowest operating frequency. On the other hand the predictor should not arbitrate for the communicator because the antenna radiation patterns may preclude the one and two hop modes.

It was stated that the system suggested in the paper by Petrie and Warren gives a good approximation to the overall MOF and LOF on the Winnipeg-Resolute Bay circuit. The results presented for this circuit indicate that the low angle ray LUFs can be calculated in terms of geometrical screening or cut off by the lower layer. In the further discussion it was pointed out by Hatton that practical communication also requires field strength predictions.

Several participants stressed the need for real time information of the variation of the MOFs about the monthly medians for appropriate design and operation of HF circuits. Considering the behaviour of transauroral oblique paths Petrie maintained, however, that even if large day to day and hour to hour variations of the MUFs are observed, the average values of the MUFs and LUFs can be predicted reliably. He held that long term predictions are required for optimum system design.

It was pointed out that minor blackouts result in a decrease in radio noise levels on HF systems along with signal strength decrease. This results in a constant S/N. A receiver with adequate gain characteristics can take advantage of the constant S/N and function quite well. For circuits with one terminal outside the polar regions the outside terminal cannot take advantage of the noise decrease and suffers in performance.

An aspect of importance for the system designer was brought up by Vincent. The communicator is required to send traffic with an acceptable error rate. This necessitates that fading rates and fading depths are taken into account in the design procedure.

Questions were raised on the meaning of "solid" communications as used by Wright. After some brief discussion it was agreed that, to the military, it may mean reliability approaching 99.99% in some cases and the ability to pass a single message at any time and with minimum delay in other cases. It was noted by Hatton that a really precise statement of requirements of the user is often lacking and that system costs can be tremendously affected by the last few per cent as reliability approaches (but never quite equals) 100 per cent.

With reference to the limitations on power and frequencies available for Air-Ground-Air communications Jull raised the question of possible use of relay ground stations. Wright replied to this that some use is actually made of existing HF ground stations when the base cannot be contacted. It appears that telephone companies have used unattended relays for some time. Experience of Stanford Research Institute with fairly complex, but largely automatic equipment such as oblique sounders, was frequently better during unattended operations. Piggott added that BBC also uses unattended relays to monitor a number of transmissions. Palmer, however, noted the problems of physical security in military unattended relay operations.

In a general discussion of the FOT prediction improvement portion of Petrie's paper Vincent raised the point that the internationally agreed definition of FOT (i.e. "Optimum Traffic Frequency", defined as the lower decile of the variation of the actual MUF about the monthly median MUF) is badly in need of revision since frequencies near the FOT are often not the optimum for traffic use. Piggott stated that, for short paths particularly, the best signal to noise ratios can frequently be obtained near the LUF where distant interferring signals have been highly absorbed. Probst added that operation well above the FOT, when these frequencies are actually supported, is desirable since it reduces the serious congestion at the lower HF frequencies. He made reference to figures 2, a to d, of Folkestad's paper as examples of cases where operation well above the predicted MUF should be encouraged. Prediction of MOF is often difficult, particularly when it is determined by high ray propagation on modes for which the low ray is earth masked or by composite modes, FEF,

EF or FE, none of which are predicted by most existing prediction techniques. Off-great-circle ground scatter modes can often be used above the on-path MUF but are seldom predicted. Vincent cited evidence from Thule to Palo Alto data that indicates high occurrence of FEF modes particularly during auroral disturbances. This path is ideally situated to provide the middle reflection from the top of the layer directly in the auroral zone, thereby avoiding penetration of the D-region and consequent absorption. He suggested that such ideal locations might be deliberately sought out for communication paths from polar to temperate latitudes.

With reference to a remark with regard to off-great-circle path ground side scattered modes, Jull held that such modes are an additional possible source of non-reciprocity due to the favouring of the ordinary wave in one direction and the extraordinary wave in the other as a consequence of the elliptically polarized character of a rhombic antenna's radiation in other than the prime direction. Vincent raised a point of disagreement on the performance of rhombics, but noted observed brief non-reciprocity in the observed MOF as the day-night line passes over the F layer reflection points of east-west paths. Jull replied that this too could be explained by slight off-path propagation but Vincent expressed continuing doubt.

Communication System Techniques

MODELING COMMUNICATION SYSTEMS

W.R. Vincent, R.F. Daly and B.M. Sifford

Stanford Research Institute Communications

Laboratory, California, USA.

Abstract: Modeling techniques have proved to be powerful tools
for investigating communication system performances and for comparing the behaviour of different systems. This report is devoted
to a description of results obtained in simulating modulation and
demodulation devices and the HF-path on an electronic computer.
The discussion has been restricted to digital communication.
Emphasis is placed on results and system implication rather than
on details of the techniques employed.

1. INTRODUCTION

Communications facilities and capacities are expanding
throughout the world at an amazing rate. Increased performance
levels are being demanded by the users, who are also quite ingeni-
ous in creating new uses for trunks. Satellites and wideband cables are being installed, and systems with ever greater capacities
are being proposed. Tropospheric scatter and microwave terminals
are being sold at a brisk rate. Also, computers and other digital
devices are being tied together in increasingly useful ways, and
there is an ever-increasing demand for telephone and other voice
service. While all this expansion is taking place, HF is still
holding its own and even seems to be expanding in most areas of
the world.

Practical growth restrictions such as crowding and the limited spectrum began to affect HF use years ago. These same restrictions are now being felt in the higher VHF, UHF, and microwave
frequencies. The crowded spectrum and limited spectrum availability led to a search for more efficient modulation and transmis-

321

sion processes. A number of new techniques have been demonstrated
and used recently. Not only have new modulation and demodulation
processes been developed, but pressures have been high to increase
transmission speeds. From the original use of 60-wpm teletype,
there was a short-lived increase to 75 wpm, and now 100 wpm is
used almost universally.

These developments have not always produced the anticipated
advances in performance. For example, early attempts to use phase
modulation on polar region HF trunks resulted in unexpectedly high
error rates. Many modulation-demodulation techniques have appeared on the marked, designed to solve the HF high-bit-rate problem. A fair amount of engineering by the seat-of-the-pants approach has gone into the creation of a number of the devices.
This has been coupled with a decreasing interest in HF expenditu-
res and a serious lack of transmission-path parameters.

Several years ago, a serious study was made on our ability to
simulate modulation and demodulation devices and to simulate the
HF path on a digital computer. This paper discusses progress in
this matter, summarizes the techniques used, and gives examples
of the results obtained. Emphasis has been placed on system results and systems implications rather than on details of the modeling techniques. To limit the scope of the problem, only digital communication systems are discussed, and noise and other modulation inputs are not considered.

2. HF PROPAGATION PATH PARAMETERS

After initial studies, the important parameters of the propagation path that limit the performance of communication systems
were defined. They are signal strength, noise or interference,
time dispersion (multipath), and frequency dispersion. Signal and
noise or interference levels have been investigated for the past
two decades. Although the literature contained considerable information on path loss and noise, it lacked adequate data on HF
path time and frequency-dispersion statistics. Therefore, instrumentation was developed and placed in the field to measure all
parameters simultaneously on selected paths. Special emphasis was
placed on time dispersion and frequency dispersion, since these
were the least known parameters. Polar paths were included, as
shown in Figure 1.

The concept of a channel-scattering function as shown in
Figure 2 was developed. This concept relates amplitude, time dispersion, and frequency dispersion, and gives some insight into the
perturbation on an HF signal by the propagation path. For convenient display of the dispersion of signals in the frequency domain,
techniques were devised to measure and plot the power spectrum

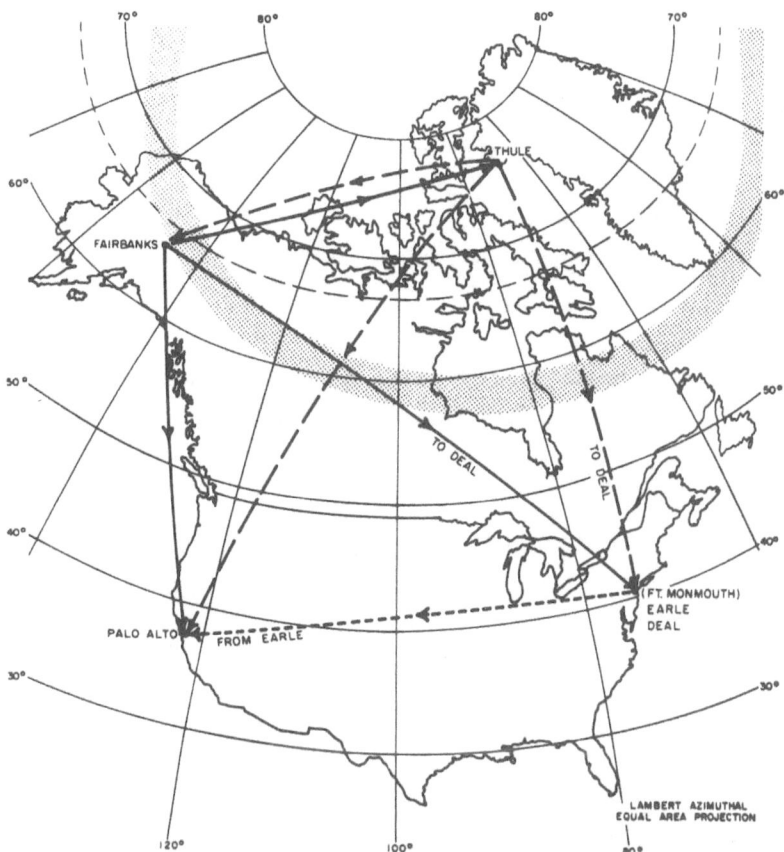

Figure 1. Locations of sounder stations and great-circle paths
           in the simulation experiment.

rapidly. Frequency dispersion is then a measure of the width of
the power spectrum. We have chosen the 2σ width as the most convenient measure to compare power spectra and as a measure suitable
for later modeling chores. Examples of power spectra are given in
Figures 3 through 5.

   Time-delay profiles have also been measured. Examples are
shown in Figure 6. The measure of time dispersion used is the 2σ
value because it is convenient in later modeling tasks.

   A sizeable library of signal-to-noise-ratio (S/N), time-dispersion, and frequency-dispersion measurements has been accumula-
lated on a variety of paths. This library permits placing practical bounds on these parameters when analyzing the performance of

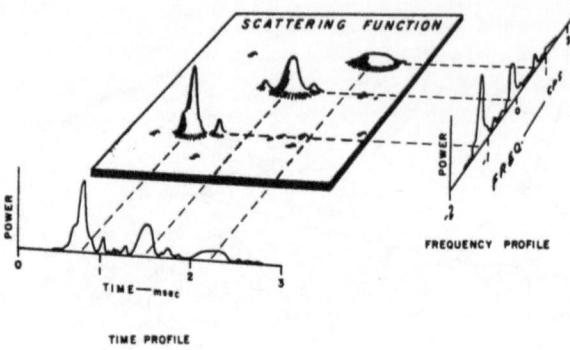

Figure 2. The time-frequency plane of the channel scattering
function.

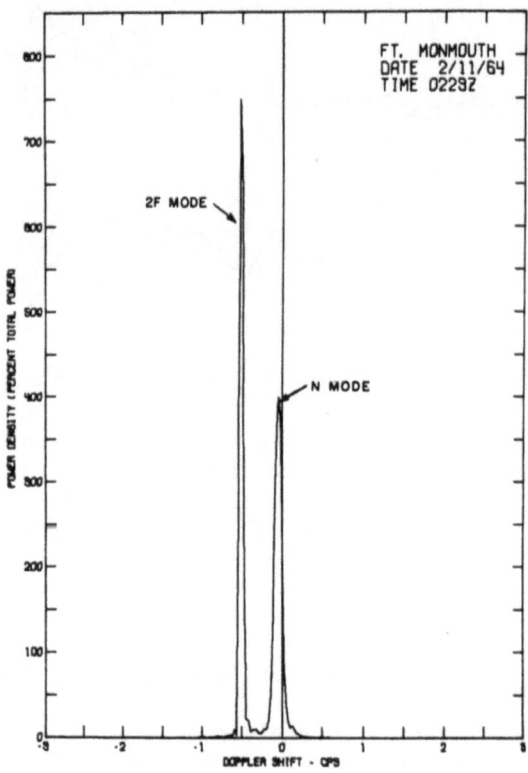

Figure 3. Power spectrum.

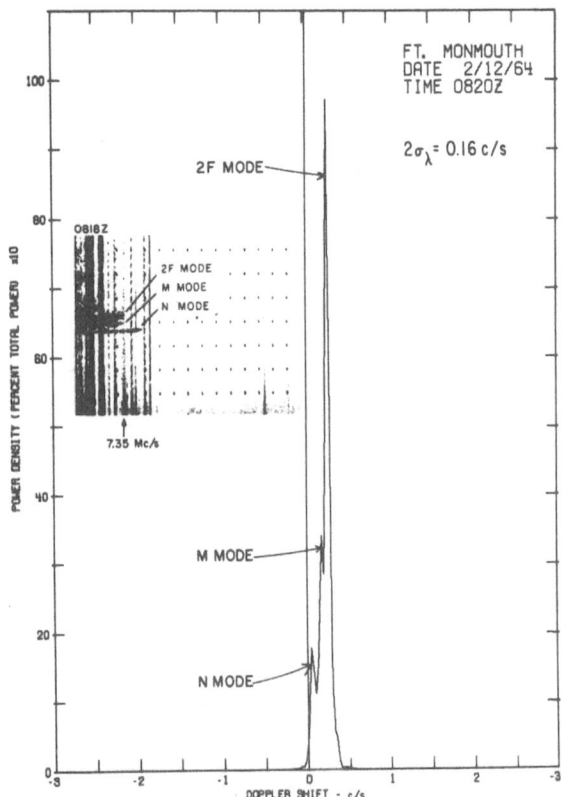

Figure 4. Three modes - power spectrum and ionogram.

communications systems subjected to these perturbations.

    Using the physically measureable concepts of time dispersion,
frequency dispersion, and S/N, a scattering model of the HF propagation channel was developed. The channel model has been described in previous Stanford Research Institute reports. This model assumes that the randomly time-varying linear channel is composed of a discrete number of independent propagation paths. Each
path is described by a time delay, a Doppler shift, and a random
complex gain. The model can be considered as an idealized version
of the HF channel when the time-delay spread and Doppler spread of
the individual paths are small compared to the time-delay differences and Doppler-shift differences between the paths. Alternatively, as the number of paths becomes large, this channel approaches one with a continuum of time delays and Doppler shifts.
The channel model accounts for the important signal-distorting

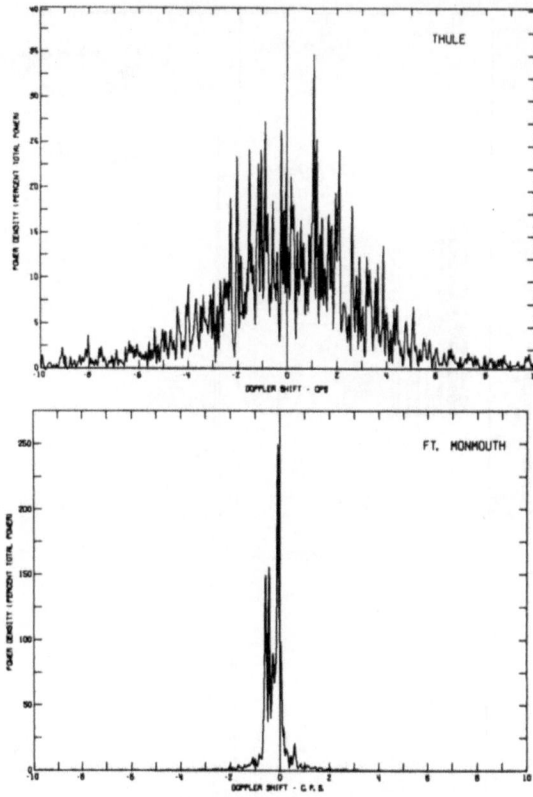

Figure 5. Frequency profiles of a transauroral propagation chan-
           nel and a mid-latitude propagation channel.

effects caused by time- and frequency-selective fading, such as
intersymbol interference and fast fading of the signal amplitude
within a signaling-element duration.

3. SYSTEM MODELING

3.1 Review of Error-Rate Models

Computational models of systems using error rate as an indi-
cator of performance must be based on statistical principles. As
such, they do not attempt to predict the exact number of errors
that will occur but only to answer the question: What is the
probability that a receiver will decide that State A was transmitted during a particular signaling interval when State B was actually transmitted? To answer such a question, probabilistic concepts must be introduced and with them certain random variables

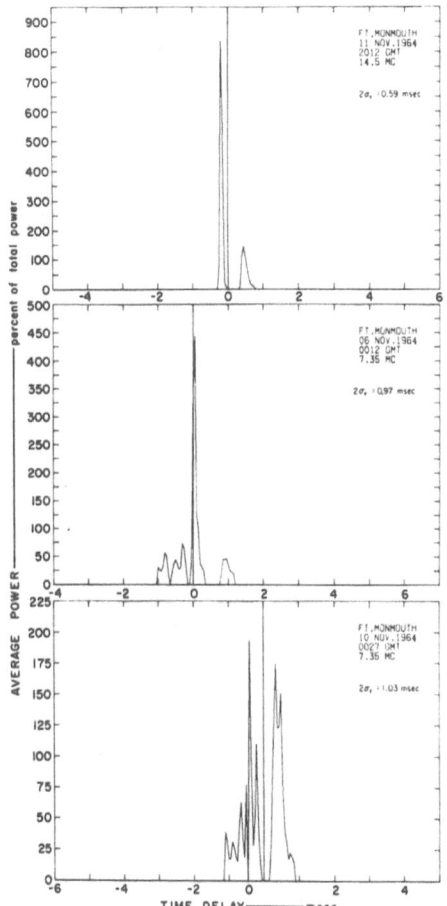

Figure 6. Multipath time-delay profiles.

that can be described only in a statistical sense.

The first attempts to derive an error-rate model for communications channels were based on the following assumptions:

(1) An ideal matched-filter receiver

(2) All signals subject only to fixed attenuation and a fixed time delay in their passage through the propagation medium

(3) Errors in reception caused solely by narrow-band Gaussian noise added to the received signal.

The ideal matched-filter receiver has knowledge of all the
possible signal waveforms that can be transmitted by the transmitter for each time interval. Coherent matched-filter detection requires that the receiver have exact knowledge of all possible
phase states of the transmitted waveforms when they arrive at the
receiver. Incoherent matched-filter detection does not utilize a
priori phase information of the received signal. The assumption
of an ideal matched-filter receiver implies exact time synchroni-
zation with the transmitted data stream and ideal integration and
sampling of the received waveform. This is followed by complete
energy dumping or quenching of all integrating filters.

The next evolutionary stage in the development of error-rate
models was to allow the received signal envelope to vary slowly
with respect to a signaling element (i.e., to fade) and to assume
a distribution of this variation in calculating the effect on the
probability of error. It was also assumed in these models that
fading over the transmission bandwidth of the signal was correlated. Such a channel has been termed flat-flat by Bello and Nelin,
because the time autocorrelation function of the channel is constant over a time difference on the order of a signaling-element
time interval, and the channel frequency-correlation function is
constant over a frequency difference comparable to the bandwidth
of the transmitted signal. The frequency correlation is directly
related to the time-delayed "multipath" propagation modes which
are observed at HF and can be predicted by propagation theory.
The time autocorrelation is directly related to the difference in
Doppler shift associated with these modes. Diversity improvements
were modeled by assuming the availability at the receiver of independent fading signals and specifying various combining rules for
the receiver. The flat-flat models show a decreasing error rate
with increasing S/N.

More recent channel models include the effects of Doppler and
delay spreading of the transmitted signals. The use of these models, however, requires additional statistical parameters to describe the spreading of the signal by the channel in time and frequency. These models show asymptotically constant error rates
with increasing S/N. The asymptotic values depend critically on
interaction of certain system parameters and on the statistical
channel spread measures.

At HF, the simpler flat-flat models have generally been used
for present binary FSK systems. These are adequate, however, only
when gross estimates of performance at moderate error rates are
needed. For normal 100-wpm teletype channels predominately handling messages with English-language redundancy, a binary error
rate less than 10^{-3} has usually been considered acceptable. With
the present increase in less redundant data traffic and the corresponding increase in reliability requirements, the simpler models

are becoming inadequate. As newer modulation systems and higher data-transmission rates are introduced, time and frequency spreading of the channel, even at HF, cannot be ignored in a performance model. But estimating or predicting these channel scattering functions from geophysical parameters presents a problem. Generally the present programs can predict individual signal strengths for each propagating mode, together with propagation-time differences, but they cannot adequately predict Doppler effects.

3.2 The FSK System Model

For the purpose of this paper, FSK modulation and demodulation can serve as an adequate example of the system model. For FSK the system model assumes a single-channel matched-filter receiver with filters matched to ideal FSK transmissions. A quadratic detector obtains the squared envelope of the output of each matched filter; the receiver then decides which signal was transmitted, on the basis of the larger squared envelope.

With the exception of a linear variation of frequency in the transition intervals, the FSK transmission is assumed to be ideal. If f_m is the mark frequency and f_s is the space frequency, then Figure 7 displays the assumed linear transition when the sequence SPACE-MARK-SPACE is transmitted. In this figure, W is the signaling element duration, and 2L is the transition time. Observe that in the transition interval the transmission is a chirp signal that spans all frequencies between the two signaling frequencies.

As the transition time, 2L, is increased, the performance of the FSK system, operating through a time-varying and frequency-selective radio channel, deteriorates. For typical ionospheric time and frequency dispersion conditions and small transition times, the deterioration appears as an increase in the binary error probability for high S/N. The randomly selective radio chan-

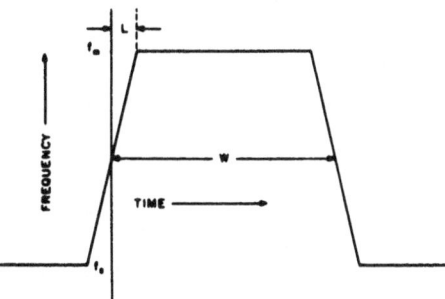

Figure 7. Linear frequency variation at transition intervals.

nel transforms the chirp transmissions at the transition time to
noise-like signals which impair the detectability of the FSK
transmissions.

3.3 FSK Model Results

Using the channel scattering model concept and the FSK system
model, the performance of the test system can be explored on a digital computer for a variety of propagation conditions. The
available catalog of S/N, frequency dispersion, and time dispersion can be used to make the modeling conditions realistic.

The output of the program is a computer-plotted curve of binary error probability versus S/N for the desired channel scattering function. Each plot displays two curves, one for no diversity
(labeled FSK1) and one for independent dual diversity (labeled
FSK2). Each plot displays a channel-scattering-function diagram
in the lower left-hand corner. In this diagram each path is represented by an X with size proportional to the strength of the
path and with position determined by the time delay and Doppler
shift of the path. Time delays relative to the average time delay
of the channel are read along the logarithmic horizontal scale and
are normalized to the signaling-element duration (T = the time delay of the path divided by the signaling-element duration).
Doppler shifts relative to the average Doppler shift of the channel are read along the logarithmic vertical scale and are also
normalized to the signaling-element duration (D = Doppler shift of
the path multiplied by the signaling-element duration). For example, Figure 2(a) displays the error-rate curves for an FSK system operating through a propagation medium composed of two equal-
strength paths spaced at a time-delay difference equal to 0.2 of
the signaling-element duration and with no Doppler spreading in
frequency.

The flexibility of digital computers permits the exploration
of many different conditions. Figure 8 is an example of FSK system performance being explored as time dispersion is increased.
For this example, the S/N values are very high, and the frequency
dispersion is zero. Additional cases are explored in the next
section.

3.4 Validation of FSK Results

It is frequently desirable to check the validity of the computed results from complex models. The FSK model has been checked
in a series of full-scale field validation tests. The purpose of
the tests was to make a direct comparison between the measured

Validation of the theoretical error-rate analysis required simultaneous measurements of the channel scattering function, S/N, and binary error probability. The scope of the experiment (the available time and rate of effort) made it necessary to limit the measurement of the channel-scattering function to the time-delay and Doppler-shift profiles. It was also necessary to concentrate the major portion of the effort on obtaining data for the no-diversity case, since the dual-diversity case would add another dimension (spacial correlation) to the classification of data. The amount of data required to obtain statistically reliable results for dual-diversity operation at all possible states of S/N, time delay, Doppler shift, and spacial correlation far exceeded our single-channel capability for collecting data, and the available rate of effort for data processing. However, a limited amount of data was obtained for dual-diversity operation.

Although time-delay and Doppler-shift profiles were measured on the channel throughout the experiment, it was not feasible or necessary to use all of the information that these functions contain for correlating the observed error rate with the state of the dispersive channel. Various computer programs were run, investigating the sensitivity of the error rate displayed by the theoretical model to the shape of the scattering function. It was found that if the scattering function is approximately symmetrical and of moderate width in the time-delay and Doppler-shift directions, then the error rate is strongly dependent on the time-delay and Doppler-shift second moments and relatively insensitive to the scattering-function shape. Hence, symmetrical scattering functions with moderate time-delay and Doppler spreads can be approximated with two equal-strength paths spaced in time delay at the time-delay spread of the scattering function, and spaced in Doppler at the Doppler spread of the scattering function. This approximation introduces a major simplification in the analysis of the data; when valid, it allows one to use only the second moments associated with the time-delay and Doppler profiles.

An example of the comparison of measured data with model results is shown in Figure 9. FSK1 is the no-diversity case and FSK2 the dual-diversity case. Measured error data are available only for the FSK1 case. All measured error-rate data, without regard to time-delay or Doppler spread, are plotted as a function of S/N in Figure 9. The theoretical error-rate curve of Figure 9 was computed for two equal-strength paths with a time-delay spread of 1 msec and a Doppler spread of 1 cps, which are typical of the spreads observed on this channel. The relatively smooth variation of measured error rate with S/N is indicative of the degree of statistical reliability achieved by the experiment. Figure 9 illustrates the important and interesting "bottoming-out" characteristic generally displayed by the measured error-rate data: the

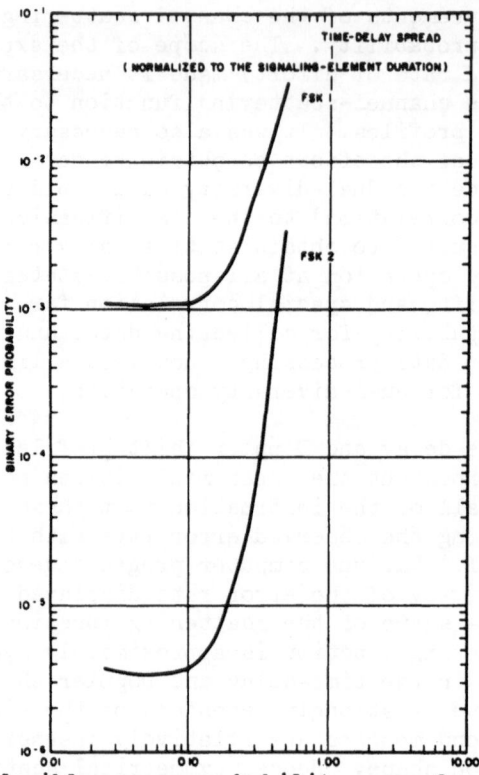

Figure 8. Irreducible error probability as a function of time-
delay spread, $2\sigma_\lambda = 0$, $2L = 0.02$.

error rates obtained from an FSK communication system operating
over a selected path and the error probabilities predicted for
that path. Measurements of the channel scattering functions were
made simultaneously with the error rate measurements and used as
inputs in the model computation.

The measurements were performed on a single binary data channel using two-stage modulation. The input binary data frequency-
modulated a sub-carrier. The resulting FSK tones single-sideband-
modulated the HF carrier, with a substantial portion of the carrier being retained. The data rate was 100 baud, which is slightly higher than that of a single 100-wpm teletype channel. The
operating frequencies were 14.360 and 7.366 Mc/s used for approximately the same amount of time. These frequencies were chosen to
maximize the amount of data collected in the scheduled experiment
time and to obtain data under a variety of propagation conditions.
The transmitter was located at Fort Monmouth, New Jersey; the receiver site was near Palo Alto, California. The great-circle distance between the sites is 4100 km.

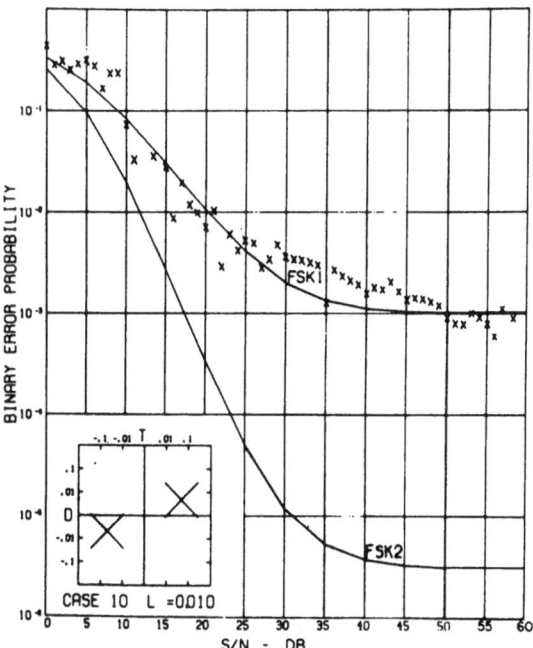

Figure 9. Data Class A - all data.

measured error rate ceases to decrease with increasing S/N and essentially maintains a constant irreducible value at high S/Ns, as predicted by the theoretical model.

The close agreement between the measured error rate of Data Class A and the theoretical error rate indicates that the theoretical model yields a good approximation to the average performance of the experimental system over the range of time-delay and Doppler spreads observed on the channel. It is to be stressed that Figure 9 is concerned with comparing the measured performance of the experimental system, averaged over all dispersive states of the channel, with a theoretical curve based on the average state of the channel. Close agreement between measured and theoretical results in this case does not necessarily imply that close agreement will be obtained when measured and theoretical error rates are compared for a particular dispersive state of the channel. Although a Doppler spread of 1 cps was the maximum observed, the theoretical error-rate curves of Figure 9 are essentially determined by the time-delay spread and transition time and would not display a discernible difference when computed for a zero Doppler spread.

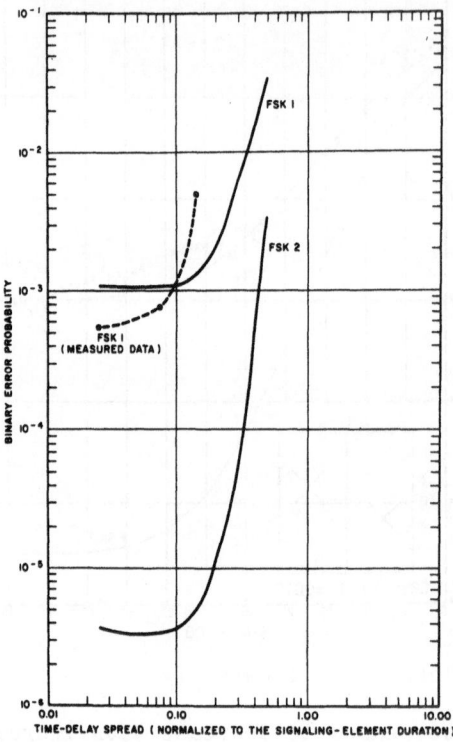

Figure 10.   Measured irreducible error probability as a function
of time-delay spread.

The measured error data were broken down into categories representing values for various time- and frequency-dispersion cases. Using only the very high S/N data the sensitivity of the
channel to time dispersion and to frequency dispersion has been
examined. Figure 10 shows the measured and calculated curves of
error probability vs. time dispersion. Figure 11 shows the measured and calculated curves of error probability vs. frequency
dispersion. Both curves clearly illustrate the irreducible-error-
rate situation and give values of dispersion beyond which system
degradation is serious.

4.   APPLICATION TO PHASE-MODULATION SYSTEMS

A number of phase-modulation techniques have been proposed to
improve HF communication systems performance. It is common practice to construct development models of each system and field-test
these models along with others in comparative tests. Modeling is

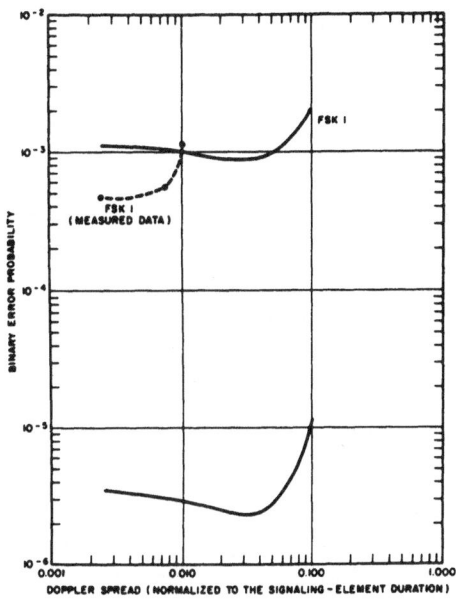

Figure 11.   Measured irreducible error probability as a function
             of Doppler spread.

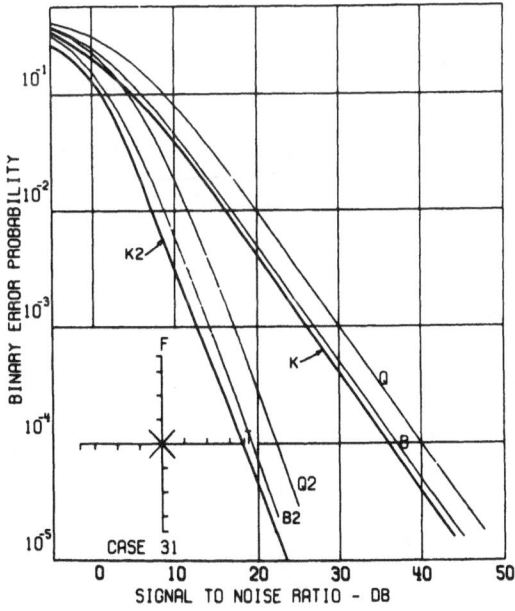

Figure 12.   Curves showing probability of error as a function of
             signal-to-noise ratio.

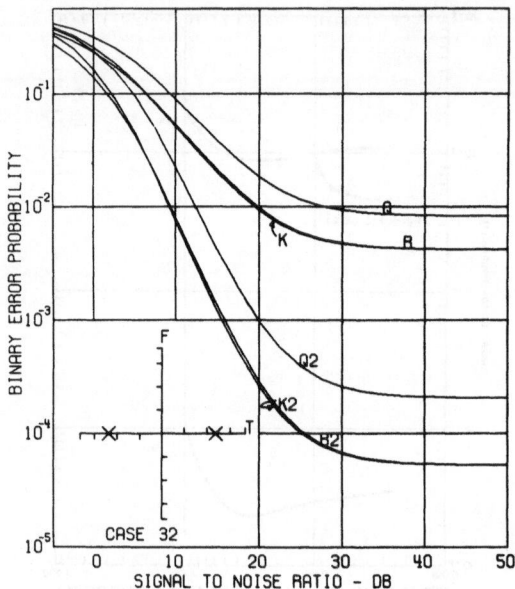

Figure 13. Curves showing probability of error as a function of signal-to-noise ratio.

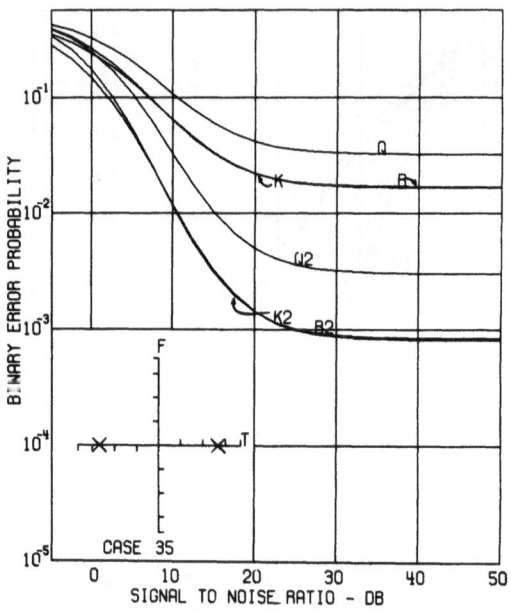

Figure 14. Curves showing probability of error as a function of signal-to-noise ratio.

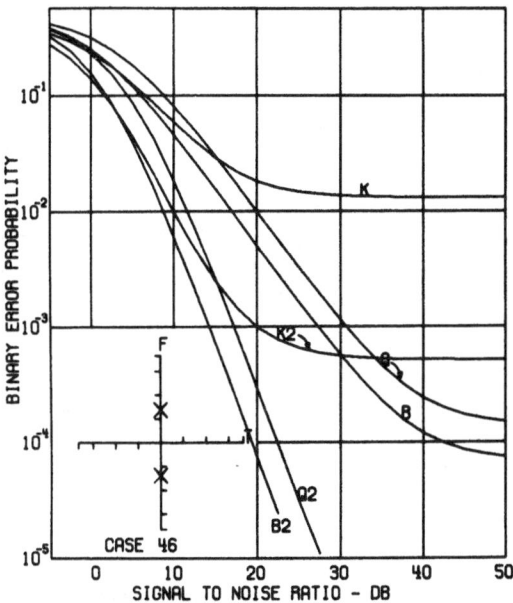

Figure 15.   Curves showing probability of error as a function of
             signal-to-noise ratio.

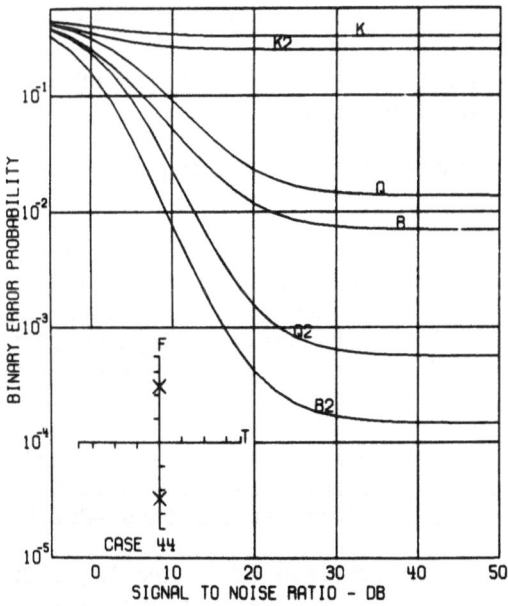

Figure 16.   Curves showing probability of error as a function of
             signal-to-noise ratio.

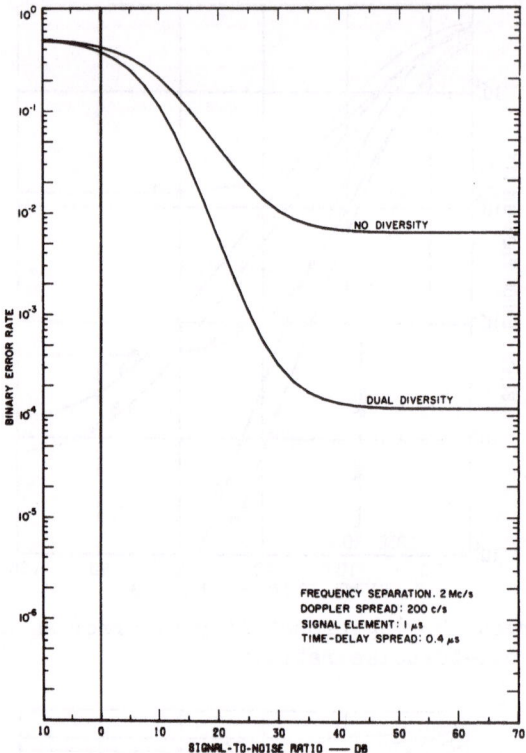

Figure 17.   TDM-FM error rate as a function of SNR with and
             without diversity.

an attractive alternate to field testing when investigating systems performance.

Using the scattering-model concepts developed under the FSK
study, selected basic phase-modulation techniques have been modeled and their performance determined for a variety of propagation conditions. Results are shown in Figures 12 through 17. The
following list relates terminology of the figures to the type of
system studied:

        K--BAPSK (Binary-Adaptive-Phase-Shift Keying)
        B--BDPSK (Binary-Differential-Phase-Shift Keying)
        Q--QDPSK (Quaternary-Differential-Phase-Shift Keying)

For the non-dispersive case of Figure 12, differences between
systems are minimal. The non-diversity and diversity cases are
shown with the same notation previously employed in the FSK results. Figures 13 through 15 show error rate as time dispersion

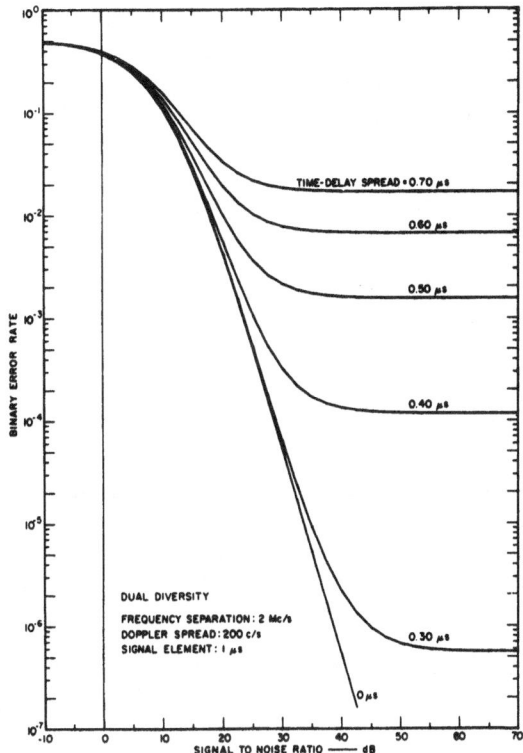

Figure 18.   TDM-FM error rate as a function of SNR with dual
             diversity.

is increased. Figures 16 and 17 illustrate the sensitivity of the
systems to frequency dispersion. It is quite clear that the particular BAPSK system (System K) should not be chosen for polar
regions where frequent high levels of frequency dispersion occur.

    Care must be taken in the generalization of results shown in
this paper. They are for particular basic systems with specific
characteristics. The BAPSK results cannot be applied to all possible BAPSK configurations, and each separate system must be modeled with its own specific characteristics.

        5. APPLICATION TO TROPOSCATTER SYSTEMS

    Modeling techniques have been extended into the analysis of
tropospheric scatter systems. The tropospheric-scatter channel is
modeled as a zero-mean, time-varying, Gaussian random field which
includes both additive noise distortions and random time and fre-

quency dispersive effects. A simple time division FM (TDM-FM) system has been examined where the system model considered no diversity and prediction combining dual diversity.

For the case where the f_m and f_s separation is 2 Mc/s, the signal element 1 μsec in duration, the Doppler dispersion 200 c/s, and time dispersion 0.4 μsec, the error-rate vs. signal-to-noise is shown in Figure 17. For this case the frequency dispersion is not a significant value affecting results appreciably. Note the bottoming effect of the error-rate curves at high signal levels.

The sensitivity of the TDM-FM system to time dispersion has been investigated and results are shown in Figure 18. Very small values of time dispersion affect error-rate values significantly.

6. CONCLUSIONS

The work described in this paper was accomplished over a period of several years and the techniques have been widely used in the analysis of communication systems. A number of conclusions are apparent from the work and are summarized in the following items:

(1) Time- and frequency-dispersive effects must be considered in analyzing the error-rate performance of radio communication systems.

(2) Time- and frequency-dispersive effects result in an irreducible error rate at high signal-to-noise levels, thus limiting maximum system performance.

(3) Modeling has been shown to be an effective tool to explore communication-system performance under a variety of propagation-path conditions.

(4) Modeling has been shown to be an effective tool in comparing systems and in exploring the result of changes in the design of a given system.

(5) Realistic inputs to the models are a key factor in their success. For example, realistic time- and frequency-dispersion values are essential to an understanding of a given communication system.

ADAPTIVE RECEIVERS FOR MULTIPATH CHANNELS

D. C. Coll

Defence Research Telecommunications Establishment

Ottawa, Canada

Abstract: This paper reports the development of a system for reception of time-dispersive PAM signals. Properties of an optimum
receiver for stationary multi-path channels are discussed and the
practical implementation of this receiver is described. Techniques for adapting the class of systems concerned to changing
channel characteristics are considered.

1. INTRODUCTION

The classical model for theoretical communication system studies contains a memoryless channel in which the transmitted signal
is linearly combined with gaussian noise. The signal itself is
not distorted during transmission through the channel, and the received signal consists of the transmitted signal corrupted by additive noise. In the classical model the channel is memoryless
and stationary; and this, of course, excludes the possibility of
intersymbol interference being considered as a factor in the design of communication systems.

On the other hand, practical channels are neither memoryless
nor stationary, and ionospheric radio channels are prime examples
of time dispersive, time-varying channels. Time-dispersion, or
multipath, causes intersymbol interference that severely limits
the rate at which information can be transmitted and received in
systems that are based on the classical channel model.

The effect of the intersymbol interference on digital data
transmission is illustrated in Figure 1. Here, the probability of

341

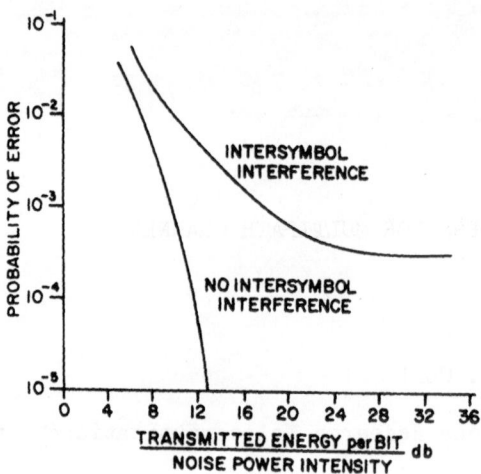

Figure 1. Effect of intersymbol interference on probability of
           error.

error decreases monotonically with increasing transmitted energy
if there is not intersymbol interference. However, when intersymbol interference is present, the probability of error cannot
be reduced beyond a certain level. This level is determined by
the perturbations of the signal pulse caused by overlaps from
pulses transmitted at other times. Because these pulses are modulated, the overlap tends to be noiselike and to behave in the
same manner as additive noise.

    Another effect of intersymbol interference is the reduction
of the information transmission rate. One of the chief advantages
of a system that is designed to operate in the presence of intersymbol interference is the increase in rate that it allows. It is
worth noting that linear time-dispersion does not impose any limitation on transmission rate in the absence of noise. It is noise
that limits the corrections that can be made to overcome time-dispersion.

    This paper traces the development of a class of adaptive re-
ceivers for quasi-static multipath channels. Several stages in
the development are described, viz.,

(1) the communication-theoretical optimum receiver for stationary
     multipath channels,

(2) the practical implementation of this receiver,

(3) a technique for adaptively controlling the response of the
     practical receiver,

(4) a general procedure for the automatic adjustment of this
     class of receivers, and

(5) a discussion of different adaptive procedures that span the
     possibilities from (3) to (4), above.

2. COMMUNICATION THEORY APPROACH TO STATIONARY MULTIPATH

     The techniques of statistical communication theory have been
applied to the problem of transmitting pulse-amplitude modulation
(PAM) signals through stationary, linear, time-dispersive channels
in the presence of additive noise, (1, 2, 3). When the additive
noise and the intersymbol interference are considered to be independent and are jointly minimized with respect to the signal, the
form of the optimum linear receiver is as shown in Figure 2, (1).
This receiver, which maximizes the output signal-to-noise ratio,
consists of a filter matched to the received pulse shape, followed
by a sampled data filter or compensator. In general, the receiver
is unrealizable.

CONFIGURATION OF THE OPTIMUM FILTER

Figure 2. Configuration of the optimum filter.

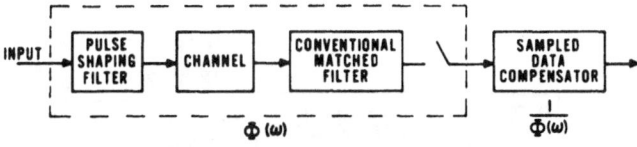

LOW LEVEL BACKGROUND NOISE SITUATION

Figure 3. Low level background noise situation.

The presence of the sampled data filter arises from the fact that the data are transmitted at regular intervals and that decisions at the receiver are made accordingly. The signal-to-noise ratio need be maximized only at the times when decisions must be made, and the fact emerges that it is sufficient to work with samples of the matched filter output.

When the noise is white, the optimum receiver has the transfer function

$$H(\omega) = \frac{S(-\omega)}{N_O + \Phi^*(\omega)}$$

where

$S(\omega)$ is the Fourier transform of the received signal,
N_O is the noise power spectral density, and
$\Phi^*(\omega)$ is the Fourier transform of the sampled matched filter output.

The receiver has the property, common to all Wiener filters, of attenuating those frequencies at which the noise or interference dominate the signal, and passing the others. When the noise is much larger than the interference, the resulting receiver reduces to the familiar matched filter. When the interference dominates the noise, the receiver reduces to that shown in Figure 3, in which the transfer function of the compensator is the inverse of that of the entire proceeding system. The total system is then an all-pass system.

3. THE PRACTICAL IMPLEMENTATION OF THE OPTIMUM RECEIVER

The ideal receiver is unrealizable, not only because of the fact that the impulse response of the matched filter is the time-inverse of the received signal (which may be overcome by the introduction of sufficient delay), but also because the sampled-data-system must process all received signals from the infinitive

EXPERIMENTAL RECEIVER

Figure 4. Experimental receiver.

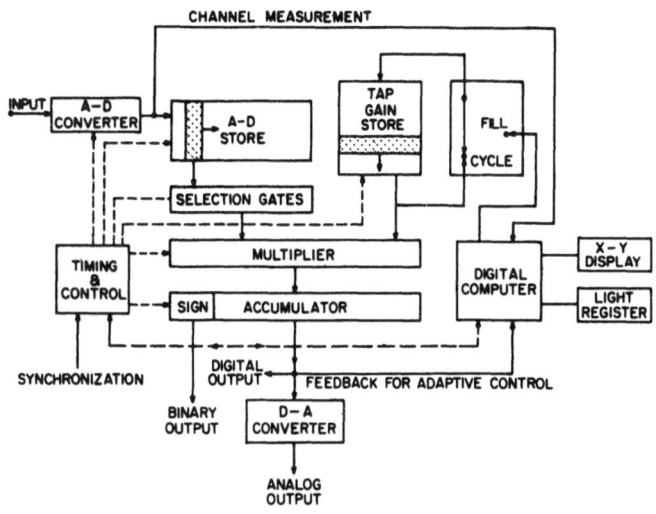

Figure 5. Sampled data compensator.

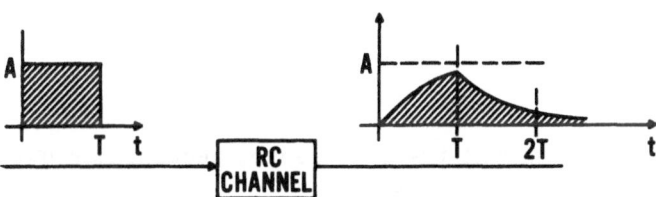

Figure 6. Rectangular pulse - single pole channel.

future. However, the ideal receiver provides a guide to a practical realization which has the form shown in Figure 4, (4, 5).

The matched filter is synthesized by straightforward techniques, using a tapped-delay-line filter. A step-function approximation is made to a reasonable portion of the desired impulse
response. The taps on the delay line are separated by intervals
that are small enough to provide a good approximation and the tap
gains chosen to minimize the distortion.

The sampled data compensator is also synthesized as a digital
version of a tapped delay line, with the distinction that the taps
are separated by the interpulse period. The output at any sample
time is the weighted sum of a set of samples of the matched filter

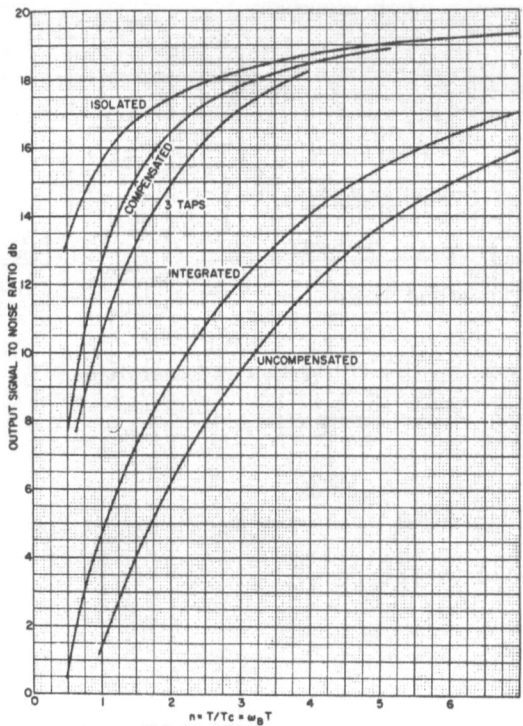

Figure 7. Output signal-to-noise radio, rectangular pulse,
           single pole channel.

output. The weights are the tap gains, which are calculated to
maximize the output signal-to-noise ratio. A block diagram of the
compensator is shown in Figure 5. Digitized samples of the matched filter output are stored in shift-registers and convolved
with the compensator impulse response, which is represented by a
set of binary numbers stored in the tap-gain store.

     To operate this receiver, the shape of the received pulses
and the additive noise level must be known. The matched filter
must be set up, and the optimum taps computed and set. The tap
values depend on the received pulse shape, the noise level, and
the transmission rate; and must be recomputed if any of these
parameters change. Likewise, the matched filter must be reset if
the received pulse shape changes.

     The performance of the practical system may be illustrated by
an example. The effect of a single pole (RC) channel on a rectangular pulse is shown in Figure 6. Calculated output signal-to-

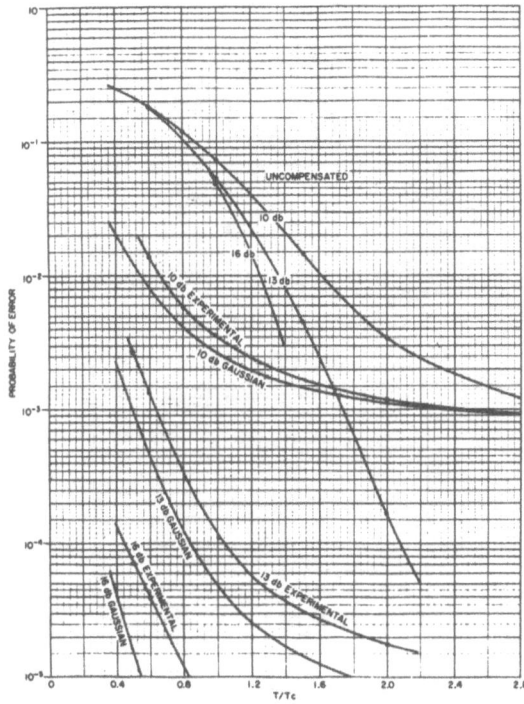

Figure 8. Probability of error vs pulse separation for exponen-
           tial pulses - experimental.

noise ratios for this situation are shown in Figure 7 as a function of the pulse length relative to the channel time constant,
for a number of different receivers. The 'Isolated' curve shows
the signal-to-noise ratio at the output of a filter matched to a
single, isolated pulse when one is received. The 'Uncompensated'
curve refers to the output of the same filter when a train of pulses is received. The 'Compensated' curve shows the output of the
ideal receiver; '3 taps' the output of a 3-tap practical receiver;
and 'Integrated' the output of a finite-time-integrator. The advantages of a compensated system are readily apparent.

    In an experimental evaluation of the practical receiver, ex-
ponential pulses were received with different additive noise levels. The probabilities of error at the matched filter and compensator outputs are compared in Figure 8 as a function of the
pulse separation. The 'gaussian' curves would obtain if the ideal
receiver were used, and if the noise in the output were gaussianly
distributed additive noise.

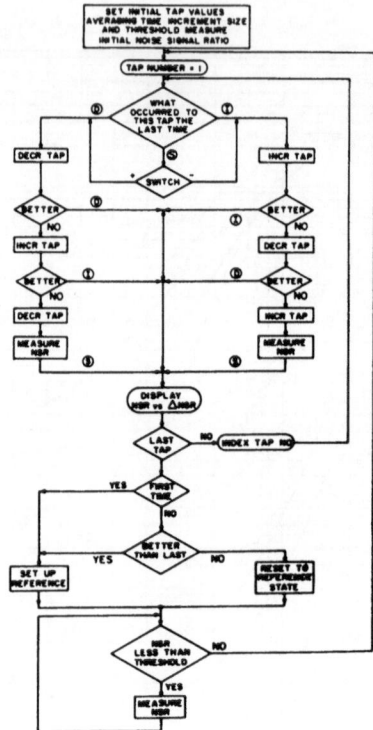

Figure 9. Flow chart of the simple search procedure.

4. ADAPTIVE CONTROL OF THE PRACTICAL RECEIVER

As mentioned above, the matched filter and optimum taps must be reset each time the channel, noise level, or transmission rate changes. On the assumption that the ultimate performance of a system with the configuration shown in Figure 4 is relatively insensitive to errors in the matched filter (at least for the purposes of the following discussion), it may be advantageous, i.e., faster, to consider a systematic procedure whereby the compensator tap values may be altered continuously to maximize the signal-to-noise ration (or mean square error) at the output. The matched filter may be adjusted by a similar technique since it is a tapped-delay-line as well, or it may be set periodically using test pulses.

The signal-to-noise ratio, or mean-square-error, may be considered as a surface in a multidimensional space whose axes are the compensator tap values. A simple iterative procedure for searching this space is outlined in Figure 9. Since the tap values are represented by finite accuracy numbers, it is natural to

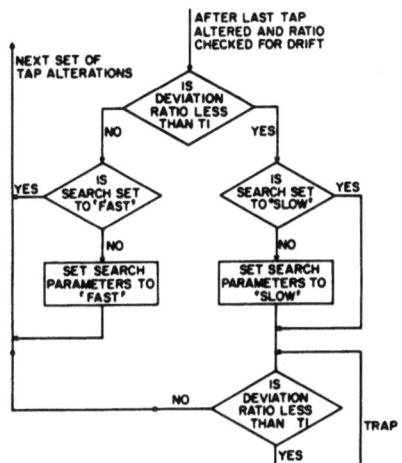

Figure 10.   Modification of iterative search procedure to include parameter changes.

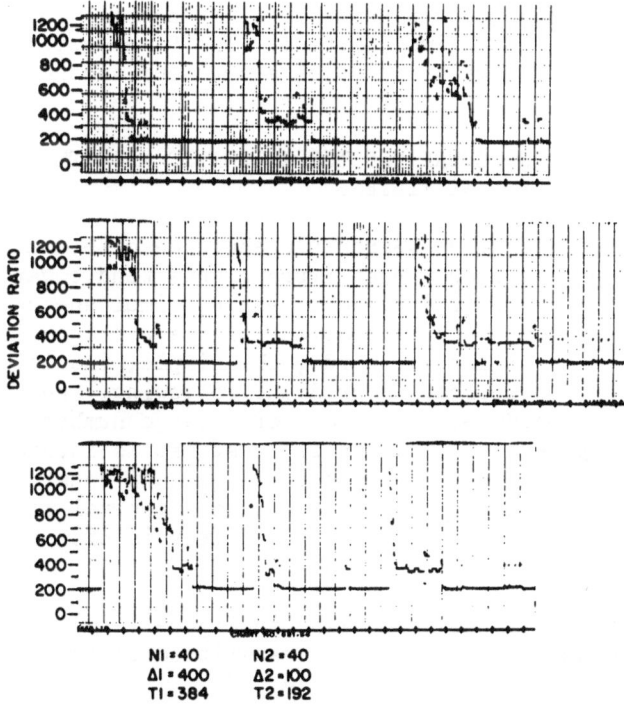

NI = 40   N2 = 40
ΔI = 400   Δ2 = 100
TI = 384   T2 = 192

Figure 11.   Examples of iterative search.

consider altering them by a simple fixed increment or decrement.
Depending on whether an increment or a decrement to a particular
tap improved the performance during the last iteration that tap is
incremented or decremented. If an improvement in performance en-
sues, the tap is left; if not, the alternate change is made. If
the performance improves, that change is retained; if not, the tap
is reset to its original value. Each tap is altered in turn, and
the process continues in this manner until the performance is satisfactory, at which time the search is suspended until conditions
deteriorate.

The basic search may be modified, as shown in Figure 10, to
allow the magnitude of the tap alterations and the observation
times to be changed during the search. This allows large alterations to be used with short observation times during the beginning
of the search, and small alterations to be used with long observation times during the end of the search.

The performance of this simple iterative procedure is demonstrated by the records shown in Figure 11. The deviation ratio,
which is related to the noise-to-signal ratio, is shown as a function of time for nine separate searches, each of which starts from
'scratch' with only one non-zero tap. In the experiment, a five-
tap compensator operated on noise-free exponential pulses separated by one-half of a time constant. The time scales are marked
in seconds; the average search took about four seconds.

5. A GENERAL PROCEDURE FOR THE AUTOMATIC ADJUSTMENT OF
TAPPED-DELAY-LINE FILTERS

The simple iterative search procedure described above can ob-
viously be improved upon. Most improvements tend towards the introduction of more memory into the search, simultaneous alteration
of more than one tap, and alterations that are proportional to the
improvements made. In other words, the simple, intuitive procedure would tend to approach a gradient following, steepest ascent
hill-climbing procedure., or any one of a large number of possible
optimum seeking methods (6). In contrast, it is interesting to
leave the extension of the simple technique and consider the problem of adjusting a tapped-delay-line from a general point of view.
The adjustment of an arbitrary linear filter is discussed in
Appendix A.

Consider the tapped-delay-line in Figure 12(a), with an input
$x(t)$ and an output $y_1(t)$. If the transmitted information is $r(t)$,
an error signal, $e_1(t)$, may be found. Now,

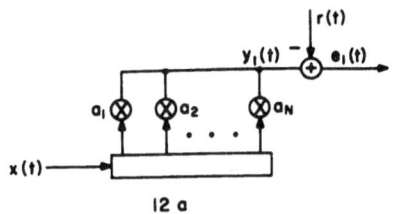

12 a

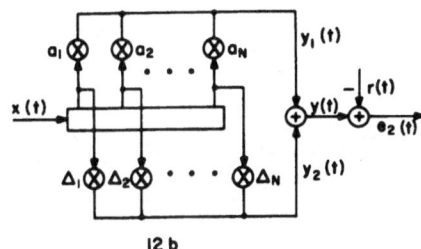

12 b

Figure 12. (a) Tapped delay line
(b) Modified tapped delay line.

$$y_1(kT) = \sum_{i=1}^{N} a_i x_{k-i}$$

where a_i is the gain of the ith tap, and $x_{k-i} = x\big((k-i)T\big)$.

Let us consider the effect of an arbitrary change to the response of the tapped-delay-line on the mean-square-error. An obvious change to the filter is a change of magnitude Δ_i to each tap gain, as shown in Figure 12(b). The output of the modified system at t = 0 is

$$y(0) = y_1(0) + y_2(0)$$

$$= \sum_{i=1}^{N} a_i x_{-i} + \sum_{i=1}^{N} \Delta_i x_{-i}$$

The mean-square-error is

$$\overline{e_2^{\,2}} = \overline{\big(y - r\big)^2}$$

$$\overline{e_2^{\,2}} = \overline{\big(y_1 - r + y_2\big)^2}$$

$$\overline{e_2^{\,2}} = \overline{e_1^{\,2}} + 2\,\overline{e_1 y_2} + \overline{y_2^{\,2}}$$

and the change in error is

$$E = \overline{e_2^{\,2}} - \overline{e_1^{\,2}} = 2\,\overline{e_1 y_2} + \overline{y_2^{\,2}}$$

The best choice of the Δ_i's is that which makes $\overline{e_2^2}$ as small as possible, i.e., which makes E as negative as possible.

Now,

$$E = 2\,\overline{e_1 \left(\sum^N \Delta_i x_{-i} \right)} + \overline{\left(\sum^N \Delta_i x_{-i} \right)^2}$$

Differentiating E with respect to a_k (k = 1, ..., N) and setting the result equal to zero yields a set of simultaneous linear equations, whose solutions are the changes to be made to the existing taps to yield the least squares estimator, i.e., to minimize the mean square error.

$$\frac{\partial E}{\partial \Delta_k} = 0 = \overline{e_1 x_{-k}} + \sum_{i=1}^{N} \Delta_i \, \overline{x_i x_{-k}}$$

The optimum Δ_i's are seen to be functions of the crosscorrelation between the error signal and the delay-line outputs, and the covariance function of the input signal. If the original system is the null system, i.e., if $a_i = 0$ (i = 1, ..., N), then $e_1 = r$ and the equations are for the taps of the least squares estimator, and have the form

$$\overline{e x_{-k}} = 0 \quad \text{for} \quad k = 1, \ldots, N$$

which is a discrete expression of the Wiener-Hopf condition for optimum linear mean-square estimators.

The method that has just been described has been implemented experimentally with the setup shown in Figure 13. The reference signal may be derived from the decision device as shown, or a 'hard-wire' reference may be used. The tapped-delay-line enclosed by the dashed lines in the upper left hand corner is the digital compensator described previously. All other functions, including the correlation measurements, are performed in the general purpose computer associated with the equipment.

With the relatively slow computer used (a Digital Equipment Corporation PDP-5) it takes about 6 milliseconds to process a set

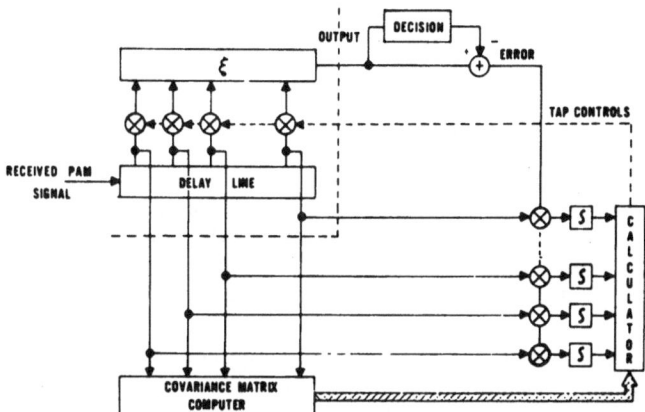

Figure 13. Implementation of the adaptive procedure.

of new samples during calculation of the correlation coefficients, for a five-tap delay line. Thus, it requires about 3 seconds to average 500 samples or about 0.4 seconds to average 64 samples. The subsequent solution of the equations requires about 0.2 seconds for five taps.

The accuracy of the solutions depends to a large extent on the accuracy of the correlation coefficient measurements. Errors in the reference will likewise be reflected in poorer performance. Research is presently underway to determine the adaptive behaviour of the system when there are errors in the reference, that is, when the filter is set up so poorly that many decision errors are made.

6. OTHER ADAPTIVE PROCEDURES

The adaptive procedures described above range from an almost 'blind' iterative trial-and-error search to a direct computational determination of the optimum receiver response. In between are many directed search procedures that might be used for various reasons: speed, accuracy, stability, equipment simplicity, and so on (6).

Examples of some searches are:

(1) Uniform Step Search, in which steps of uniform size are made or not depending on the gradient.

(2) Variable Step Search, in which the step size is chosen to minimize the error along each coordinate.

(3) Parallel Tangent Search (6), which is a technique that terminates at a minimum after 2N steps.

(4) Steepest Descent Search, in which the local gradient is followed.

Each of these searches requires knowledge of the gradient of the surface after each step. With the well behaved surface involved in the mean-square-estimator, a single measurement is sufficient: the gradient may be updated computationally.

These techniques have not been compared to either the iterative search, the computational procedure or the Lucky-Rudin 'directed' iteration (7).

7. INVERSE FILTERS AND TIME-DOMAIN EQUALIZATION

The consideration of multipath channels from a statistical communication theory point of view has led to a practical receiver that contains a tapped-delay-line as an approximation to an unrealizable sampled data filter. Other approaches to the reception of PAM signals lead naturally to the use of tapped delay lines, or transversal filters as they are sometimes called (7, 8). These approaches have been called inverse filtering or time-domain equalization. The distinguishing factor about inverse filtering and the majority of time-domain equalization is that they are based on the reduction of signal distortion, and not with the simultaneous reduction of additive noise. In low-noise situations, time domain equalization (which is the use of a tapped-delay-line to reduce distortion due to time dispersion) and the techniques described in this paper are similar. They differ only in the methods used to choose the tap values, and in the criteria of performance used.

8. SUMMARY

This paper has described the development of a system for the reception of time-dispersed PAM signals. The rate at which such a system can be adapted to changing channel conditions is compatible with the rate of change of ionospheric paths and hence, warrants consideration as a receiver for HF ionospheric communication systems.

Studies to date have indicated that on 'channel limited' communication paths, large improvements in performance result from

the use of better channel models in the design of the systems; and
that even minimal allowance for multipath distortion yields significant rate increases. Specification of the parameters of an operational receiver for ionospheric paths would require a detailed
study of the path, the particular data transmission requirements,
and the equipment budget.

APPENDIX A

THE EFFECT OF A LINEAR SYSTEM CHANGE ON MEAN SQUARE ERROR

It is possible to determine how any arbitrary modification to
a linear system will affect the mean-square-error. Further, for a
given change, it is possible to determine what magnitude that
change should be to minimize the mean-square-error.

Consider the linear system shown in Figure A-1. The input to
the system is $x(t)$. The output of the initial system H_1, is $y_1(t)$.
If the desired output is $y(t)$, the error $e_1(t) = y_1(t) - y(t)$. If
an arbitrary change were made to the initial system H_1, and were
represented as a system ΔH_2 placed in parallel with H_1, the new
error would be $e_2(t) = y_1(t) + \Delta y_2(t) - y(t)$. Δ is an arbitrary
gain factor, and $y_2(t)$ is the output of H_2 when $x(t)$ is the input.
The change in mean-square-error is

$$E = \overline{e_2^2} - \overline{e_1^2}$$

$$= 2\Delta \overline{e_1 y_2} + \Delta^2 \overline{y_2^2}$$

which may be expressed as

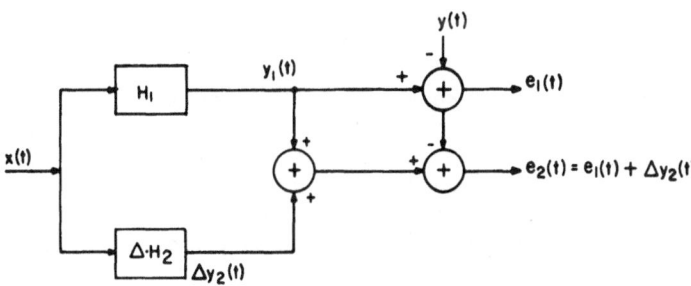

Figure A-1. Modification of an arbitrary linear system.

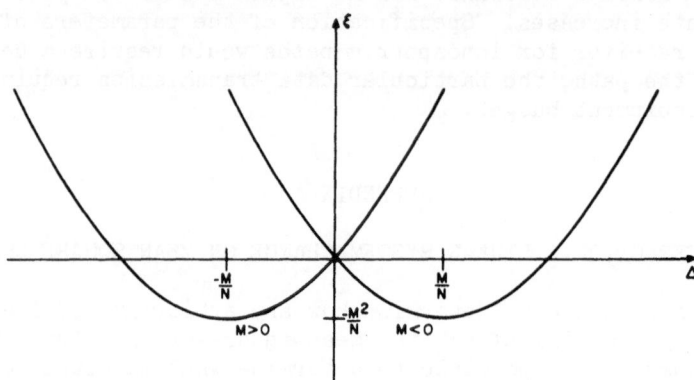

Figure A-2. Error vs modification.

$$\left(E + \frac{M^2}{N}\right) \;=\; N\left(\Delta + \frac{M}{N}\right)^2$$

where

$$M \;=\; \overline{e_1 y_2}$$

and

$$N \;=\; \overline{y_2{}^2}$$

i.e., as a parabola with its apex at

$$(E_o, \; \Delta_o) \;=\; (-M^2/N, \; -M/N)$$

as shown in Figure A-2. Note that if M is positive the left hand parabola holds, if M is negative the right hand one holds. The slope of E at the origin is $2M = 2\,\overline{e_1 y_2}$.

It is desirable to make E as negative as possible. This is accomplished by choosing Δ so as to move down the parabola. That is, if M is positive, Δ should be made negative, and vice-versa. The direction in which to change the system is given by the cross-correlation of the output of the proposed modification and the error signal. This is the principle behind the operation of the Lucky-Rudin system (7).

A great deal more information may be derived from this approach than just the sign of Δ. The optimum value is, in fact,

$$\Delta_{opt.} \ = \ - \ \overline{e_1 y_2} \ / \ \overline{y_2^2}$$

APPENDIX B

INVERSE MULTIPATH CHANNELS AND THE EXTENSION OF THE PRACTICAL RECEIVER

Linear systems whose tranfer functions are the inverses of those of multipath channels are shown in Figure B-1. A comparison of the inverse filter for discrete multipath shown in Figure B-1 (c) with the practical receiver suggests the extension of the practical compensator to the form shown in Figure B-2. Neither the determination of the optimum values of the taps, nor the specification of an adaptive scheme have been developed for this compensator as yet.

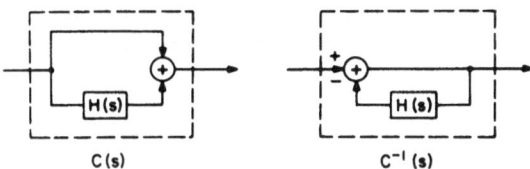

(a) TWO PATH MULTIPATH AND INVERSE

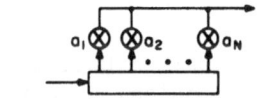

(b) DISCRETE MULTIPATH CHANNEL

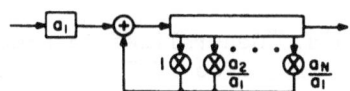

(c) INVERSE OF DISCRETE MULTIPATH CHANNEL

Figure B-1. Inverse multipath channels.

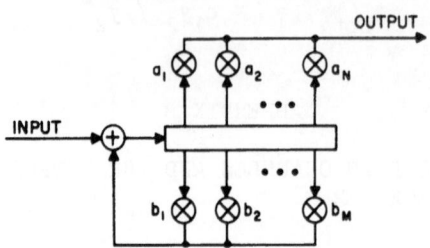

Figure B-2. Extension of the practical tapped-delay-line.

REFERENCES

1. D.A. George, "Matched Filters for Interfering Symbols'. IEEE Trans. on IT, vol. IT-II, No. 1, January 1965, pp 153-154.

2. J.W. Smith, "The Joint Optimization of Transmitted Signal and Receiving Filter for Data Transmission Systems". BSTJ, vol. XLIV, No. 10, December 1965, pp 2363-2392.

3. E.W. Tufts, "Nyquist's Problem – The Joint Optimization of Transmitter and Receiver in Pulse Amplitude Modulation", Proc. IEEE, vol. 53, no. 3, March 1965, pp 248-259.

4. D.C. Coll and D.A. George, "A Receiver for Time-Dispersed Pulses", Record of the 1st IEEE Communications Convention, June 1965, pp 753-758.

5. D.C. Coll, "A System for the Optimum Utilization of Pulse Communication Channels". DRTE Technical Report No. 1168, December 1966.

6. D.J. Wilde, Optimum Seeking Methods. Prentice-Hall, 1964.

7. R.W. Lucky and H.R. Rudin, "Generalized Automatic Equalization for Communication Channels", 1966 IEEE Int. Comm. Conf. Digest, June 1966, pp 22-23.

8. H.R. Rudin, "Automatic Equalization Using Transversal Equalizers", IEEE Spectrum, vol. 4, no. 1, January 1967, pp 53-59.

THE CHEC SOUNDING SYSTEM

E. E. Stevens

Defence Research Telecommunications Establishment.

Ottawa, Canada

Abstract: The CHEC sounding system was developed to enable operators of air-ground links to communicate with optimum effiency.
This report describes the basic system concepts, operational assumptions, and instrumentation features. Results of evaluation
trials are discussed together with established calling procedures.

1. INTRODUCTION

The reliability of high frequency radio communications can be
improved if real-time propagation information for the transmission
path is known and can be applied operationally (1). This is particularly so with Arctic communications due to the large variations that occur in the real-time MUF with respect to a monthly
median value. This is illustrated in Figure 1 which is a plot of
the hourly maximum observed frequency for the transmission path,
Ottawa to Resolute Bay, for the month of December 1960. It will
be noted that the median of the observed values, and the predicted
monthly median, are in fairly close agreement despite the large
variations in the actual observed values.

The instrumentation that is required for oblique incidence
ionospheric sounding has now reached the state-of-the-art where it
is operationally feasible to employ this technique, as a means of
providing real-time propagation information, in parallel with a
high-frequency communications system.

Channel sounding is one method of obtaining propagation information concerning the transmission path. While this technique

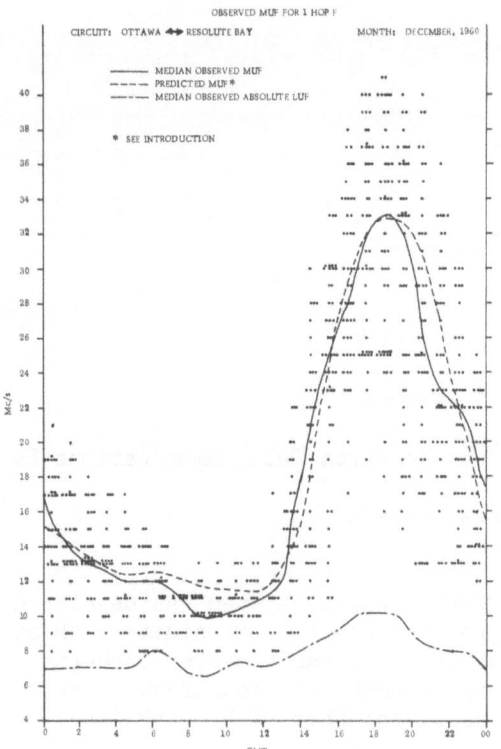

Figure 1. Comparison of predictions with observations on the
circuit Ottawa-Resolute Bay, December 1960.

does not provide the more detailed information that is possible
with the conventional oblique incidence ionosonde, the advantages
are; that the relatively simple instrumentation reduces installation cost by at least one order of magnitude, and as sounding
transmissions are made on the actual assigned frequencies, the
performance of the sounding link can be related directly to the
communications link.

    The CHEC Sounding System, C-H-E-C for channel evaluation and
calling, was designed at DRTE to assist long range patrol aircraft,
operating in and north of the auroral belt, to maintain reliable
high frequency communications with their base communications terminal. The basic system concept can be applied to ship-shore communications, and with modifications, to point-to-point circuits.

    This paper will describe the CHEC system as it applies to an
air-ground-air communications environment and will discuss some of
the results obtained during evaluation trials made under actual
operational conditions.

2. SYSTEM CONCEPT AND INSTRUMENTATION

In applying the CHEC system to air-ground-air communications the collowing is assumed;

a) Sufficient frequency diversity is available, both in frequency assignments and communications equipment, that efficient use of the propagating medium can be realized under all anticipated operational conditions.

b) The air-ground transmission path is the sensitive link; if propagation conditions are satisfied for this path the ground-air link should present no problem, providing the same channel is used to communicate.

The CHEC sounding system provides the communicator in the aircraft with the most essential information concerning his communications link with the ground. This is the signal-to-interference ratio his transmissions can expect to have at the ground receiving station, on any of his assigned channels over which propagation is possible. In addition, the calling sub-system provides the facility for,

a) the aircraft communicator to call the ground station, indicating automatically the channel that will be used for communications and,

b) the ground station to call a particular aircraft, or a general call to all aircraft in flight, and have this indicated automatically in the aircraft.

The CHEC system concept is based on the premise that, if the level of the desired signal and the undesired interference on a particular channel, is measured or sampled for a period, short compared to the time interval between measurements, an accurate estimate of the signal-to-interference ratio is possible, providing the levels measured are consistent over several time intervals (2).

Reference is made to Figure 2. This shows the configuration of the CHEC system integrated with a typical air-ground-air communications complex. The ground communications center, with its conventional transmit and receive facilities, is also equipped with the CHEC stepped-frequency ground interference receiver, for measuring and encoding ground interference levels, (GIL), and the stepped-frequency sounding transmitter, for the transmission of coded sounding signals. The aircraft, in addition to its communications transceiver, is equipped with the CHEC airborne stepped-frequency receiver for the reception and evaluation of the sounding transmissions. Both air and ground units are maintained in

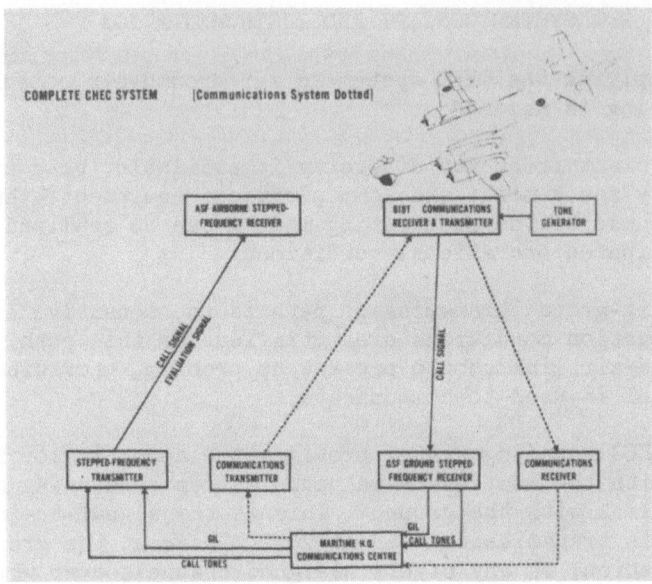

Figure 2. CHEC system integrated with typical air-ground-air
           communication complex.

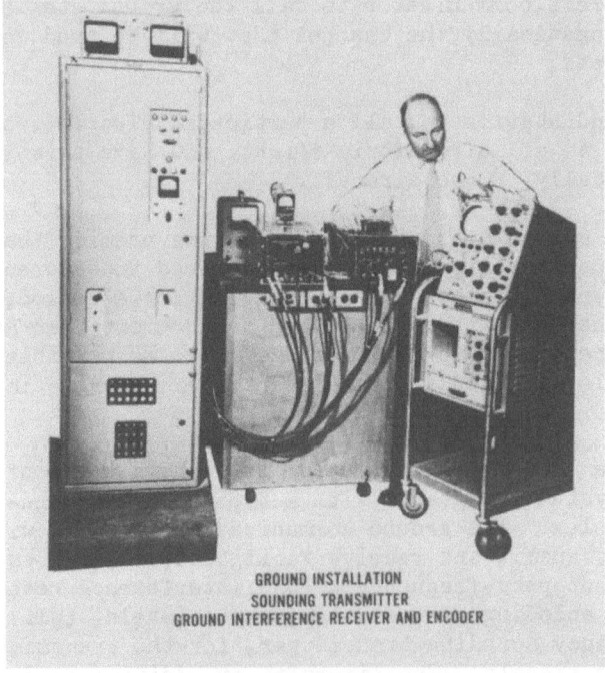

Figure 3. Ground installation, sounding transmitter and ground
           interference receiver and encoder.

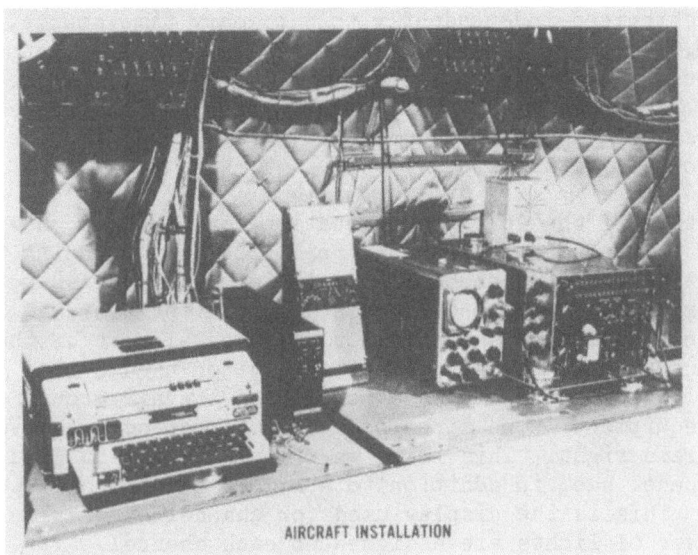

Figure 4.   Aircraft installation.

Figure 5.   Airborne stepped-frequency receiver.

frequency and time synchronism by internal crystal controlled clocks synchronized independently to a primary time source, e.g., WWV or CHU. The tone generator shown is used for air-ground calling and its function will be described later under the section "Calling Sub-systems". The CHEC signal transmissions are shown in solid lines while those of the air-ground-air communications system are shown in dotted lines.

A picture of the CHEC ground station equipment is shown in Figure 3. This consists of the stepped-frequency sounding transmitter on the left, the encoder and modulator for the sounding transmitter, and the stepped-frequency ground interference receiver. The test equipment shown is not an integral part of the system. Figure 4 shows the aircraft installation used during the evaluation trials and includes a portion of the air-ground communications equipment. The airborne stepped-frequency receiver is at the extreme right. This is shown in more detail in Figure 5. The front panel has, in addition to a number of controls, two rows of lights. This is the display used for channel evaluation. A vertical pair of lights are assigned to each channel. If the bottom one lights, propagation is possible between the ground station and the aircraft. If both light, a preset level of signal-to-interference ratio has been satisfied for the channel.

The equipment is designed to accomodate sixteen frequencies in the off-route aeronautical mobile bands between 3 and 21 MHz. This is probably more than would be required for the usual operational requirement. Four seconds are required on each frequency for channel evaluation, and for the 16 frequencies the sweep sequence requires 64 seconds. Sweeps can be repeated as often as every two minutes.

To evaluate a channel, during the first two seconds the background interference level at the ground receiving terminal is measured, quantized, and encoded. During the third second this information is transmitted to the aircraft in the form of an appropriately coded signal. During the fourth second an additional transmission is made for the purpose of measuring the signal strength at the aircraft for transmissions between the ground station and the aircraft. If propagation is possible between the ground station and the aircraft, these transmissions will be detected by the airborne stepped-frequency receiver. The ground interference level is automatically decoded and stored for subsequent comparison with the signal level measurement at the end of the fourth second. The receiver logic determines if the ratio of the signal level measured at the aircraft, to the ground interference level, is equal or greater than a preset value. The result is indicated on the receiver display.

As the system gains for the air-ground communications link
and the ground-air sounding link have been normalized, by assuming
reciprocity the communicator in the aircraft can predict the performance of this air-ground communications link on the basis of
the displayed information. The output of the receiver display is
recorded on a paper chart to provide a history of channel performance which can be used as a ready reference for channel evaluation.

3. EVALUATION TRIALS

The CHEC system was evaluated operationally during the spring
and early summer of 1966. The necessary ground terminal was set
up at DRTE, Ottawa, equipped with both sounding and communications
facilities. A series of flights using long range patrol aircraft,
equipped with the airborne installation, were made into regular
operational areas, out to distances of several thousands of kilometers and as far north as 64° latitude. During the flights it
was only possible to devote the first half of each hour to CHEC
evaluation tests. This usually afforded acceptable continuity
although continuous testing would have been preferable.

The dynamic accuracy of channel evaluation that is possible
with the CHEC system is shown in Figure 6. At the beginning of
each hour, on the basis of the information displayed from four
consecutive sweep sequences, four channels were selected and rated,
one as good, one poor, one fair and one as above the MOF. The ac-

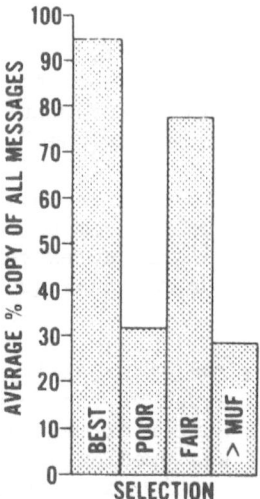

Figure 6. Communications performance on frequencies selected
using channel evaluation information.

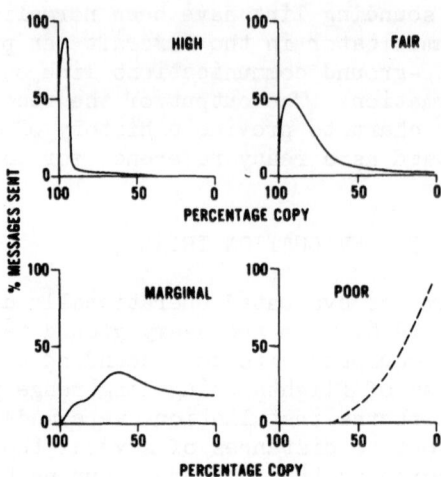

Figure 7. Teletype copy received for indicated channel perform-
           ance CHEC system.

curacy of these ratings was checked by sending a series of teletype messages on each of the channels over the aircraft communications system, and noting the percentage copy received at the
ground terminal. The histogram shown in Figure 6 is an average
for all data for the trials, and shows the percent copy received
on all channels rated as good, poor, fair, and above the MOF.
These results would confirm that channel evaluation is reliable.

Figure 7 illustrates the analysis that has been made of the
channel selected as "best", with respect to quality of communications, based on channel evaluation information available to the
aircraft communicator. Operationally the "best" channel selected
would be used for communications. During a large portion of the
trial period it was possible to predict that the "best" channel
selected would provide high quality communications performance.
Occasionally, however, particularly at extreme ranges, it was necessary to predict only fair or marginal quality, and on several
occasions, poor.

The distribution shown for each quality predicted is based on
the average percent copy received at the ground station, for each
category, against the total number of messages sent. When high
quality channel performance was predicted, i.e., where the indicated signal to interference ratio was consistently in the order
of 10 db or greater, for 90% of the messages sent the percentage
copy was 90% or better. The distributions for fair, marginal, and
poor quality are as would be expected.

The minimum accuracy considered acceptable for high quality communications during these trials was 90% copy. This may appear to be a rather low standard. However, with a transmitter power of 100 watts, and the usual low gain aircraft antenna system, this figure is, perhaps, not unrealistic.

A comparison was also made between the performance of two experienced communicators in the same aircraft, one using standard frequency selection techniques and the other operating on the CHEC-determined "best" frequency. During these trials the ionospheric propagation conditions were excellent, and, as might be expected under these conditions, there was no substantial difference between the quality of messages received at the ground terminal from the two communicators. Even under these excellent conditions, however, there was a marked difference between their performance with respect to the time taken to establish radio contact with the ground on a suitable frequency. The CHEC-aided communicator transmitted his message without preliminaries, on the channel indicated as "best" by the CHEC display and without acknowledgement from the ground. The unaided communicator, on the other hand, took an average of several minutes to establish contact with the ground before sending his message.

It should be mentioned that on other trials carried out by DRTE (3) with sounder equipment somewhat similar to the CHEC system, it has been shown that during periods of ionospheric disturbance the sounder assisted communicator can achieve very significantly better time of contact and message quality than can the unassisted communicator.

4. CALLING SUB-SYSTEMS

In many communications systems there are neither enough receivers nor personnel to monitor all assigned frequencies at the ground station. Since the CHEC system permits the aircraft communicator to choose his best frequency, it is necessary to provide a means to ensure that the ground terminal personnel monitor the correct frequencies. This requires an air-ground calling system, which can be provided for in the CHEC system by the addition of the following very simple equipment:

a)   a call-tone generator in the aircraft

b)   a call-tone decoder at the ground terminal.

The call can be sent in the following manner. First the aircraft communicator selects his best frequency for communications and sets his transmitter on that channel. He then uses the call-

tone generator to modulate his communications transmitter on that
channel during the regular sounding sweep sequence. The call is
received and decoded by the stepped-frequency ground interference
receiver, which, as part of the evaluation system, is sequentially
monitoring background interference at the ground terminal on all
assigned frequencies. A call-tone decoder, connected to the output of this receiver, operates a display which indicates the channel on which the aircraft communicator is calling and on which he
wishes to communicate. This air-ground call facility is an essential part of the CHEC system and can be implemented with only very
minor addition to the basic channel evaluation system.

A ground-air calling facility can also be added to the system
and again this can be done with little increase in system complexity and cost. By means of tone generators on the ground, call
tones can be added to the evaluation transmissions. These can be
added to the evaluation transmissions. These can be received and
decoded in the airborne stepped-frequency receiver and used to
operate a simple call display in the aircraft. Specific aircraft
can be called in this way if signature tones are used. Perhaps
the most important advantage of this facility is that a call can
be sent on all assigned channels and will be registered in the
aircraft if at least one channel is open to the aircraft.

5. SUMMARY

The CHEC Sounding System has been demonstrated to be operationally feasible in an air-ground-air communications environment.
The system is now being modified for application to a 2800 km
point-to-point communications link.

The application of CHEC to a high frequency communications
link permits all system parameters to be known. The operational
advantages are obvious. For example the CHEC system permits the
communicator to select his best channel with no time delay and
transmit his traffic with confidence that it will be received. It
indicates when communications are impossible. Thus futile efforts
to establish contact, which may cause unnecessary co-channel interference, are avoided. It also provides an independent monitor
of the performance of the communications system. If, for example,
the quality of message reception is consistently lower than the
CHEC predictions, a fault can be suspected in the communications
equipment. The calling facilities are an additional bonus provided by the CHEC system at minimal cost.

6. REFERENCES

1. D.R. Nielson, K.D. Felperin, D.R. Macquivey and T. Kailath, "Communications Sounding as an Aid to Frequency Management", Final Report Phase 1, Control 50-189 prepared for D.C.A. S.R.I. Project 4554, June 1965.

2. G.W. Jull, "Prediction of Communication Reliability using an HF Channel Sounder", DRTE Technical Memorandum 459, March 1966.

3. G.W. Jull, D.J. Doyle, G.W. Irvine and J.P. Murray, "Frequency Sounding Techniques for HF Communications over an Auroral Zone Path", Proc. IRE 50, 7, July 1962.

THE CURTS CONCEPT AND CURRENT STATUS OF DEVELOPMENT

S. E. Probst

Defense Communications Agency, Washington D.C., USA

Abstract: The CURTS concept envisions the use of oblique synchronous ionospheric soundings, noise and interference measurements on the assigned communications frequencies, and current performance monitoring of the communications frequency in use, as inputs to a centralized computer. The computer memory will contain
all the necessary information of the available frequency resources
and the computer logic will utilize the inputs to predict the operational performance of assigned frequencies, select the frequencies for use on all of the controlled trunks, and provide, as output, directions for frequency change in a manner to optimize the
continuity of the communications system being controlled. This
paper reviews the concept, reports on the current stage of development, and provides some of the measured performance results to
date.

1. THE DEFENSE COMMUNICATIONS SYSTEM

To provide the necessary background against which the CURTS
concept can be examined and evaluated, it is necessary to briefly
discuss the nature of the Defense Communications System. Prior to
1960, the three U.S. Military Departments had each developed large
communications systems tailored to their own needs with only a
small degree of interconnection between these systems.

In 1960, the Secretary of Defense issued directives establishing the Defense Communications System (DCS) and the Defense
Communications Agency (DCA). The DCS was defined as consisting of
the long haul point to point circuitry previously contained in the
three separate systems and the subsequent evolutionary development

of the newly integrated system. Excluded from the DCS were tacti-
cal communications and wholly dedicated communications in support
of specific weapons systems. The three Military Departments continue to operate their respective portions of the DCS. The DCA
was established to manage the DCS, plan its development, establish
design and performance standards, and to maintain information on
the system status of the DCS.

The DCS now consists of over 32,000,000 channel miles of
leased and government owned communications and is truly a world-
wide system. It contains submarine cable, landline, microwave relay, tropospheric scatter, ionospheric scatter, rapidly expanding
communications satellite and, of course, extensive high frequency
radio trunks. The high frequency (HF) trunks of the DCS constitute what is undoubtedly the largest HF network in existance consisting of more than 100 two-way radio trunks.

The experience of DCA in monitoring the performance of the HF
portions of the DCS has clearly shown that a major limitation, if
not the major limitation, on the reliability of the system is imposed by the problems of congestion and frequency sharing in the
HF bands. The necessity for widespread sharing of frequency assignments within the DCS is best illustrated by the following
facts. If it were possible to assign six completely clear and un-
shared frequencies in each direction on each HF trunk of the DCS
with the required emission bandwidth available on each frequency,
the DCS by itself would consume the entire HF fixed band allocations in the International Table of Frequency Allocations. Of
course, the DCS must occupy these bands along with all of the
other U.S. Military, other nations' military, the world's commercial and all other users of the HF fixed bands. The net result of
these facts is that DCS frequencies are widely shared within the
DCS as well as being shared in most instances with non-DCS assignments. Real-time day-to-day management of frequency utilization
in this congested environment in an efficient manner is a nearly
insurmountable task unless every possible aid can be provided.

2. THE CURTS CONCEPT

The development of oblique synchronous ionospheric sounders
was seen to present an important potential for the rendering of
positive assistance in the real time frequency management problem.
The CURTS concept is the result of several years of thinking, research, and development and can best be summarized as follows.
The aim was to develop a single set of sounder operating parameters (number of channels, pulse length, PRF, etc.) which would
satisfy the maximum number of HF communications and research interests so that the total number of sounder transmitters which would

need to be installed could be kept to the minimum. It was recognized that a proliferation of incompatible sounder systems can result in the sounders themselves becoming a significant source of additional contamination in the HF bands. If the common user approach is successful, however, the needs of some 90 per cent or more of fixed point HF communicators of the world could be satisfied with a total number of oblique sounder transmitters less than the number of vertical incidence sounders that operated throughout the International Geophysical Year, the International Geophysical Cooperation, and the International Year of the Quiet Sun. It is fully recognized that there will continue to be some research and operational applications which will require different types of ionospheric sounding and no attempt is intended to restrict the development of such sounders.

With the establishment of a world-wide common user sounder network, transmissions would be available from all of the major points of interest to HF communicators and use of these transmissions could be made by anyone possessing a compatible sounder receiver. The discussion which follows, however, describes the manner in which these transmissions would be used in the DCS application of CURTS. A sounder transmitter would be located at each major node in the DCS HF network with sounder receivers at each connected point in the network. The data from these sounder receivers would be digitized, reduced to the minimum essential information, and transmitted through the DCS (including non-HF circuitry as available) to frequency selection computers located in the major DCS area control centers (probably no more than four in number to cover the world). At each DCS receiving station, digitally tuned receivers would measure the noise and interference background on each assigned frequency. These data would also be digitized and passed to the frequency selection computer. The computer memory would contain the complete list of assigned frequency resources for each HF trunk controlled, with all of the assignment limitations specified. The frequency currently in use on each trunk path would also be known. The computer logic would be designed to consider the measured propagation characteristics of each path, the implications of sharing within the frequency resources, and the relative priorities of the communications being carried on each trunk, and to provide as its output, directive orders for all frequency changes and the times at which those frequency changes should be made in manner to optimize the continuity of the communications system. This output would, of course, be based on a prediction logic and there also would be provided inputs on existing or predicted ionospheric disturbances which would appropriately initiate changes in that prediction logic.

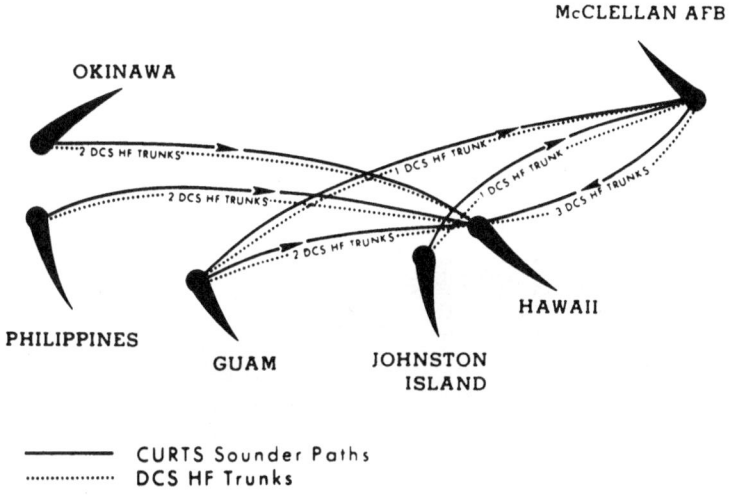

Figure 1. Test network in the Pacific Area.

3. THE CURTS R&D TEST NETWORK

The research and development program for CURTS has progressed
through the implementation of an R&D test network in the Pacific
Area as illustrated in Figure 1. Granger sounder transmitters in
the network are located at McClellan AFB, California, Johnston
Island, Guam, the Philippines, and Okinawa. Receivers in the test
network are at McClellan AFB and in Hawaii. The sounders are operating on 120 frequency channels between 4 and 32 MHz. Four 1 ms
pulses are transmitted on each frequency channel at a PRF of 20
pulses per second. Each sounder scan requires 24 seconds and each
path is sounded once each ten minutes. Figure 2 shows a typical
ionogram for illustrative purposes, however, it should be noted
that the normal ionogram display is utilized only for a visual
check by the operator to determine that the sounders are function-
ing properly. The sounder signals on each frequency channel are
digitized and passed directly to a computer. The output of the
sounder receiver for each received pulse is sampled each 50 μs beginning prior to the receipt of the transmitted pulse. Signal
amplitude in each 50 μs is digitized, with the levels measured
prior to receipt of the sounder signal providing the background

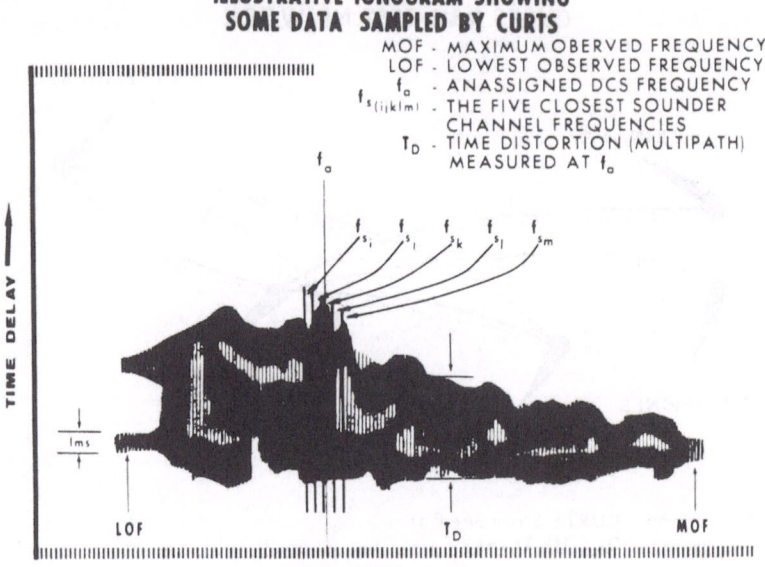

Figure 2. Typical ionogram.

noise level against which subsequent values are compared to determine when sounder signals are actually received. The data from
each sounder scan are compared with a time history record maintained on a peripheral memory tape and are then added to that time
history for use on subsequent days at the same time.

The computer memory contains the lists of assigned frequencies for each of the DCS radio trunks controlled (9 trunks at
Hawaii and two at McClellan). The sounder signal amplitude received on the sounder channel frequency (f_{s_k}) closest to each assigned frequency (f_a) and the signal for each of the two lower
(f_{s_i}, f_{s_j}) and higher (f_{s_l}, f_{s_m}) sounder channel frequencies are
utilized to provide a prediction of the signal energy that would
be received on the assigned frequency (f_a). This method of averaging smooths out the effects of fading from one received sounder
pulse to the next in a manner to provide an adequate representation of the median signal. The sounder data samples are also utilized to provide a measure of the time dispersion (multipath delay, T_D) that will be experienced on the assigned frequency. The
sounder receiver is automatically switched to the appropriate receiving antenna multicoupler so that the communications receiving
antenna for the appropriate path is utilized for the sounder
transmissions related to that path.

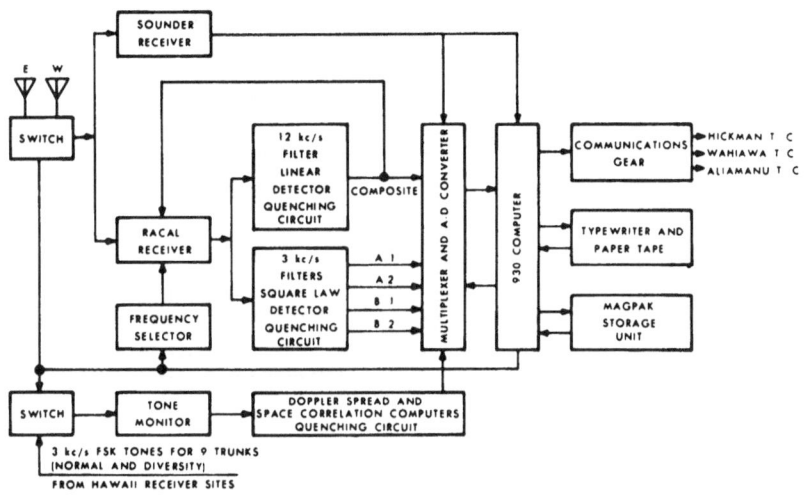

Figure 3.   Block diagram of receiving system.

Each of the DCS trunks involved carries independent sideband modulation consisting of four 3 kHz slots in an over-all emission bandwidth of 12 kHz. One of these 3 kHz slots on each trunk normally contains the separate tones for 16 channels of 100 word per minute teletype while the remaining three 3 kHz slots each contain one voice channel. Figure 3 is a block diagram of the receiving system configuration and shows a digitally tuned Racal receiver which is tuned successively to each of the assigned frequencies on each radio trunk (again utilizing the communications receiving antenna) and a measurement of the combined noise and interference in the 12 kHz emission bandwidth and in each of the four 3 kHz slots on each frequency is taken. These levels of interference plus noise are also digitized and passed to the computer.

The teletype tone package on the currently-in-use operating frequency from each of the two diversity receivers on each HF trunk is passed through a diversity correlation computer where the correlation of fading is measured. In addition, the deviations of the AFC voltage are observed and the zero crossings are counted to provide a measure of frequency dispersion (doppler spreading F_D) on the operating frequency. These data are also passed to the computer. Each assigned operating frequency which is greater than

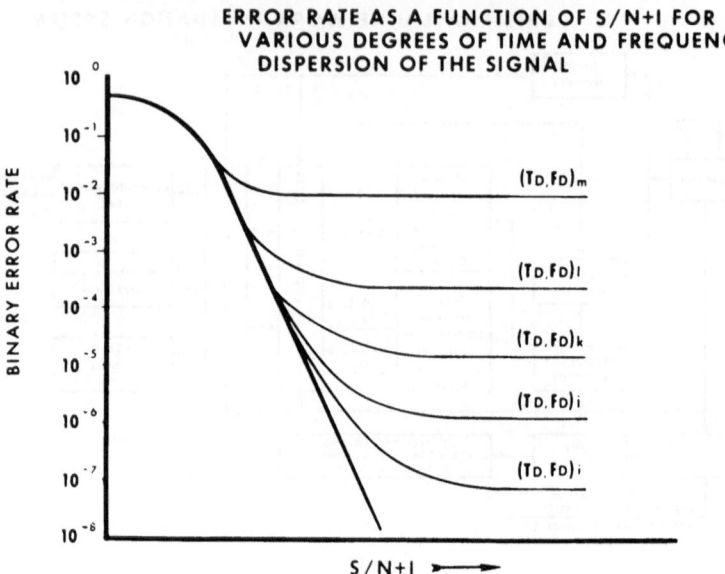

DEFENSE COMMUNICATIONS AGENCY

**ERROR RATE AS A FUNCTION OF S/N+I FOR
VARIOUS DEGREES OF TIME AND FREQUENCY
DISPERSION OF THE SIGNAL**

Figure 4.   Error rate as function of S/N+I for various degrees of
time and frequency dispersion of the signal.

the LOF and less than the MOF is evaluated by the computer in the
following manner. The predicted value of median received signal
as determined above is compared to the measured noise plus interference to produce the predicted S/N+I ratio. The value of time
dispersion (T_D) measured from the sounder data and the value of
frequency dispersion (F_D) measured from the operating frequency
are assigned.

   The performance of the modulation system utilized on the DCS
trunks has been modeled and a set of binary error rate versus
S/N+I curves for various values of time and frequency dispersion
(T_D, F_D) as illustrated in Figure 4 have been generated and stored
in the computer. From these curves, the predicted binary error
rate for each operating frequency is determined and the exponent
of the binary error rate is assigned as a quality figure for each
frequency. That is to say, a frequency assigned a quality figure
og 6 is one for which the predicted binary error rate is one in
10^6. A subsidiary quality figure is also assigned to each of the
four 3 kHz slots. The list of assigned frequencies for each controlled trunk is ordered in terms of the descending quality figure

Defense Communications Agency

CURTS PHASE II NETWORK
EVALUATION SYSTEM
FREQUENCY SELECTION DATA PROVIDED TO
FACILITY CONTROL

NPN		78EB01/78BB01				2300
22990	7	3	3	3	3	
23050	7	3	3	3	3	
19947.5	7	3	3	3	3	
25440	7	3	3	3	3	
14612.5	5	3	2	3	3	
19020	2	3	3	1	1	
10283	1	1	2	1	1	
20725	U					
	S					
11100	U					
15835	S					

Figure 5.

and is provided on an output teletypewriter to the communications technical controller as illustrated in Figure 5. This ordered list of frequencies is utilized in determining frequency change decisions, while the subsidiary quality figures are used to determine which 3 kHz slot should be used for the teletype tones. The entire process described above is repeated each ten minutes and an updated output is provided each ten minutes.

During periods when none of the assigned frequencies on a path is propagating, which occurs on some days particularly during the predawn transition period, the computer logic has been modified to provide a prediction, based on the time history information and measured noise and interference of each assigned frequency, of the frequency that will be first returned to a usable level. This capability has proven invaluable in shortening the periods of experienced outage.

4. OBSERVED RESULTS

Experience has shown that, when left to their own devices,
the Technical Controllers will usually request a frequency change
whenever the error rate on the operating frequency is approximately in one to 10^3 or worse. On the other hand, again when left to
their own devices, the Technical Controllers will rarely request a
frequency change even though frequencies capable of providing
quality 7 or 6 performance are available so long as the operating
frequency has not dropped below 10^3. During periods when the ordered frequency lists are actively utilized in determining frequency changes, a significant reduction in printed errors and in
circuit outages is consistently observed. Extensions of the computer logic to consider the availability of shared assignments and
to produce directive frequency change orders are now in prepara-
tion. Several significant impacts of availability of this system
have been observed. Some of the more significant of these are as
follows:

(1) 98 per cent of the frequency changes based on the CURTS out-
     put lists provide immediately acceptable traffic. This figu-
     re compares to approximately 70 per cent for frequency chan-
     ges ordered by technical controllers without the availability
     of the CURTS output.

(2) Repetitive unsuccessful frequency changes during periods of
     actual propagation outage are eliminated. The Technical Con-
     troller's confidence in the CURTS output lists has reached
     the point that they will change to the frequency first pre-
     dicted to recover and restore their traffic at the earliest
     possible moment.

(3) Confidence in the predicted quality figures has increased to
     the point that unsuccessful circuit performance on a frequen-
     cy assigned a high quality figure is promptly identified as
     equipment maladjustment difficulty. The practice of charging
     poor performance to propagation problems, when in fact propa-
     gation is not the cause, has been significantly reduced. In
     fact, the observed performance of the Okinawa to Hawaii path
     when compared with the Japan to Hawaii path which is not yet
     included in the CURTS network raised suspicion with regard to
     a long continued period of difficulty on the Japan-Hawaii
     trunk which had been charged to propagation. Investigation
     as a result of this suspicion disclosed equipment difficul-
     ties which were corrected restoring normal circuit performan-
     ce.

(4) The almost continuous availability of frequencies with quali-
     ty figures of at least six indicates that higher data rates

on many HF paths may indeed be possible. It is currently planned to conduct a test in the near future utilizing the data rate of 2400 bits per second, a rate which could certainly not be maintained under past practices of frequency selection.

The CURTS development program is nearing the end of its research and development phase. Significant and persuasive demonstrations of its potential impact on future reliability of HF circuitry are accumulating daily. While approval for the full implementation of the CURTS concept within the DCS has not yet been obtained, it is hoped that this decision will soon be made.

NOTES ON A MULTI-MODE COMMUNICATION SYSTEM

R. A. Kulinyi

Communications/ADP Laboratory, U S Army Electronics

Command, Fort Monmouth, New Jersy, USA

Present design philosophy for ionospheric scatter and meteor burst systems tends to optimize these facilities for a single mode of propagation. Typically, observations of signals over these circuits reject identifiable periods of sporadic E (Es) mode propagation and the equipment is not designed to operate efficiently during these occurrences.

Based on the above facts and certain other capabilities described below, a communication system has been designed, fabricated and tested by a group at the Communications/ADP Laboratory, U S Army Electronics Command and their contractor, Melpar, Inc which has the inherent ability to operate with great efficiency at both low and high data rates, signal strengths and duty cycles. Weak scatter modes, either tropospheric or ionospheric, depending on path length are used to operate at low data rates approximately 500 bits/second with very high or continuous duty cycle. Automatically, upon occurrence of signal enhancements due to meteor trails or, optionally, E_s clouds, bursts of 50 kilobits per second data from a magnetic memory file of up to 10^6 bits of lesser priority traffic can be transmitted and received. Error detection and correction techniques are used. Operation is in the 50-150 MHz range at a power level of 5 kW. Simple 9 db log-periodic antennas are used.

In addition to the scatter, burst and E_s modes given above, a capability has been developed to generate an electron cloud, equivalent to an over-dense meteor trail or E_s occurrence by use of a simple two-stage rocket, chemical-explosive payload and very simple launch facilities. Nike-Cajun rackets with 20 to 40 lb payloads of a caesium (Cs) salt plus pure aluminium (Al) powder and

explosive yield a chemical release at heights from 95 to 105 km.
This produces an excess of electrons with durations of useful extent both in sunlight and darkness. Without solar radiation, propagation times of enhanced signal strengths are of much shorter
duration but still of great use for long bursts of high-priority
data.

This system can interoperate in both continuous and store-
and-forward modes as part of switched digital communications systems at distances from trans-horizon ranges to between 1200 and
1400 miles. It is constrained to operate from a static location
but is mounted in vans and trailers. It can be deployed and set
up quickly and can be packed and transported to a new location
easily and with little delay.

A NOTE REGARDING THE PERFORMANCE OF VARIOUS COMMUNICATIONS SYSTEMS IN A NOISY BACKGROUND

A. D. Watt

DECO Electronics, Dept of Defence and Space Centre

Boulder, Colorado, USA

It is possible to compare the performance of different communications systems in a noisy background by plotting the performance described in terms of the percentage of character errors or percentage of errors received correctly as a function of rms carrier to rms noise in a given bandwidth. Figure 1 shows the results of measurements taken with typical frequency-shift keying teletype receivers employing a start stop teletype printer operating at 60 words per minute and also a good and a fair operator receiving random five letter groups at 16 wpm. The noise background for the solid curves was typical VLF noise which contains rather large impulses. The amplitude probability distribution for this noise has a very wide dynamic range. The dashed line was taken in a thermal or gaussian type of noise background and shows that the performance varies much faster in terms of a changing signal to noise than is true for the atmospheric noise background. It is important to note that when human operators are involved, there is an upper limit of performance which appears to vary from individual to individual as well as with the rate of transmission and the general state of well being of the operator. This upper level of performance is in the order of 1% character errors for a good operator and something in the order of 10% for fair to poor operators.

It is important to note that the systems shown are operating at different data rates. As a result, a direct comparison of system effectiveness in data transmission can not be made on these curves. A method for intercomparing systems has been described by Watt, Coon, Maxwell and Plush in the 1958 Proceedings of the IRE. The systems performance factor employed in this paper normalizes results such as shown here in terms of data rates so that direct

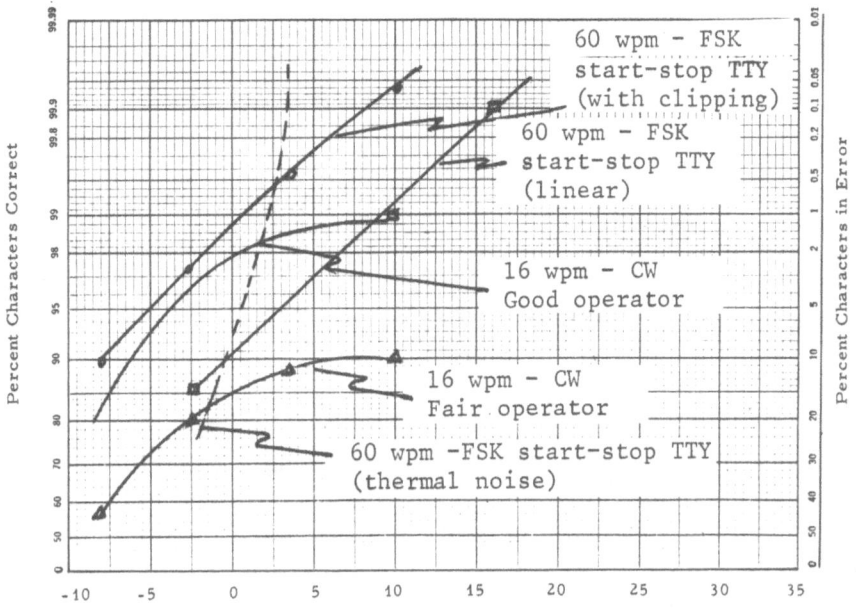

Figure 1. $C/N_{1\ kc/s}$, rms carrier to rms noise in a 1 kc/s band, db.

intercomparison of system effectiveness can be made. The system performance factor is similar to and in fact related through appropriate conversion formula to the E/N_0 employed in information theory papers.

It should be noted that the system performance will vary not only with the noise background but also the characteristics of the transmission path. In the referenced paper, information was given for tropospheric fading signals and also some for ionospheric fading signals. The characteristics of the multiple path transmission circuits prevalent at HF can influence different communication systems in a widely varying manner. Details of the way in which high frequency multipath affects the performance of some communications systems has been described in a very thorough manner by Mr. Vincent during this conference.

SUMMARY OF TECHNIQUES AVAILABLE FOR OVERCOMING PHYSICAL PROBLEMS OF HIGH-LATITUDE COMMUNICATIONS

N. H. Knudtzon

SHAPE Technical Centre, Den Haag, Netherlands

Abstract: This paper surveys techniques available for combating the effects of attenuation, time and frequency dispersion in ionospheric radio propagation. The techniques described are not peculiar to transmissions in the polar areas. However, their properties may be of particular significance in these regions where the channel parameters often vary in a critical and unpredictable manner.

1. INTRODUCTION

This symposium deals with two main subjects:

- geophysics of the ionosphere
- communications techniques

and I have been asked to give a summary of the latter. This will be rather short, since many of the papers already presented have been surveys.

The discovery and exploitation of new means of <u>transmission</u> have led to much progress in telecommunications. However, as the possibilities for radically new means of transmission are getting exhausted, further progress will depend on better use of the means now available by improved <u>techniques</u>.

The latest mode of transmission which is given much attention these days is by means of <u>satellites</u>, but it should be noted that the synchronous satellite cannot cover regions north of a latitude of 81.5°N, in fact the practical limit would be, say, 73°N for a minimum ground station elevation angle of 7.5°.

However, the present symposium is to deal with systems depending on <u>ionospheric propagation</u> only.

2. INFLUENCE OF THE IONOSPHERE

To the communications engineer who wishes to make use of the ionosphere - rather than the <u>geophysisist</u> who wishes to explain its behaviour - signals and noise are affected in the three following ways:

- attenuation
- time dispersion (multipath)
- frequency dispersion (Doppler spread)

These effects vary with time.

VHF signals propagated by ionospheric and meteoric scatter modes have quite different characteristics. The first type exhibits rapid variations superimposed on a fairly stable median value, and the latter consists of bursts.

Basically, the techniques available for overcoming the communications problems caused by the influence of the ionosphere, are:

- antenna optimization
- change of frequency
- relaying
- diversity
- modulation and coding
- error detection and correction.

None of these techniques are peculiar to high-latitude communications. However, the techniques may have to be particularly carefully exploited in this geographic area, because propagation conditions at high latitudes often are critical. Two approaches are possible:

- a <u>specific</u> design based on statistical data for the signal and noise characteristics to be expected for the path concerned

- a flexible design which enables the system to <u>be adapted</u>, or to <u>adapt itself</u>, according to a given criterion, such as signal strength or actual communications performance, e.g. error rate, measured at the actual time of transmission.

3. ANTENNA OPTIMIZATION

Antennas, both transmitting and receiving, are important components of any radio system. The only special feature which should be taken into account at high latitudes, apart from the need for physical ruggedness, is the low ground conductivity which may occur in certain areas. Techniques exist to meet many different requirements. However, the physisists must provide guidance on the desired radiation pattern and its direction, both in the horizontal and vertical planes, which are desirable in each case, such as, for example, to account for tilted ionospheric layers or to strike meteoric "hot spots". Present-day techniques allow rapid switching among antennas.

4. CHANGE OF FREQUENCY

The two classical ways for obtaining guidance on the choice of frequency for HF systems are:

- predictions prepared from statistics of past observations
- oblique or backscatter soundings performed at the time communication is to be established.

Both methods have been the subject of the previous papers and discussions at this symposium.

Sounding can be made by special apparatus which scan the relevant frequency band covering the pertinent frequency assignments, or in a more crude way by operating the actual communications equipment on a set of selected frequencies, see Reference (1).

The criterion for selecting the frequency to be used can be signal strength, interference level or communications performance, and the operation can be made more or less automatic.

Reference 2 gives a typical demonstration of the improvements achieved by sounding techniques. It describes sounding experiments on an air-ground link of less than 2200 km in length at high latitudes. It was found that loss of contact during a two-day period of geomagnetic storms could be reduced from 77% for frequencies selected from predictions to 18% for frequencies selected from sounder observations. This important gain was due to the ability to determine the optimum frequencies for communications via auroral sporadic E which persisted during the geomagnetic storm.

Oblique sounding is generally preferable to backscatter sounding. However, in cases where one terminal is unable to transmit, e.g. due to the requirement of radio silence for naval ships, backscatter sounding may prove worthwhile.

With regard to VHF, one great advantage of ionospheric and meteoric scatter systems is that no frequency changes are required.

5. RELAYING

HF relaying is a well-known communication expedient used:

- to avoid parts of the ionosphere which are frequently subject to serious disturbance

- to reduce the length of difficult individual paths and so increase the time for which the most favourable propagation mode is usable

- to minimize the effects of low transmitter power and antenna gain as in ship-shore communications, where a message may be relayed via several ships.

Relaying is often performed by a quasi-random path selection, i.e. the originating station sends the message to any other station willing to pass it on.

6. DIVERSITY

Diversity takes advantage of the lack of correlation between signals:

- separated in frequency
- **received** on antennas separated in space
- transmitted and received at different angles
- having different polarization.

The three forms of diversity combining systems in use are:

- maximal-ratio, also called variable-gain or ratio-squared, combining which adjusts the gain in each branch to deliver the maximum signal/noise at the output,

- equal-gain combining which adds the signals in the branches linearly,

- selection combining which selects the one branch which has the highest signal-noise ratio.

The diversity gain which can be achieved depends on which of these techniques is used and on the number of diversity branches, see Reference (3). Of the three techniques maximum-ratio combining provides the highest diversity gain, which for two-fold diversity theoretically amounts to 3 db in average signal/noise ratio. However, the reduction in telegraph error rate may be appreciably greater than indicated by the gain in average signal/noise ratio, because the amplitude distribution of the combined output signal is becoming flatter as the number of diversity branches increases, see References (4) and (5).

7. MODULATION AND CODING

Modulation and coding is a field which is receiving much attention, in particular for the transmission of digital information.

Repeated attempts have been made to provide models and guiding theories for the synthesis of communications systems for various propagation characteristics including their bandwidth limitations. Although there are limits to how far it is useful to develop an all-embracing generalized model and theory, these attempts have led to a much better understanding of the problems involved, and gradually also to improved techniques.

No attempt is made here to present a survey of these modern modulation and coding techniques. However, it is pointed out that the propagation mechanism is complicated and that a particular system which is able to counteract the effects of time dispersion may well be vulnerable to frequency dispersion, and vice versa. One may, therefore find that, on the average, a simple system such as frequency shift keying (FSK) compares very favourably with more elaborate modulation and coding schemes.

If real improvements are to be expected, it seems that adaptive systems, i.e. with automatically time-varying parameters, will be required. These systems become rather sophisticated because they must be able to probe the channel at any time and immediately adjust the receiver in accordance with a given criterion.

8. ERROR DETECTION AND CORRECTION

Error detection and correction systems have been in common use for digital transmission during the last two decades. Modern technology which provides small switching and memory components have led to the development of various compact systems, operating with or without return paths.

These systems can achieve a considerable increase in reliability in low signal/noise or high interference conditions at the expense of a relatively small reduction in the average information rate. However, the advantages to be gained by this technique are generally greater on circuits where deterioration is not as sudden and severe as often is the case with high-latitude paths.

9. CONCLUDING REMARKS

In spite of many new developments in the field of transmission, ionospheric propagation continues to provide a particularly important mode of communications at high latitudes.

Many techniques are already available for overcoming the physical problems of high-latitude communications, and modern technology makes it possible to implement reliably methods requiring increasing degrees of sophistication.

In designing a communications system for meeting specified operational requirements, we - the communications engineers - in selecting the techniques to be used must take due account of such factors as operational conditions, maintenance, security, radio silence, frequency assignments and costs. This is, after all, what engineering is about.

10. REFERENCES

1. K.W. Blake and N. Knudtzon, "Comparison of propagation predictions and measured performance for two paths over the Norwegian Sea", Record of NATO Advanced Study Institute on Ionospheric Radio Communications in the Arctic, these proceedings.

2. G.W. Jull, D.J. Doyle, G.W. Irvine and J.P. Murray, "Frequency sounding techniques for HF communications over auroral zone paths", Proc. IRE, July 1962, pp 1676-1682.

3. D.G. Brennan, "Linear diversity combining techniques", Proc. IRE, June 1959, pp 1075-1102.

4. J.R. Pierce, "Theoretical diversity improvement in frequency-shift keying", Proc. IRE, May 1958, pp 903-910.

5. N. Hveding, "Comparison of digital modulation and detection techniques for a low-power transportable troposcatter system", Conference Record of First IEEE Annual Communications Convention 1965, pp 691-695.

SUMMARY OF DISCUSSION ON COMMUNICATION TECHNIQUES

With reference to Vincent's paper discussion took place on the method used to determine the 2σ spread in doppler frequencies. This was done in practice by determining the average frequency received in an AFC circuit, and then using the received doppler-spread signal to count zero crossings of the average frequency. The 2σ spread in doppler frequencies was stationary with time, and in practice valid over a useful area.

It was also pointed out that in principle it would be possible to obtain a "scatter funtion ionogram", which may have application to fine structure studies of the ionosphere. In particular, travelling disturbances, and other changes in reflecting layer characteristics could be determined.

A plea was issued for more of this type of information to be made available to the scientific community, and in reply it was stated that a limited number of copies of this study would be available.

In the discussion on "The CHEC System" by Stevens it was stated that the operator in the aircraft could determine which channel was best, as well as all other channels which could support reliable transmissions.

It was also pointed out that the doppler frequency measurement equipment could be added to make the determination of channel characteristics more reliable. It was mentioned that the CHEC system represents a relatively simple method to determine important channel characteristics, and that if users wish to have short-term reliable information, "will the ends" then it is necessary for them to "will the means".

Discussion after Coll's paper, "Adaptive Receivers for Multipath Channels" established the following points about the receiver:

a)   The system now under study operated only on low-pass signals. A quadrature receiver would be required in practical application of the technique to radio transmission systems.

b)   The hill being climbed is smooth.

c)   The technique is applicable to quasi-static multipath channels, and is not concerned with channels showing frequency dispersion.

In the discussion concerned with the presentation of R. Kulinyi on "A Multi-Mode Communication System" it was stated that propagation via an artificially ionized patch at 100 to 105 km had the following properties:

a) Durations of useful transmission times were up to one hour in
    the day and about 20 minutes at night.

b) Asymmetrical path geometry is possible because the ionized
    patch acts as a scatterer.

c) Frequencies in the VHF band would be the most useful; this
    is the band in which the system would normally work using
    scatter and meteor burst modes of propagation. Effects have
    been noted above and below the VHF.

In addition to propagation aspects of a communication system
other factors must be considered. Some of these factors are advanced technology, frequency allocation problems such as the crowded spectrum and interference, reliability, weight, size, primary
power supply, man-made noise, economic considerations, and a variety of human problems, etc. Frequency allocation considerations
are particularly serious.

It became abundantly clear that the technical aspects of communications design can not be separated from economical factors.

Survey of Existing Communication Facilities and User Problems in the Artic

EXISTING COMMUNICATIONS FACILITIES AND USER PROBLEMS IN THE

CANADIAN ARCTIC

T. Ringereide

Telecommunications & Electronics Branch

Department of Transport, Ottawa, Canada

Abstract: This paper reviews the telecommunications systems in use in Northern Canada against the background of population density, economic development of the region and known natural resources.

The activities of the Department of Transport in the North are reviewed briefly, and a discussion is given of some main problems faced by users of northern radio systems.

Finally, the expected future development of telecommunications in Northern Canada is discussed with particular reference to the planned establishment of a domestic satellite communications system providing service to remote communities.

1. SOME BACKGROUND DATA ON CANADA'S NORTH

1.1 Area

In order to understand the special communication problems existing in the Canadian North, it may be useful to review very briefly some basic geographical facts. The total area of Canada i.e. land area and fresh water, is around four million square miles which gives Canada the second largest area in the world, second only to that of the Soviet Union, but larger than that of the United States of America or Brazil.

1.2 Population

For various historic and economic reasons Canada's population
of 20 million is very largely concentrated in the southern part of
the country with about two-thirds of the population living in large urban centres. On Figure 1, which is a map of Canada, a heavy
broken line has been drawn to show the northern boundary of the
densely populated portion of Canada. The population living north
of this "population frontier" is only 1.5% of Canada's total population or 300,000 people. It is also worth noting that out of
this total "northern" population only about 38,000 live in the
Yukon Territory and the Northwest Territories. The Yuon has a population of only 7 people per 100 square miles, while the Northwest Territories has a population of 2 people per 100 square miles.
The total Eskimo population in the North is about 13,600, while
approximately 8000 Indians live in the Yukon and Northwest Territories.

1.3 Transportation

Most of northern Canada is accessible only by air for 9-10
months of the year but the Yukon Territory and the western-most

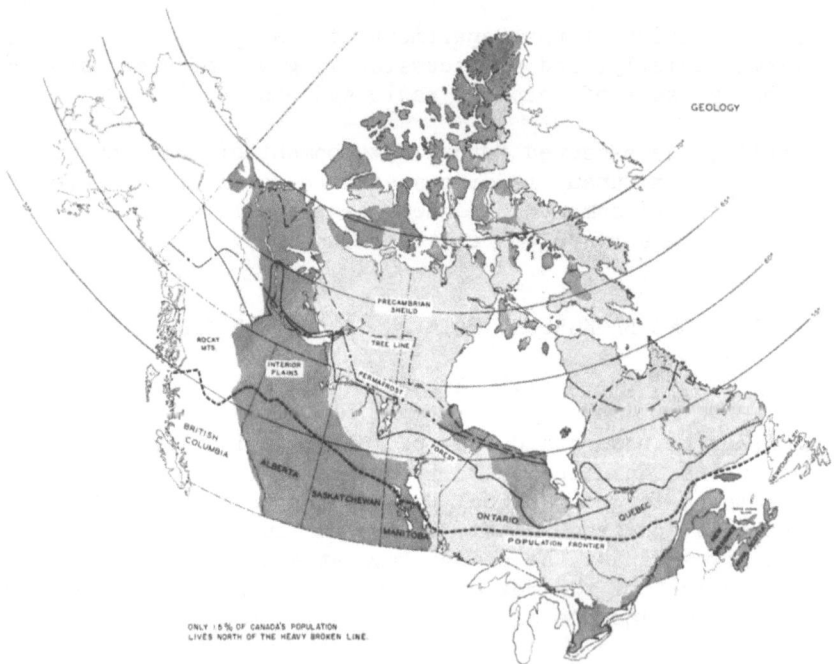

Figure 1. Map of Canada.

Figure 2.   Transportation routes.

part of the Northwest Territories have more developed transporta-
tion facilities than other portions of northern Canada. Figure 2
is a map on which the main roads and railway lines extending into
Canada's North have been shown. You will notice that the Alaska
Highway runs from Dawson Creek in British Columbia through the
Yukon Territory into Alaska, and that the towns of Dawson City,
Mayo and a number of other points in this general region are served by roads. The network of roads also extends up to Great Slave
Lake connecting such places as Yellowknife and Hay River to
southern points. Similarly you will see from the map that railway
communications extend into the Yukon and certain western points in
the Northwest Territories. Further east, roads are much scarcer
but you will note that the town of Churchill on Hudson Bay and
Moosonee on James Bay are served by railway lines. The North
Shore and Labrador Railway connects Schefferville in Quebec and
parts of Labrador to Sept Iles on the Gulf of St. Lawrence.

The Mackenzie River which is navigable along its entire
length, is an extremely important transportation route in the sum-
mertime. Coastal shipping during the all too brief summer season
is vital to the resupply of all communities along the arctic coast.
In recent years the town of Churchill on Hudson Bay has become an
important port for the shipping of grain to overseas markets.

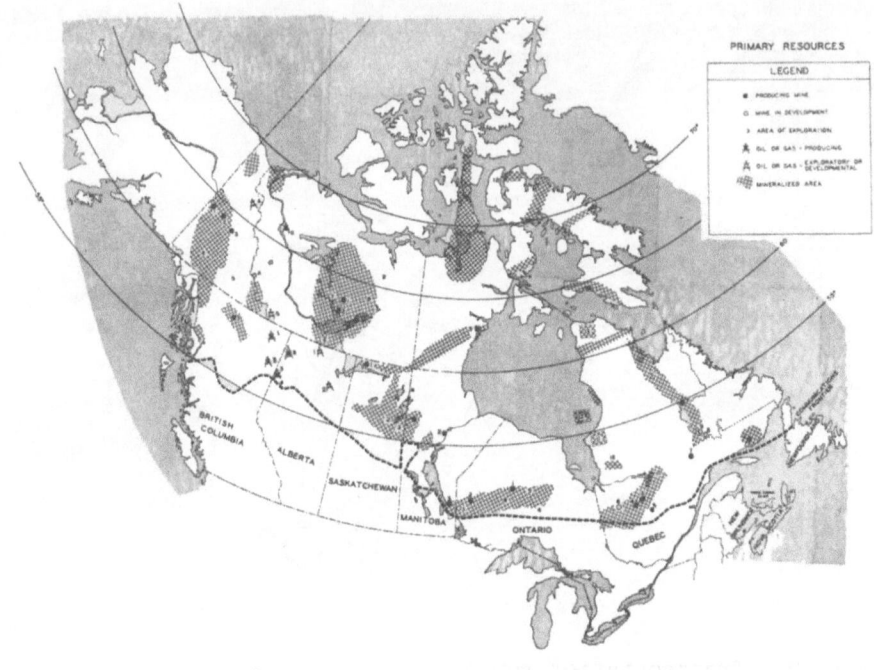

Figure 3. Natural resources in northern Canada.

It should also be mentioned that most of the established communities in the North have airstrips and larger centres have improved airport facilities.

The reason for showing this particular map is to indicate the
general development of the North, and it will later become apparent that the main communication routes to a large extent follow
the transportation routes.

1.4 Resources

To further illustrate the communications problems of the
Canadian North, Figure 3, which is a rough map of developed and
potential resources in northern Canada, is shown.

Northern Canada and in particular the western part of the
Northwest Territories contains very large potential mineral resources as well as oil. The hydro electric resources are also
very considerable. Perhaps the major reason why economic development of the Canadian Far North has not proceeded at a more rapid

pace is because more accessible natural resources have, in the
past, been readily available at more southern latitudes. While
mining development and oil exploration in parts of the Yukon and
Northwest Territories as well as in northern Quebec and in Labrador are being accelerated, the economic development of other known
resources is largely a matter of transportation economics. The
expected doubling of Canada's population in the next 15-20 years
should also have an important effect as regards the development of
northern resources.

2. TELECOMMUNICATIONS

2.1 Backbone Routes

Until recently the "population frontier" shown in Figure 1
also corresponded fairly closely to what might be called the "communications frontier", i.e. the regions north of this line would
mainly be dependent on HF radio communications for internal and
external communications. Due to developments of so-called "backbone" routes, mainly within the past 10 years, a number of the
more important northern communities have now been linked to the
nation's long distance telephone and telegraph network by means of
radio relay systems, cables or wirelines. Most of the northern
regions are still, however, served by HF radio systems generally
operating to a number of base stations which have been located
along the so-called "backbone" or fixed communications routes.

Figure 4 shows the main Canadian long distance communications
trunk routes. These circuits are mainly provided by means of
radio relays. Also shown in Figure 4 are some spurs or extension
routes linking key points in the North to the main national communications routes. Some of these systems are wireline systems as
the one linking Inuvik on the Mackenzie River to Hay River and
southern points, and a wireline system linking Churchill on Hudson
Bay to the telecommunications network of southern Manitoba. The
radio relay system along the Alaska Highway connects the town of
Whitehorse and other points in the Yukon Territory to the main
national networks and carries civilian/military communications to
Alaska. Similarly Hay River on Great Slave Lake and Moosonee on
James Bay are served by radio relay systems. The military Dewline
system is not shown on this particular chart as it has little civilian significance but it is pointed out that the tropospheric
scatter system from Lady Franklin Point on Victoria Island to Hay
River provides both a connection to the Dewline system and circuits for civilian use. Similarly the Quelab troposcatter system
connecting Goose Bay in Labrador to the Quebec telephone network
serves a dual military/civilian purpose.

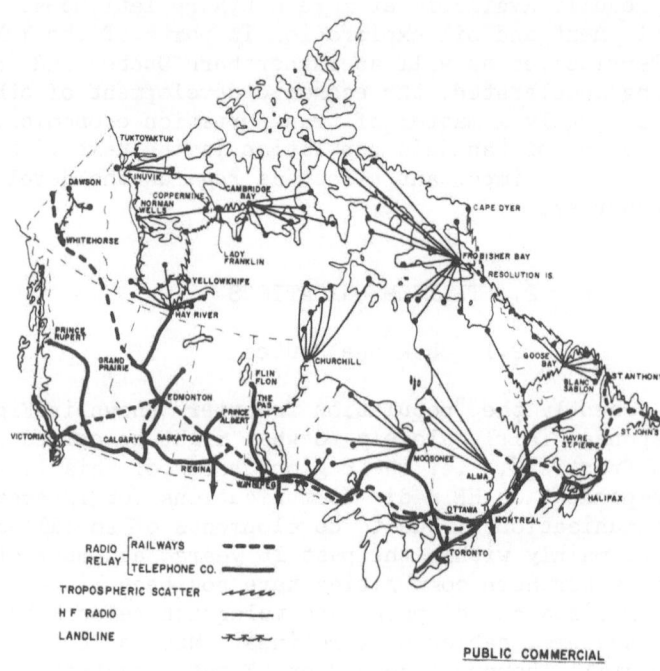

Figure 4. Canadian long distance communications trunk routes.

The Polevault North or BMEWS Rearward Communications System from Goose Bay in Labrador northward is operated by the USAF and interconnects with the military dewline and the commercial Quelab systems. Polevault North also carries civilian traffic between Goose Bay in Labrador and Frobisher Bay in Baffin Island. By agreement between the USA and Canada, the Polevault South System extending from Goose Bay in Labrador into the Island of Newfoundland, is being upgraded by Canadian common carriers to carry additional circuits for civilian use.

VHF relay systems are used for some routes with up to 24 voice channels. A system of this type is used along the coast of British Columbia between Vancouver and Prince Rupert where it is also paralleled by a troposcatter system extending into Alaska; another VHF system links Inuvik at the northern end of the Mackenzie River wireline system to Tuktoyaktuk on the Dewline, and a 65 mile hop links Coppermine to Lady Franklin Point on Victoria Island. Several other VHF links are used over short distances in the Yukon and western region of the Northwest Territories.

A VHF relay system is now being built along the North Shore of the Gulf of St. Lawrence from Havre-St-Pierre to Blanc Sablon on the Strait of Belle Isle where it will be interconnected with the networks of Newfoundland.

2.2 Commercial HF Systems

In eastern Canada the main commercial base stations are located at Alma in Quebec, Frobisher Bay in Baffin Island, Goose Bay in Labrador, Moosonee on James Bay and at Churchill, Manitoba. These base stations are usually all compatible for SSB and AM. Most base stations in eastern Canada operate on a simplex basis but some stations also provide duplex service to outstations. Because of the need to operate on many frequencies and to outstations on many different bearings, it is difficult to employ high gain antennas. The trend now is to use several log-periodic antennas at each base station, but rhombics, half-wave dipoles, etc. are in wide use. Gradually, the base stations are being equipped with extra monitoring receivers to keep a listening watch on all assigned frequencies.

It has been pointed out by the common carrier companies that available ionospheric propagation data are not always reliable for use at high magnetic latitudes. Thus Churchill is known as a "dead spot" for radio communications on HF.

In western Canada major commercial base stations for northern communications are located at Le Pas in Manitoba, Hay River, Norman Wells, Inuvik and Cambridge Bay in the Northwest Territories and at Whitehorse in the Yukon. This listing is by no means complete but covers the main base stations. Most of these stations are compatible for AM/SSB. It is noted that most of the western commercial stations are operated on a half-duplex basis.

2.3 Private HF Systems

The very extensive private HF networks in northern Canada serving the Hudson Bay Co., the Royal Canadian Mounted Police, various church missions and the mining and oil industry, are shown in Figure 5. The Department of Transport encourages the use of common carrier facilities as such public systems provide good service and interconnection to the national telephone and telegraph networks, and they also need a strong traffic base to operate on an economical basis. There is, however, a strong desire on the part of many organizations to operate their own systems, using their own base stations in the South. The DOT will assign shared frequencies for such purposes when frequencies are available, and will usually not prevent the establishment of private HF systems. Our experience is that when the common carrier companies are able to provide a better service in an area, they will gradually per-

Figure 5A. Private radio networks.

suade the users of private HF systems to make use of public HF
facilities and networks.

3. DEPARTMENT OF TRANSPORT ACTIVITIES IN NORTHERN CANADA

3.1 General

     All radio communications in Canada are under the control of
the federal Department of Transport. Altogether about 2000 HF
radio stations have been licensed for use in northern Canada. The
Department of Transport will assign a set of frequencies for the
use of such stations on the basis of a careful analysis of ionospheric data, availability of frequencies and other factors and is
now in the process of introducing a computer system to be employed
for this purpose. Because of the peculiar conditions of propagation existing in the Canadian Arctic, the trend so far has, however, been to rely largely on user experience and experimental
data for the assignment of frequencies.

Figure 5B.   Private radio networks RCMP.

The DOT does not normally subsidize telecommunications in the North but has, in some instances, guaranteed a minimum revenue for the first 5-10 years of operation for some systems installed in the Yukon and Northwest Territories. The two common carrier companies serving most of Canada's North are the Bell Telephone Co., serving Labrador, northern Quebec and Ontario as well as that part of the Northwest Territories east of 102° West longitude; and the Canadian National Telecommunications, which is essentially a railway telegraph company, provides all public telecommunications in the Northwest Territories west of 102° West longitude and in the Yukon. Certain private organizations have very extensive private systems such as the Hudson Bay Co., some mining companies and Government agencies, etc. Most HF outstations used in Canada have a power output of less than 100 watts, while the commercial base stations mostly have a power of around 1 kw with a few more powerful stations. Very few multiplexed HF systems are in use in Canada's North although a few routes are served by multiplexed VHF systems.

The DOT operates a certain number of LF and HF radio systems in northern Canada for meteorological, marine and air traffic communications. For many years the DOT stations also used to handle an important volume of civilian message traffic, and still do this

HF ROUTES _____
MAIN LF ROUTES --------

Figure 6. Department of Transport northern HF & LF radio commu-
           nications networks.

in some locations prior to the establishment of commercial communications. The Department is normally willing to close down its
own radio systems in favour of the use of commercial leased facilities whenever reliable systems become available. In addition
the DOT maintains some ionosonde stations for measurement of ionospheric data. More about these will be said later.

3.2 Department of Transport HF & LF Systems

Figure 6 shows the main DOT LF and HF systems used for RTT
meteorological circuits, air traffic control, navigational aids
etc. Most of the aeronautical radio beacons used in the North are
in the frequency range of 200 to 415 kHz and it may be of interest
that many of these beacons can be modulated to provide back-up
circuits for our HF systems.

The Department operates the following major arctic LF radio
circuits:

     Resolute - Cambridge Bay,
     Inuvik - Norman Wells (back-up system),

Figure 7.   Canadian ionospheric stations.

    Churchill - Coral Harbour, and
    Coral Harbour - Frobisher.

    These LF stations all have a 3 kw transmitter output and pro-
vide a reliability of well above 90%. The DOT HF radio transmitters used in point-to-point service are usually more powerful than
those employed in commercial HF systems, many having a 5 kw output
power. Both AM and SSB equipment is used as well as CW and FSK
for message traffic.

3.3   Ionospheric Program

    A brief orientation may be in order about the ionospheric
propagation measurements carried out by the Department of Transport. From time to time additional work in this field is also
undertaken as special projects of the National Research Council,
the Defence Research Telecommunications Establishment or by one of
the universities, but my remarks here will be limited to the continuing program carried out by the DOT. Figure 7 shows the location of the main ionospheric stations.

All Canadian ionospheric data can be considered "Arctic Data" since ionospheric conditions peculiar to high latitudes begins to be evident at 45 degrees North latitude. Our most southerly ionospheric station is located at Ottawa which is 45.4 degrees North.

Three of our stations (Resolute, Churchill and Kenora) are located along the 95th meridian to monitor gradients related to latitude. Our Ottawa station provides a N-S link along the 75th meridian with Thule in the North and Washington, San Salvador, Bogata, Huancayo and Concepcion (Chile) to the south. The St. John's station is the most easterly North American Station and therefore provides measurements related to the mid-North Atlantic. It also completes a chain of stations along the 45th meridian extending from Godhaven (Greenland) through Paramaribo, Buenos Aires and Port Lockroy in the Antarctic.

About a dozen additional stations have operated for periods of at least a year in order to provide measurements along North to South or East to West lines. Altogether, stations have been established at 25 locations in Canada.

Canadian data is obtained through conventional pulsed type vertical incidence ionospheric recorders. Pulse lengths are 50 µsec. and most of our stations employ a pulse power of 10 KW. Ionosonde output (Ionograms) are on 35 mm film in the form of a height (time) versus frequency graph which is linear along the height coordinate and progresses logarithmically along the frequency coordinate.

Ionograms are reduced in conformity with rules prescribed by the World Wide Sounding Committee (WWSC) into daily "f plots" and monthly numerical charts. Numerical values are placed on punched cards which are machine processed to produce synoptic data. Punched cards and numerical charts conform to WWSC formats.

Automatic equipment has been in use on most of our stations since 1948, which has provided six ionograms per hour as recommended by the WWSC. St. John's, Churchill and Ottawa data produced hourly by manual equipment dates from early 1943.

Data accuracy is limited initially by the definition of the photographic process which is in turn limited by distortion from scattering of the pulses returned from the ionospheric layers. Layer height measurements are made to the nearest 10 km and measurements related to frequency are within 5 kHz below 5 MHz, 10 kHz from 5 to 10 MHz and 20 kHz above 10 MHz.

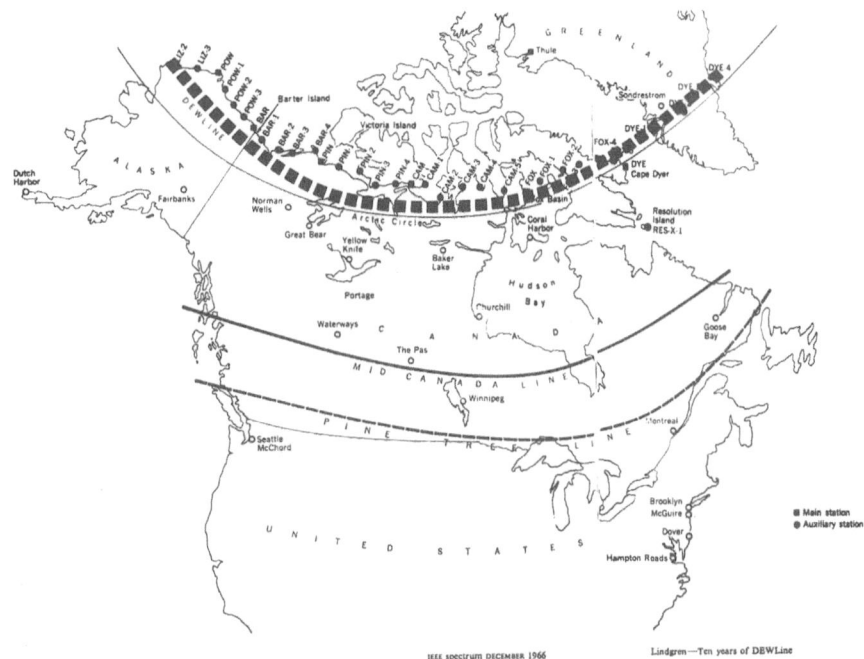

Figure 8. The Dewline communications system.

4. DEWLINE SYSTEM

The Dewline communications system is shown in Figure 8. The Mid-Canada line also shown in this figure has now been closed down. A good orientation about this system was given in the IEEE Spectrum for December 1966, so I will confine my discussion of this system to a bare minimum. Figure 9 shows a schematic representation of the system which should be self-explanatory.

The system which started out as a first line of defence against manned bombers ten years ago, has since changed its character to become essentially an early warning system for ballistic missiles (BMEWS).

As previously observed, the Dewline system has little civilian significance, and it is expected that it will, in the future, become more economical to provide civilian communications to remote communities by means of a communications satellite system rather than through the use of troposcatter systems. The satellite communications system would also have the advantage of being able to carry television channels.

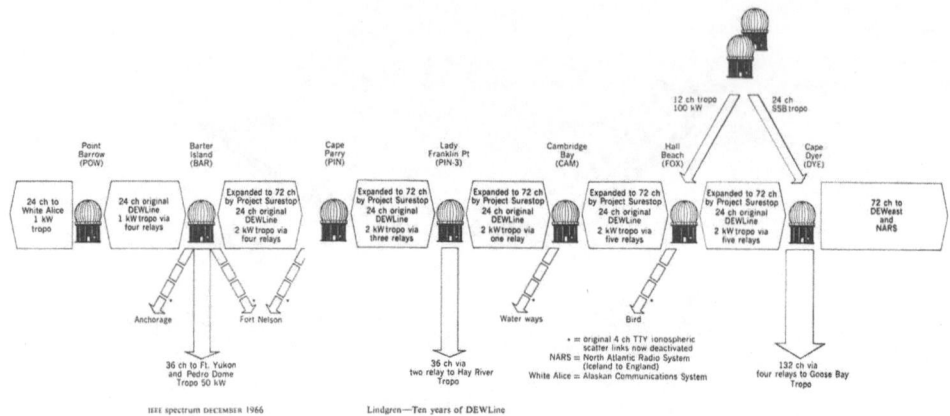

5. USER PROBLEMS

5.1 Power

The high cost of electric power (12-16¢ per kwh) is a major reason why most outstations use fairly low power transmitters (mostly around 100 watts). The introduction of transistorized equipment has been of great benefit in this regard. The DOT is vitally interested in the development of new power sources for use under arctic conditions. At the present time this Department is testing new thermoelectric power facilities at Wabush Lake. The thermoelectric generator used is manufactured by the 3M Company, while the shelter and the installation have been engineered by DOT. If anyone should be interested in further details regarding our experience with this project, I have brought along some copies of reports dealing with the project.

5.2 Frequency Assignment

The original choice of sites and optimum frequency for new HF outstations is normally made following field testing over a period of 3-4 weeks. It has been the experience of most user organizations that a frequency assignment based on use of published ionospheric data only, is not always satisfactory. All of these stations are assigned a number of frequencies for both day and night use. For circuits over fairly short distances (less than 500 miles) it has been found that the assignment of frequencies for day time use usually also is satisfactory for night time operation, and many of these stations therefore tend to use one frequency for 95% of the time. Circuits used over greater distances are more dependent on the use of a variety of higher frequencies for 24 hour, year-round operation. Such long circuits are also affected to a greater extent by different propagation conditions between day and night operation, the season of the year and the sun spot cycle. Prediction charts are particularly useful for setting up the optimum frequency for long distance circuits for use during different seasons of the year.

5.3 Some Notes on Propagation

The base station at Churchill, Manitoba often receives signals of good strength from outstations but is unable to contact the calling stations. This occurs most often when the calling station is located south of Churchill.

Arctic flutter affects HF only but is not very often a serious problem. Some users claim that this normally occurs during strong and quickly changing auroral activity. Blackouts in HF communications become more severe and occur more often during periods of high sun spot activity. Under such conditions it is found that north-south circuits provide a better path than east-west circuits, thus the HF circuit Resolute Bay-Edmonton will become useless long before the Resolute Bay-Churchill circuit. In many cases there would be no interruption to the Resolute Bay-Churchill circuit. It is also found that our LF circuits are not seriously affected by blackouts caused by high sun spot activity.

During periods of poor transmission CW signals will be intelligible during periods when voice transmission is impossible. Error correction equipment is employed on some of the more important circuits used for teleprinter and data transmission.

5.4 Modulation

Most of the new HF radio stations employ USB single side band equipment, and the use of temperature controlled crystal oscillators or frequency synthesizers allowing more accurate control of transmitter frequencies, is becoming common. The trend is now to equip all commercial base stations with monitoring receivers on all assigned frequencies and to assign additional frequencies to outstations allowing them to operate on more than one base station. Automatic signalling systems are not presently used on commercial HF systems, although some such systems are being evaluated by the common carriers. Operators at base stations may patch calls through to long distance telephone circuits but always have to be present to adjust levels regardless of whether duplex or simplex operation is used. The duplex HF circuits are all of the $\frac{1}{2}$-duplex type, i.e. a separate frequency is used for transmission and re-

ception but transmission only takes place in one direction at a
time.

5.5 Maintenance

Maintenance of HF stations presents a very difficult problem
because of the high cost of transportation. Most outstations need
to be visited 2 or 3 times per year, but some improvements in the
maintenance standards are being achieved by equipping many outstations with duplicate transistorized sets. When one set breaks
down, the spare set is put into operation, and the maintenance
depot will send a serviceable set to the station without waiting
for the defective set to be returned. In many cases local personnel are trained to perform such minor maintenance tasks as re-
pairing broken cables, etc.

5.6 Expected Future Developments

Planning is now taking place for a domestic satellite communications system for Canada. The DOT is conducting a major study
on such a system and special emphasis is placed on the use of a
multiple access satellite communication system to provide telephone, telegraph and data service to remote communities and also
to provide television programs in English and French over the same
system for local retransmission on VHF channels or over cable.

Communications satellites used in such a domestic satellite
communications system will be of the geo-stationary type. A synchronous satellite, as you know, is a satellite in an equatorial
orbit travelling in the same direction as the earth's rotation and
completing a travel around its orbit in approximately 24 hours, so
that it will appear to remain stationary above a point on the
earth's Equator. Such a satellite, located at approximately 95°
West would be able to cover points in the Canadian Arctic up to
about a latitude of 78° North. A major Canadian manufacturer is
at the present time developing an especially rugged but low cost
satellite ground station for use in the Arctic. It is hoped to
start experimental transmissions in January, 1968. The first experimental station is now being built at Bouchette in Quebec, not
far from Ottawa.

It will not be necessary for many northern communities to
have a satellite ground station which is able to handle telephone
and telegraph traffic as well as the reception of television signals, as some northern communities already have a satisfactory telephone and telegraph service via wireline systems, radio relays,
etc. In these communities a less expensive receive-only satellite
ground station may be used for the purpose of receiving television

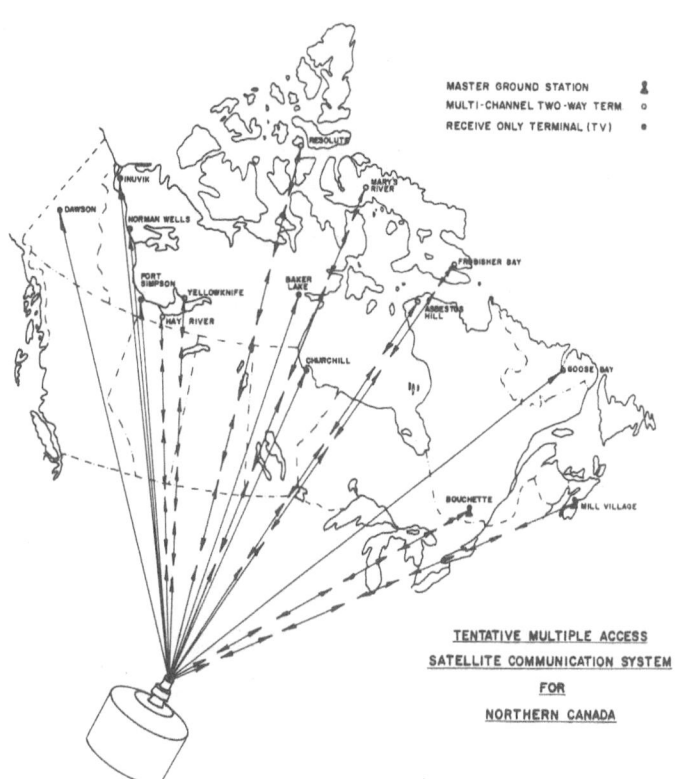

Figure 10.   Possible sites for satellite ground stations in
            northern Canada.

signals. This less expensive type of satellite ground station
could, for instance, be used in a number of communities along the
Mackenzie River.

     It is hoped the type of satellite ground station capable of
use for telephone, telegraph and television can be built for
$ 1,000,000. Due to the importance of high reliability and mini-
mum maintenance, the preferred solution is to use uncooled para-
metric amplifiers in such stations. Such stations will of course
have a much higher noise temperature than the satellite ground
stations used in international communications, but it would appear
preferable to use more bandwidth in the satellite rather than ad-
ding complexity to these satellite ground stations.

     Figure 10 shows the location of some possible sites for sa-
tellite ground stations in northern Canada. Even if, in the not
too distant future, satellite ground stations are built at all of

the locations shown, there will still be a number of smaller communities where the installation of a satellite ground station will
not be economical in relation to the size of the local population.
It is, however, expected that a number of communities in the
neighbourhood of a satellite ground station may be served econo-
mically over VHF one or two-hop systems capable of carrying television programs. It is further expected that when a number of
communities throughout the Canadian Arctic have been equipped with
satellite ground stations, most base stations for HF will be moved
farther north, and greater emphasis will be placed on the development of suitable facilities for providing reliable communications
to the nearest community equipped with a satellite ground station.
It is possible that low powered, low cost, one-hop scatter systems
may be suitable for this purpose in some regions.

MINES AND MINERALIZED AREAS

QUEBEC AND LABRADOR

1	Iron Ore Co. of Canada *	Iron
2	Quebec Iron & Titanium *	Iron, Titanium
3	Campbell Chibougamau Mines *	Copper, Gold
	Copper Rand Chibougamau Mines *	Copper, Gold
	Merrill Island Mining *	Copper, Gold
4	Opemiska Copper Mines *	Copper, Gold
5	Coniagas Mines *	Zinc, Lead
6	Mattagami Lake Mines	Zinc
7	Quebec Cartier Mining *	Iron
8	Wabush Iron Ore	Iron
9	Iron Ore Co. of Canada	Iron
10	Albanel Minerals	Iron
11	Great Whale Iron Mines	Iron
12	Belcher Iron Mines	Iron
13	Murray Mining Co.	Asbestos
14	Ungava Iron Ores	Iron
15	Tache Lake Mines	Copper
	M.J. Boylen	Zinc
	McIntyre Porcupine Mines	Copper
16	Canadian Dyno Mines	Copper

ONTARIO

1	Pickle Crow Gold Mines *	Gold
2	Campbell Red Lake Mines *	Gold
	Cochenour Willans Gold Mines *	Gold
	H.G. Young Mines *	Gold
	Madsen Red Lake Gold Mines *	Gold
	McKenzie Red Lake Gold Mines *	Gold
	New Dickenson Mines Ltd. *	Gold
3	Nickel Mining and Smelting	Nickel
4	Annaconda Iron Ore	Iron
5	Little Long Lac Gold	Gold
6	Steep Rock Iron Mines	Iron

MANITOBA

1	Sherritt Gordon Mines *	Nickel, Copper, Cobalt
2	International Nickel *	Nickel
3	Fourty-Four Mines *	Gold
	San Antonio Gold Mines *	Gold

SASKATCHEWAN

1	Eldorado Mining and Refining *	Uranium
	Gunnar Mines *	Uranium
2	Waddy Lake Mines	Gold
3	Augustus Exploration	Gold

ALBERTA

1	Caribou Mountain Area	Oil
2	Athabaska Tar Sands	Oil
3	Worsley Area	Gas
4	Premier Steel	Iron
5	Rainbow Lake	Oil

BRITISH COLUMBIA

1	Fort Nelson Area	Gas
2	Blueberry-Milligan-Boundary Lk Area	Gas & Oil
3	Cassiar Asbestos Corp. *	Asbestos
4	Granduc Mines	Copper

YUKON

1	Yukon Exploration Ltd. *	Gold
2	Yukon Consolidated Gold Corp. *	Gold
3	United Keno Hill Mines *	Silver, Lead, Zinc, Cadmium
4	Western Minerals	Oil
5	Conwest Exploration	Silver, Lead, Copper, Zinc
6	Dominion Explorers	Copper

NORTHWEST TERRITORIES

1	Consolidated Discovery *	Gold
2	Consolidated M & S *	Gold
	Giant Yellowknife *	Gold
3	North Rankin Nickel Mines *	Nickel
4	Canada Tungsten Mining	Tungsten
5	Fort Liard Area	Gas
6	Pine Point Mines	Lead, Zinc
7	Consolidated Discovery	Gold
	Consolidated Northland	Gold
8	Taurcanis Mine	Gold
9	Canadian Nickel	Gold
10	Nahanni Sixty Syndicate	Silver
11	Norman Wells	Oil
12	Texas Gulf Sulphur Co.	Lead, Zinc
13	Melville Island	Oil
14	Mary's River	Iron

* Producing Mines

Table I. Mines and mineralized areas.

In order to develop the resources of the Canadian Arctic both improved communications and transportation will be required. The prospects now appear very bright for providing high quality telecommunications to Arctic communities, and it is to be hoped that improvements in transportation technology will also soon make it possible to extend the shipping season for heavy transport on an economical basis. The future prospects for the Canadian North have never been better, and I believe we have reason to be optimistic about the outlook for development in this vast area.

Acknowledgement: The author wishes to thank the Bell Telephone Co. of Canada and the Canadian National Telecommunications as well as colleagues in the Telecommunications and Electronics Branch of the Department of Transport for assistance received in preparing this paper.

Figures 1, 2 and 3 were originally prepared by the Trans-Canada Telephone System. Minor additions have been made by the author to update the illustrations.

Figures 8 and 9 have been reproduced without change from the IEEE Spectrum for December 1966.

SOME EXPERIENCES WITH MILITARY COMMUNICATIONS IN THE CANADIAN ARCTIC

P. J. Pratley

Canadian Forces Headquarters

Ottawa, Canada

Abstract: With the expanding civilian and military activity in Canada's Northland there has been an increasing demand for radio communications in this area. This contribution reports on the development of the military network in the regions concerned. Results from various experiments conducted to test antennas, equipment and optimum use of allocated frequencies are displayed. In particular the experience with three circuits, influenced in different manners by the auroral belt, is discussed.

1. INTRODUCTION

It is a distinct privilege, and a pleasure, for me to be able to attend this symposium, and to have the opportunity to hear so many papers of interest on so wide a range of subjects, having so great a bearing on the future of the business in which I am bound up, namely, communications of one sort or another. I am pleased, particularly to have been asked to present a paper recounting some of our operational experiences with communications activities in Canada's north, and hope that they will prove interesting, or at least diverting, to some of you.

Mr Ringereide has given you some background on our north, touching on the area it covers, its population and its resources, and describing the development of its transportation systems and what has been, in general terms, the attendant growth of its communications.

It has, of course, been the vast expanse of Canada's North-

land, together with the unpredictable behaviour of radio waves in
the transauroral area which have posed the principal problems in
the development of military communications in the Arctic. Difficult terrain and severe weather conditions coupled with short periods of two or three months in the year when construction work
can be carried out have not made the job of keeping abreast of
technological advances any easier. However, the need and tenaci-
ous initiative have overcome many almost insuperable hurdless and
in the past four decades military communications across the tundra
have made substantial advances in both quality and quantity.

2. THE NORTHWEST TERRITORIES AND YUKON SYSTEM (NWT & Y)

The Yukon and Northwest Territories (Figure 1) include all
the mainland and islands north of the 60th parallel from Hudson
Bay to Alaska, an area of one and a half million square miles,
comprising about two-fifths of Canada. After the first World War
Canada became very interested in opening this immense but sparsely
populated region. The Territories at that time depended largely
on river transport, infrequent coastal vessels, and dog teams for
their communications. Commercial development and communication go
hand in hand, but development in that era had not progressed to a
stage where commercial companies could afford to install a communication system. Accordingly, it became necessary for some branch
of the Canadian Government to undertake the task.

In 1923 the Canadian Army commenced construction of an extensive radio telegraph network in Western Canada, the Yukon and

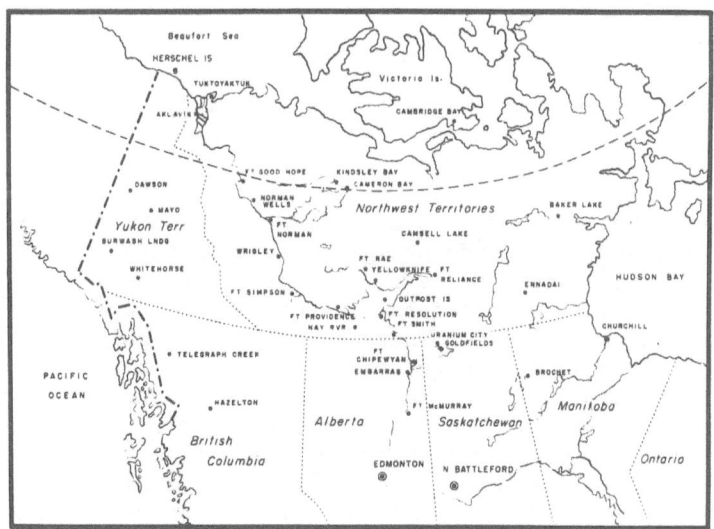

Figure 1. The Northwest Territories and Yukon Radio System.

Northwest Territories. The undertaking originally called for the
erection of stations at Dawson City and Mayo Landing with the
southern terminus at Edmonton. Originally this network was intended to provide a basic administrative link to the outside and,
in addition, was utilized to gather meteorological information.
Gradually, as the northland opened up to the advancing civilian
population, the NWT & Y system took on a certain commercial aspect
with the handling of civilian traffic. The network expanded rapidly and stations were installed at Herschel Island, Fort Smith
and Fort Simpson. The volume of traffic continued to increase and
further expansion included stations at Fort Resolution, Fort Norman, Lindsley Bay and Port Radium. Stations at Camsell River,
Fort Chipewyan, Fort Rae, Whitehorse and Burwash Landing were added later. At the peak of its operations the NWT & Y system com-
prised 24 stations throughout the north. The first transmitters
had a power of 120 watts, but these were gradually replaced with
more powerful equipment until the manual morse outstations were
using 500 watts in the 2 to 16 MHz range for short haul communications, and the main stations at Edmonton, Fort Smith, Fort Simpson
and Norman Wells were equipped with ten kilowatt transmitters for
radio teletype operation on L.F. frequencies in the 117 to 124 kHz
range for distant communications. During the first year of operations the NWT & Y handled 3360 messages. When the network, then
consisting of 19 stations, was transferred to the control of the
Department of Transport in 1957, the annual quantity of traffic
handled was in excess of 2,523,000 messages which included some
300,000 messages a year for the Distant Early Warning or DEW Line
pending completion of the DEW Lines own permanent communication
facilities which occurred at about that same time.

3. AERIAL SURVEY

During the late nineteen-forties and early nineteen-fifties
the Royal Canadian Air Force conducted an aerial survey of Canada.
Every square mile of the country from coast to coast and from the
southern boundary to the great uncharted frontier of the arctic
islands was photographed. To accomplish this mammoth undertaking
it was necessary to provide navigational aids throughout the entire area so that accurate navigation could become a matter of
routine. Thus were born dozens of temporary HF radio stations
throughout the wilderness. These communication stations, operating in the 2 to 16 MHz range, were flown in to predetermined remote sites by amphibian aircraft set up and operated from portable
auxiliary power units. They were connected in an interlaced network to assist in the survey by reporting weather conditions and
by providing fix locations for the airborne crews. Most of the
operations were conducted by manual morse but some voice modulation was used.

Canada began to realize the enormous potential of its north and elaborate plans were developed to encompass the area with a network of communications such as had been pioneered by the NWT & Y system many years before. Hitherto obscure locations like Aklavik, Cambridge Bay, Alert, Frobisher Bay, Resolute Bay, Fort McMurray and a host of others suddenly became household words. The development of the Canadian North had begun and with it the trials and tribulations, and the excitement and satisfaction of building permanent communications throughout the north had become an economic necessity and a reality.

4. PROPAGATION

By this time Radio Teletype and automatic equipment had come into common use, but morse operators were still required - they alone could keep traffic moving when high frequency blackouts and heavy atmospheric conditions on low frequencies rendered radio teletype circuits useless for hours, and sometimes days, at a time. By this time too, it was realized that a study of propagation and atmospheric phenomena must be undertaken and the prospects for new methods and systems explored if efficient military communications were to exist in the arctic. Thus the Canadian Armed Forces acknowledged the need for the experimental stations to examine these problems and to obtain data on which to base decisions as to the most suitable systems to develop to provide the desired standard of communications for strategically located military installations.

5. SUPPLEMENTARY RADIO SYSTEM

The Canadian Armed Forces maintain several radio stations throughout the arctic. In addition to their research into arctic communications these stations play an important role in the Search and Rescue operations of the North American continent and its environs. Bases are located at Whitehorse in the Yukon Territory, Inuvik in the Mackenzie River delta, Alert at the northernmost tip of Ellesmere Island, at Frobisher Bay on Baffin Island on the west coast, in Hudson Bay and on the east coast. (Figure 2). These stations form a High Frequency Direction Finding net in support of Search and Rescue. However, the role that I will discuss here is that of Arctic Research.

6. ARCTIC RESEARCH

In this field of endeavor we concern ourselves primarily with the conducting of tests between various locations to determine from practical experience, the best types of antennas, equipment

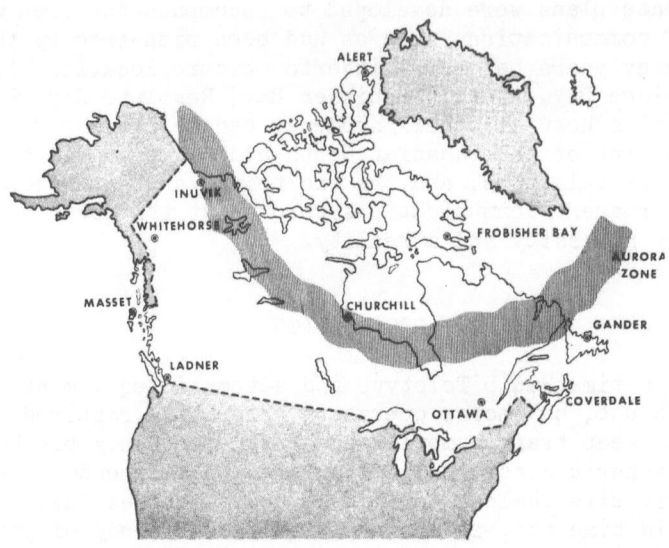

Figure 2. Canadian Armed Force Radio Stations supporting Search
           and Rescue.

configurations and the most efficient portion of the frequency
spectrum, within our assigned frequencies, for use in high
northern latitudes at given times of the year. Prior to conducting major tests, consulting engineers are often called upon to
conduct surveys in the light of known data. Their recommendations
are considered, the test programme drawn up and subsequently the
practical results are compared with the engineers' predictions.
Careful analysis of these results often leads to reconfiguration
of various systems and to a consequent improvement in communications throughout our various networks.

    Radiated power varies from 750 watts to 5 kilowatts utili-
zing both single frequency transmission and simultaneous transmissions on two frequencies. Rhombic and folded dipole antennas predominate in our work with distances varying from 700 to over 4000
kilometers. Some of our rearward communications utilize troposcatter facilities with amazingly good results. Evaluation of a
circuit from Alert to Ottawa, a distance of 4200 kilometers, using
Low Frequency, Troposcatter and Microwave, indicates very little
effect by transauroral hop and atmospheric conditions. This circuit maintains a fairly constant efficiency of 97 percent regardless of seasonal changes or fluctuations in atmospheric phenomena.
Auroral conditions apparently have no effect on the portion of the
circuit covered by microwave although it was originally suspected
that there may be some reasonably significant disruption on the

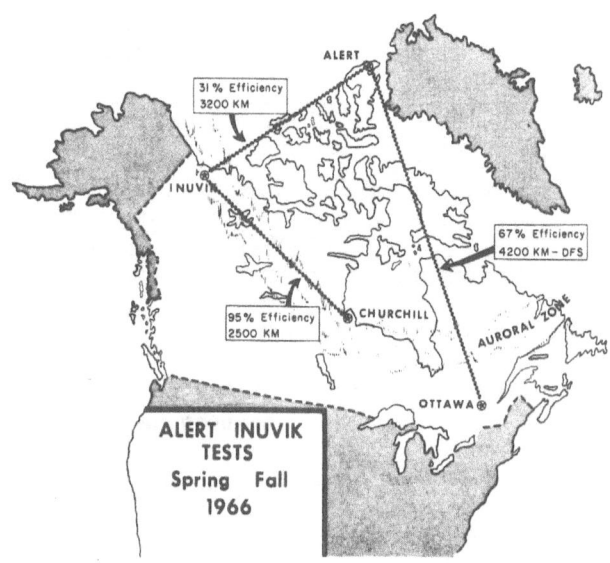

Figure 3.   Location of Test Circuits.

microwave circuitry due to auroral activity and thermal inversion.

7.   TEST RESULTS

I intend now to discuss briefly three particular cases of
circuits we have tested, or operated through necessity, in recent
years in attempting to establish reliable and high speed communications links through the Arctic. These three, displayed in Figure 3, have been selected as interesting examples because each was
conducted within a distingly different influence from the Auroral
Zone.

Of the many tests conducted by the Armed Forces not all have
been successful. An illustration of this lack of success is provided by this first case which was an attempt to establish a link
between Alert and Inuvik, a hop of 3200 kilometers in a Northeast-
Southwest direction, chiefly inside the Auroral Zone but penetra-
ting it at the Inuvik end. Power at 5 kilowatts was diffused by
rhombic antennas suitably oriented at Alert, and folded dipoles
were used for reception at Inuvik. On the return leg similar antennas were used but Inuvik radiated a power of only 1 kilowatt.
The attempt to establish this long haul radio teletype link was
basically unsuccessful but some interesting data was collected.

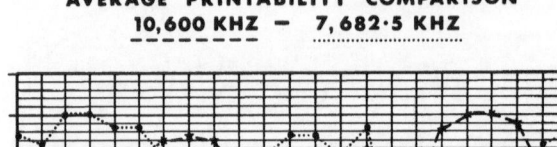

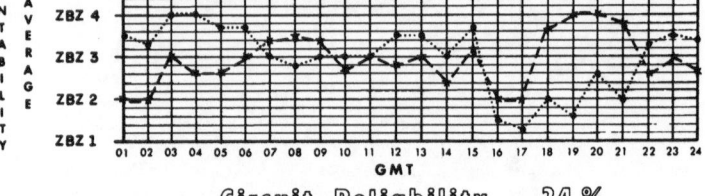

Figure 4. Alert-Inuvik Test March-April 1966.

It was found that there were extensive periods of both circuit ex-
cellence and unpredictable blackout. This resulted in a variable
reliability with peak efficiency periods occurring between 0300 -
0600, 1200 - 1500 and 2200 - 2400 Greenwich Mean Time on 7 MHz
while between 0600 - 0900 and 1800 - 2100 were prime times on 10
MHz (figure 4). The overall efficiency of the circuit was a somewhat dismal and, from a practical standpoint, unsatisfactory 24%.
In the fall of the same year a further attempt was made for a period of sixty days. This time simultaneous frequency operation
was used on frequencies in the 4, 7, 9 and 11 MHz bands (figure 5).
The same transmitters and antennas were used. The peak period of
reliability on this occasion occurred between 0600 - 0900 GMT,
however, the readability varied constantly as compared with the
spring tests during which extended periods of printable signals
punctuated with blackouts had been experienced. Overall circuit
reliability of the fall test was 31%, a slight but, not really
significant, improvement.

     In contrast to that case is the link between Inuvik and
Churchill a distance of some 2500 kilometers in a Northwest -
Southeasterly direction and at a somewhat lower latitude. This
circuit closely parallels and transposes the auroral belt. Using
similar equipment, power and antennas with simultaneous frequency
keying this circuit, operating in the 7, 9, 11 and 13 MHz bands,
has, over a period of twelve months, sustained a circuit efficiency approaching 95%. The full significance of the disparity between the two circuits, one running parallel to and within the
auroral belt and one at an obtuse angle to it is not known.

A third High Frequency circuit to which we have devoted much

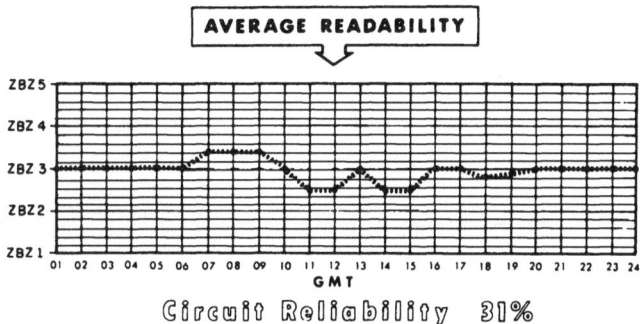

Figure 5. Alert-Inuvik Test September-October 1966.

attention is the link between Alert and Ottawa. Three wire rhombic antennas are used for transmitting and receiving at each terminal and both stations use a power of 5 kilowatts. This circuit utilizes double frequency shift radio teletype operation over a distance of 4200 kilometers. Frequencies in the 7, 10, 13 and 16 MHz bands are used. The mean efficiency over a twelve month period during 1966 was 67%. It is of interest to note that 16 MHz was in use only 5% of the time but had an average efficiency of 87% while 7 MHz was used 40% of the time but efficiency was only 52%. This is the longest circuit that is operated into the north by the Department. Although it also runs perpendicular to the auroral zone, results are much better than on the circuit tested to Inuvik.

8. CONCLUSION

We have conducted many tests similar to those described here and are preparing to conduct many more. Though changes of command and rotation of personnel will bring many new faces to the north, our job will remain the same, to man the stations of the Search and Rescue net and probe the ionosphere in a continuing effort to improve communications for the welfare and protection of the northern populace and persons in distress.

SURVEY OF THE USER PROBLEMS AND EXISTING COMMUNICATION FACILITIES IN THE NORWEGIAN ARCTIC

J. R. Veastad

Norwegian Telecommunication Administration

Oslo, Norway

Abstract: This paper surveys the arctic communication facilities funded and operated by the Norwegian Telecommunication Administration. Users problems encountered are described. Remarks pertaining to the usability of the applied frequencies are included.

1. INTRODUCTION

The following presentation is based upon three different sets of arctic communication facilities for which the Norwegian Telecommunication Administration is responsible. The circuits are:

1) The main radio connection Spitsbergen-Norway

2) An HF link to serve the satellite telemetry station at Kongsfjord

3) The HF en-route Radiotelephony Network, constituting a part of the Polar East Network.

2. GENERAL REMARKS

As the radio circuit Norway-Spitsbergen has been operated for the last 50 years, one should perhaps expect to find sound scientific reasons for the frequency usage and good explanations for the problems which must have been tackled. Unfortunately this is not the situation, for several reasons. As is well known, the problems encountered in transmission systems based upon the polar ionosphere may be of a rather complicated nature. The circuit concerned has been established and operated on restricted funds. Furthermore the population on the Spitsbergen Island, essentially miners, seldom have required urgent communication. Usually they have been satisfied with the services offered to them.

But with increasing flight traffic in the polar area, and with the establishment of an expensive satellite telemetry station on the Island, we are faced with a new situation with emphasized requirements for regular radio transmissions.

3. DESCRIPTION OF CIRCUITS AND USER PROBLEMS

3.1 The Harstad-Longyearbyen HF Link

The northern terminal is located at Longyearbyen at 78°N, 15°E (Figure 1). The station buildings are placed at the bottom of a valley shielded by some rather steep mountains. The direction of transmission to Harstad passes along another narrow valley branching off at the station area. The screening elevation angle is about 9°. Rhombic antennas are applied both for transmission and reception. Channel transmitters are used to give four pre-tuned channels, each with an output of 2-2.5 kW. For telephony amplitude modulated double sideband transmissions, A3, are used. Telegraphy is based on frequency-shift keying, F1. The receiving equipment consists of ordinary communication receivers for A3, F1 and for unmodulated telegraphy, A1.

The station at Harstad is positioned at 68°N and 16°E. The horizon is low. Radio equipment and antennas are of the same type as at the Spitsbergen terminal.

Figure 1. Northern circuit terminals.

A considerable part of the connections have been established
at daytime hours between 0900 and 2100 local time. As a rule two
transmitters at each site are started and keyed by morse-slip.
Based on the observation of the signal strength the best channel
is chosen. During the first daytime hours usually the frequencies 6995/7690 kHz are selected. At mid-day the combinations
9255/7690 kHz or 6995/9092 kHz are used. In the afternoon and
evening hours the same frequencies used in the morning usually
give a satisfactory connection. The operators state that there is
no marked seasonal variation in the frequency usage, but there is
a slight tendency to use the lower frequencies 5805 kHz and 3717
kHz during the wintertime.

In an attempt to establish the statistical distribution of
the traffic on the allocated working frequencies an investigation
of the operating reports for the last 11 years was undertaken. It
appears that the information in these reports leaves much to be
desired as to preciseness and accuracy in description of the running transmission conditions.

According to our best estimate the different groups of frequencies have carried the following percentage of the traffic:

7 MHz	60%
9 MHz	32%
3.7/5.8 MHz	7%
11-12 MHz	1%

Table 1. Distribution of traffic on allocated frequencies.

It should be noted here that until recently, when the requirement for communications became urgent, the highest frequencies, 11-12 MHz, have not been tried as extensively as the others
in the efforts to establish contact. This is partly the reason
that these frequencies have carried but a small fraction of the
total traffic.

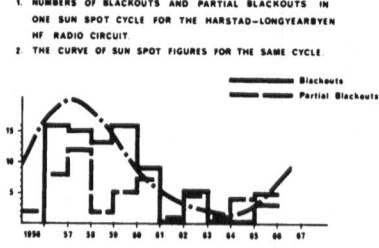

Figure 2. Occurrence of blackouts during sunspot period.

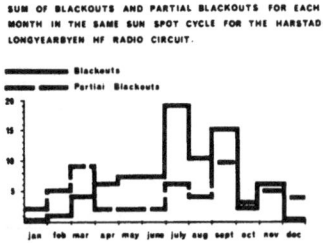

Figure 3. Annual distribution of numbers of blackouts.

Figure 2 shows "blackouts" and partial "blackouts" during one sunspot cycle for the Harstad-Longyearbyen HF-circuit. The full line indicates the number of total "blackouts" and the dashed line occasions of partial "blackouts". Partial "blackouts" in this case means that the traffic has been limited to less than three hours per day. The dashed-dotted line displays the sunspot figures for same cycle. The presentation in Figure 3 gives the sum of blackouts and partial blackouts for each month in the same sunspot cycle. It is noted that there is a pronounced maximum during the summer season.

3.2 Communications for the European Space Resarch Organization

During this year a new satellite telemetry station, designed by the European Space Research Organization (ESRO), will be established at Kongsfjord near Ny-Ålesund. The Norwegian Telecommunications Administration was asked to plan and construct the HF radio stations to perform telephony and teleprinter-connections from Ny-Ålesund to Southern Norway. From there the circuits are carried on by ordinary PTT lines to ESTEC in the Netherlands. Due to the importance of regular data transmission from the telemetering station ESRO specified the rather strict requirement of not more than three hours of blackout per year! Equipment cost should be kept at minimum. The estimate for an HF circuit, essentially similar to the Harstad-Longyearbyen link, was worked out and presented to ESRO together with a statistical data analysis of blackout periods during the last three years.

The radio station at Ny-Ålesund is situated near Kongsfjorden at a position 78°N, 11°E. Two rhombic antennas are used for "space diversity" reception with a single rhombic antenna used for transmission. The direction of transmission is somewhat shielded by nearby mountains so that the radiation characteristics are essentially based on two hop F2 layer propagation. The free angle to the horizon is 12° in the transmitting direction and 10° and 4°

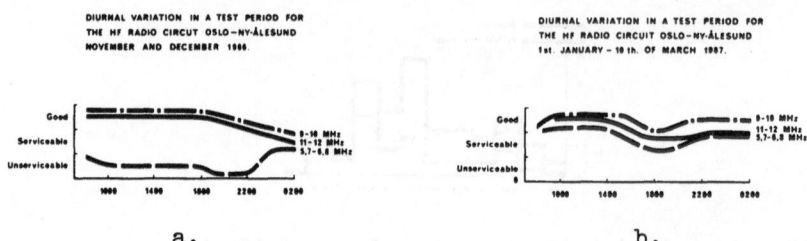

a.   b.

Figure 4, a and b.   Results from quality tests of the circuit
                     Oslo - Ny-Ålesund.

for the receiving antennas.

     The transmitting equipment constitutes a 10 kW SSB/ISB group
for 3 HF channels. The receiving equipment is designed for "space
diversity" and equipped with crystal controlled frequencies.

     Existing transmitting and receiving locations at Jeløy and
Ski, outside Oslo, had to be implemented with equipment equivalent
to that at Ny-Ålesund. The distance from Ny-Ålesund to the stations in southern Norway is about 2200 km. The circuit is orientated approximately north-south.

     As this radio circuit is still not put into regular operation,
some tests have been carried out since October last year. The
purpose of these tests is to obtain a practical evaluation of the
usability of the allocated frequencies. No exact measurements of
signal strength have been performed.

     During the tests a combination of telephony and telegraphy,
A9B, with a 26 db carrier has been used in both directions. The
lower side band carries the telephony channel, and the upper side
band a teleprinter channel.

     Figure 4a presents the derived usability of three frequency
groups, 5.7-6.8 MHz, 9-10 MHz and 11-12 MHz, based on diurnal variation for November and December 1966. A rather rough quality
division, graded "Good", "Serviceable", and "Unserviceable", was
employed. Figure 4b shows the diurnal variations in the period
Januar-March of this year. No test has been carried out during
nighttime hours from 0200-0700 local time. The curves are based
on the overall estimate of the results obtained during each month
in the test period. There are several single events, occurring at
irregular hours, departing considerably from these mean values.

3.3 HF en-route Radiotelephony Network in the North Atlantic and Polar Region

The third part of this presentation is concerned with the HF r/t network for ground-air communications in the North Atlantic Region, the Norwegian Sea, and the Polar area, involving air routes established by different companies flying from Europe to Anchorage and Greenland, as well as the Aeroflot routes Moscow-Havana and Moscow-Montreal.

Our stations in this network are located at Bodø and at Isfjord. The station at Bodø is equipped with 2.5 kW fixed-channel transmitters and ordinary communication receivers. It does not have the facility to select directive antennas for either the transmitting or the receiving side. A 63 metre ground wave antenna is used for the lowest frequency. In addition, half-wave, delta matched, electrically-tuned transmitting antennas are utilized.

According to the international frequency plan, the frequency family D2 consisting of 2868, 5626.5, 8913.5, 13324.5 kHz is allocated to this region. Based on experience during the last 10 years at Bodø air-radio, the operators state that the frequency 5626.5 kHz gives the best connections towards the west and southwest, i.e. in the direction of Iceland and the southern parts of Greenland along the auroral zone. See Figure 5.

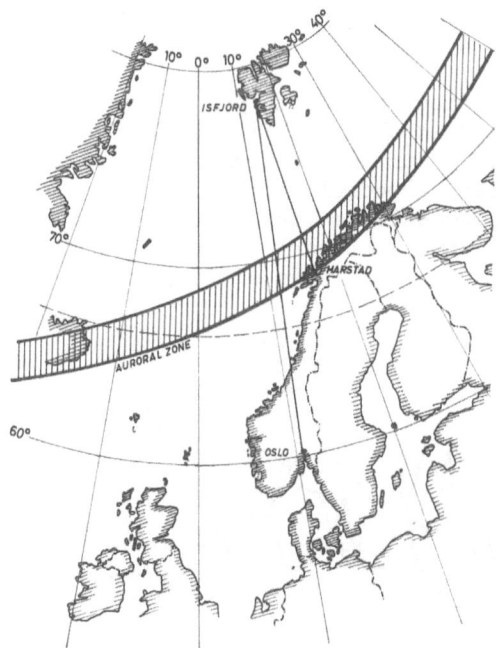

Figure 5. Area covered by Polar East Network.

The frequency 8913.5 kHz is good for covering the area beyond 75°N over the polar area and towards the west over northern
Greenland and under good conditions even as far as New York,
Tokyo and Manilla. In a restricted zone between 70 and 75°N and
from 15°E somewhat west of 0° longitude, 8.9 MHz gives no connections at all. This zone is, as mentioned above, covered by 5.6
MHz. During bad radio conditions one has to rely on the ground
wave frequency 2868 kHz which at times reaches airplanes over
Alaska.

Total blackout (in this connection defined as conditions in
which communication is impossible between stations and planes in
the air, using this frequency family) has occurred only twice
during the last 10 years.

For a period from 1964 through July 1966 the number of contacts including interceptions shows the following distribution:

5626.5	kHz	45%
8913.5	kHz	30%
2868	kHz	25%

Table 2. Distribution of established contacts on available
          frequencies.

The frequency 13324.5 kHz, which at times has provided good
contacts, has been of restricted use due to an inappropriate local
installation.

The traffic distribution in the Polar East Network for the
above-mentioned period is shown in Figure 6. 2868 kHz (the dashed
line) shows low traffic during the summer months, April-September.
The higher frequencies, especially 5626.5 kHz, represented by the
dashed-dotted line, indicates a marked increase from autumn 1965
to the present. The same statement is valid for 13324.5 kHz,
which has been used a lot more this winter. The frequency 8913.5

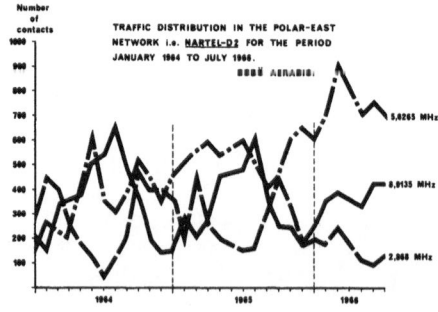

Figure 6. Traffic distribution in Polar East Network.

kHz (the full line) has a clear maximum in traffic during the summer months.

4. CONCLUSIONS

During the lecturing here at Finse it has become evident to me that the problems encountered in the field of arctic communication are not only of technical nature, but organizational and human as well.

As you have seen, a small telecommunications administration is able to establish HF radio circuits capable of meeting reasonable demands, even to a degree that some of you have to rely on it when flying over the Pole. My colleagues and I are grateful for having been invited to this symposium, and I hope that it has inspired some of us to find a common language for future collaboration.

Acknowledgement: The helpful asistance of Mr A. Lingaas, Mr. T. Øwre and Mr H. Dahl in providing technical and operational data is gratefully acknowledged.

DISCUSSION

In the discussion on Ringereides' presentation it was stated
that:

1) Mostly simplex systems are employed for communications in
   Northern Canada.

2) The "spectrum of distances" for HF communications in Canada
   falls below 500 km.

Attention was drawn to the economical aspect in the design of
arctic communications. It is of primary importance to use efficient antennas and low power transmitters. Thermal electric generators are now available, operating at a power cost of about 6
c/kWhr. With conventional equipment the power cost is about 16
c/kWhr.

Some discussion took place on the problems of reliability and
maintenance of communication equipment in remote areas. Generally
in the polar regions there is a need for rapid exchange of information of varying propagation characteristics. It was pointed out
that information delayed more than one hour is not very useful.

It has been found that FSK increases the reliability of communications. In Canada the trend is towards voice communication.

Probst commented upon procedures established by US Military
in the Pacific to forecast MUF a few hours in advance. In particular he was concerned with an arrangement worked out for the US
HF-network in the pasific area. Up to date vertical incidence
data are used. The data are processed at ESSA, Boulder.

A basic drawback of this system is the time-delay of more
than one hour inherent in the exchange of information between the
areas concerned in South-East Asia and the processing centre.

It was suggested by Wright that a pamphlet be prepared stan-
dardizing the terms used by scientists, engineers and operators in
describing the performance of communication systems.

Panel Discussion of Future Developments

SUMMARY OF SESSIONS

T. R. Hartz

Defence Research Telecommunications Establishment,

Ottawa, Canada

The more formal sessions of the NATO Advanced Study Institute on Ionospheric Radio Communications in the Arctic that have oc- cupied us throughout the past week are now almost completed. Fol- lowing this summing-up session, we will return to our homes and each of us will try to assess the essentials of this meeting and to plan and set in motion future activities concerning Arctic Communications. It is, of course, not possible now to say what most of these activities will be, but as a first step toward their formulation we will try to summarize the main points of signifi- cance that have arisen during the preceding sessions. Each parti- cipant will likely have his own opinion as to what were the high- lights of the Study Institute; here this author can only present his own biased viewpoint which, hopefully, includes a majority of the main essentials and is not too cursory or oversimplified.

Essentially, the meeting has involved three main groups of people, who can be identified as (1) ionospheric specialists, (2) technical or systems specialists, and (3) communications users or operators. There was a time when these groups could not be clear- ly distinguished because one person might be engaged in several functions, but the present situation seems to be that the diffe- rent functions are much more separated and isolated. This meeting has brought the three groups together, cut them off from outside influences, and re-established communications between them. Each group has been heard from, in formal presentations and in informal discussions; let us, then, review their contributions.

The first sessions were dominated by the ionospheric specia- lists, who tried to describe the sort of an ionosphere one has to

deal with in the Arctic. We heard about the polar-cap and auroral
disturbances that introduce anomalous ionization in certain parts
of the high-latitude ionosphere; we discussed radio wave absorption and blackouts; we dealt with the distribution of ionization
under quasi-normal times, and discussed HF propagation modes and
the predictability of modes; we discussed LF and VLF propagation,
as well as VHF scatter types of communication systems; and we had
a spontaneous session on high latitude sporadic E. Indeed, a
wealth of detail on all these physical phenomena was presented and
discussed; most of this can be found in the proceedings and will
not be gone into here. During the discussions, however, one heard
the occasional remark or query from the other groups; "What does
this mean to the communicator"? "What is the signal dispersion"?
"What about reliability"? These and other questions made it clear
that the ionospheric specialists were being challenged as to
whether their fund of empirical knowledge on the ionosphere could
be put to more practical uses. Could they predict the behaviour
of the ionosphere at some future time, or, on the basis of a measurement at one place, could they specify the ionospheric conditions at other locations?

This term 'prediction' was discussed in four different con-
texts, and it is worthwhile enumerating these:

(1) the prediction of ionization distributions, and therefore
     operating frequencies, on a propagation path for purposes of
     planning communication circuits and equipment requirements;

(2) the prediction of operating frequencies on a particular path
     so as to guide the operator in his day-to-day and hour-to-
     hour operations;

(3) short term forecasts of specific conditions on one or more
     propagation paths, (for instance, the prediction of operating
     conditions for the next hour on the basis of the current con-
     ditions rather than on the monthly statistical prediction);
     and

(4) the forecast of storms or major disturbances that are global
     in nature, along with their effects on specific circuits.

The discussion suggested that a great deal can be done in regard to these different 'predictions', but as yet much remains to
be done. The prediction of frequencies for planning purposes in
the Arctic regions does not seem to be a major problem, particularly if one takes note of the Canadian methods for high latitudes
and their new prediction scheme for the F1 layer. The prediction
of conditions for operations purposes is less certain; monthly medians can be specified fairly well but the variance is not yet

predictable, nor is the signal strength. At low latitudes, at least, it seems likely that short term forecasts of operating conditions can be made; however, a lot of effort and money is required for this, and it is debatable whether the same sort of thing could be even partially successful in the Arctic. As for the forecasts of storms, it looks as though this might be possible in many cases; there are several groups now making such forecasts with variable success, and we have had some discussions here of likely storm precursors, but the fact remains that a reliable warning service is not available.

A question that was not resolved in the discussions is, to what extent could and would the operators use the forecasts if they were made? This would seem to lie at the root of any decision as to how far to go in making predictions, and for what purpose.

Turning our attention now to the next phase of the meeting and the second group of specialists, there were a couple of very interesting sessions on communications systems and techniques. In these the speakers described the propagation path in terms that are readily identifiable with equipment functions, such as the time dispersion of the signal, rather than in terms of the actual distribution of ionization along the path. A mathematical and statistical description of the signal can permit the simulation of the HF communications circuit on a computer. The implication was that the system specialist should be at least as capable of predicting future communications on this basis as the ionospheric specialist is on the basis of his knowledge of the ionization. The advantage here is that the equipment built for the transmitter and receiver terminals can be matched to the propagation characteristics of the signal much more readily than it can to the electron density distributions in the ionosphere. From the discussion it became evident that this is a very active current field of endeavour and that already there have been very substantial developments along the lines indicated: methods are known for combating time dispersion, or frequency dispersion, but as yet these two cannot be handled simultaneously.

The potential capability exists for the communication systems of the future to adapt automatically the information to be sent to the carrying capacity of the channel under most conditions (although blackouts may still present some problems). We heard about particular systems and developments that fit into this context; the CHEC system is one example wherein there is an evaluation of the communication channel, and the system under development at S.T.C. can adapt to several propagation conditions. Details on these, and other systems discussed, can be found in the proceedings.

Yet another facet of adaptive procedures that was discussed is the possibility of adapting the ionosphere to suit the communications requirements under particular situations. We heard of developments for the purpose of producing an artificial ionosphere by means of rocket seeding of the upper atmosphere near the mid point of a specific circuit.

The sessions on systems became enmeshed with those dealing with operations, and we had a number of papers on existing communications in the Arctic; notably, in Canada, Greenland, and Norway. We were reminded of the physical situation in these regions, the sparse population, and the reasons for economy in the system installations. In assessing the problems of the users in the Arctic, one must keep in mind that they need to communicate. These requirements can be considered by grouping them into four categories according to the volume of traffic and the degree of urgency (a T/U index, with T for traffic and U for urgency) as follows:

(1) The L/L category, for which there is low traffic and low urgency. The communicator in this case can afford to wait until propagation conditions are optimum. Because the traffic is low the system is relatively simple and inexpensive. A fair amount of Arctic communications is in this category and high reliability can be readily achieved.

(2) The H/L category, for which there is high traffic but low urgency. Again the operator can wait for suitable propagation conditions, although the high traffic requirement might prohibit waiting too long. The volume of traffic calls for expensive systems, so that this category includes trunk communications between major centres or bases. High reliability would appear to be somewhat harder to achieve than in the case of the L/L category.

(3) The L/H category, for which there is low traffic but high urgency. Included here might be various military requirements; for instance, a single aircraft of Coastal Command, or a six-main infantry patrol; a number of civil requirements concerned with safety and survival, or with aircraft control may also fit in this category. The reliability for this category is much more difficult to achieve, but a number of techniques were discussed that could prove helpful in particular applications; these include HF slow-information-rate systems, and LF and VLF systems which now can be produced in manageable packages with the aid of miniature electronics.

(4) The H/H category, for which there are high traffic and high urgency requirements. Many military communication systems and some civil systems fall in this category. It will be obvious that high reliability in this instance is very diffi-

cult and very costly to achieve. Some of the complex systems
that we heard about at this meeting were the CURTS System
described by Probst and the Multi-Mode System that Kulinyi
described.

From the point of view of the user, we can summarize communications in the Arctic in terms of four different requirements.
For each of these categories reliable communications appears to be
possible, but the cost of achieving reliability is very much dependent on the T/U index. Any one who really wants a highly reliable system must be prepared to put hard limits on his traffic
and urgency requirements in order to keep within a realistic budget.

During the meeting, and particularly in the discussions, a
number of issues were raised to which more consideration might
well be given. Some of the more important are recommended to you
for your further thought and appropriate action:

(1) The matter of interference was mentioned a number of times as
     a limitation of communications, and it might be well to con-
     sider spectrum management and whether serious efforts should
     be devoted to this subject along lines different from those
     currently followed.

(2) The duplication of HF communication systems in many regions
     of the Arctic is of some concern, particularly in cases
     where relatively few people are involved. The population ge-
     nerally is sufficiently low that some optimization of traffic,
     urgency, reliability, and cost per unit of population seems
     possible and desirable.

(3) The potential of anomalous propagation modes at high latitu-
     des does not seem to have been exploited to the full. There
     would appear to be a requirement for some simple system that
     tells an operator what frequency or mode is possible, and
     which can adapt quickly and efficiently to that mode -- in-
     cluding the facility to change antenna directivity if an off-
     path mode can support propagation.

(4) In the interests of long period synoptic studies, can the
     ionospheric specialists decide on some limited number of de-
     sirable parameters that might be read out of the receiving
     equipment of an adaptive communication system, which would be
     meaningful in terms of the upper atmospheric ionization?

There are probably many other things that might be included
in a summary of the Study Institute, and each of you should be
able to add to this list those missing items that you found significant. So far, one of the most beneficial aspects of the meeting

has been the fact that the three groups of specialists have carried on worthwhile discussions on a subject of common interest.
When we leave here we will each carry with us a lot of new ideas
and, perhaps, an entirely different viewpoint on Arctic Communications from the one we brought to the conference. It remains to be
seen how our future activities are to be influenced by what tran-
spired here.

RELATIONSHIP BETWEEN THE USER, SYSTEMS DESIGNER, & RADIO PHYSICIST

W. L. Hatton

Defence Research Telecommunications Establishment

Ottawa, Canada

In this discussion I will try to present a communication system designers viewpoint of the interrelationship between the systems designers, the users, and the radio physicists.

The problems that the system engineer will try to solve depend on his environment. If his environment does not include input from the user, he may well try to solve problems that do not require solution; if input is not included from the physicist he may solve problems that do not exist. If the systems designer does not receive these inputs he will try to find solutions to the needs that he believes exist. His choice will be influenced by such factors as; what can be done, what is challenging, what hasn't been done, and what is theoretically possible. The results may not be wanted. It is therefore essential that the systems designer be influenced by a strong interjection of real problems to keep him on the right track.

At the same time, it must be recognized that many of our most important developments, such as FM radio and radar, originated in the minds of the inventor, and not as a result of a statement of a requirement by a user.

Because it is often impossible to anticipiate new developments, new techniques and new ideas, it is essential that user requirements be stated as an operational requirement and not as a specification for equipment. A requirement, rather than requesting an equipment or so many voice channels, should indicate the purpose for communication and how much communications is required. Engineering includes economic factors which can only be balanced with physical factors, and the right trade offs will be obtained, only if the systems designer has a complete understanding of the problem.

My first plea is therefore for the user to bring the systems designer completely into his confidence in establishing requirements.

Once the user requirement has been stated and explained, for example for captains of all ships to be able to talk to each other, a continuing dialogue must be maintained between the user and the systems designer. For example, if the designer accepts the above statement of requirement as absolute it may cost two orders of magnitude more than would be required if the designer went back to the user and determined that a 99% probability of being able to communicate would be satisfactory.

My second plea is for the user and designer to remain in contact throughout the development phase of a new communication system.

While the system designers progress depends on a better understanding of the problems that need solving it also depends on a better understanding of the physical environment. In obtaining a better understanding of this environment, the work of the radio physicist is essential to the systems designer.

There has been an increasing recognition, in the last few years, of the importance of understanding the propagation environment in the design of communication systems. I have been fortunate to work in the Radio Physics Laboratory in Ottawa where this was recognized at an early date and where Dr. Forsythe, a radio physicist, invented JANET, a meteor burst system, which was the first predetection feedback communication system. This concern with the propagation environment was a direct result of our concern with communications under difficult propagation conditions in the Auroral zone in Canada.

The dialogue between the radio systems designer and the radio physicist is equally important as the dialogue between the user and the designer. Even in such an establishment as the Canadian Defence Research Telecommunications Establishment this can prove to be difficult. At the same time as JANET was being readied for testing in the Canadian Arctic, the Polar Cap Blackout phenomena was being discovered. This fact did not really sink in until the JANET system was blacked out in its first week of trials between Edmonton and Yellowknife.

The radio physicist should be able to tell the communications engineer much about the physical characteristics of the ionosphere

as well as about the physical processes found there. Often measurements which are made to obtain a better understanding of the
ionosphere could, by small modifications, be used to obtain a better understanding of the transmission characteristics of the media.
Greater use could be made of the physicists knowledge if he took
into account the potential use of his measurements and modified
his experiments to obtain results which could be interpreted as to
how they could affect communication systems and then passed these
results to the systems designer. Similiar modifications of systems tests by systems designers would provide much information
that would be of great help to the physicist in obtaining a better
understanding of the propagation media.

My third plea therefore is for the radio physicist and the
communication systems designer to work together to improve our
knowledge of radio propagation and our ability to use it for communications.

My last plea is for continuing cooperation between the user,
the systems designer and the radio physicist which, as has been
shown at this Study Institute, will make us all stronger and wiser.

USERS VIEWPOINT CONTRIBUTION TO THE SUMMARY PANEL DISCUSSION

S. E. Probst

Defence Communications Agency

Washington, D.C. USA

I have been requested to express the needs of the communications users in the Arctic. Any brief summarization of the needs
of all users must of necessity be a gross oversimplification, however, the papers and discussions of the past several days have
suggested what may probably be a new approach to the expression of
such needs.

The lack of precisely agreed definitions for such terms as
"reliability", "continuity", "speed of service", "complexity", etc
has contributed to the confusion. I shall attempt, therefore, to
elaborate on levels of communication "needs" which have been al-
luded to these past few days and to present an outline which might
be of use to the systems designers particularly, with a few comments appended for the ionospheric physicist.

Three broadly defined and admittedly somewhat overlapping
categories of communications needs in the Arctic might be associated with the words "safety", "commercial" and "personal". Communications categorized as "safety" represent those in support of
the most critical needs and the title has been chosen to reflect
the fact that these communications all concern themselves in some
way with the safety of groups of individuals. In the hard light
of present day reality, the safety of single individuals does not
qualify. The groups of individuals, however, may range all the
way from a plane-load of people, through a community, a region, a
nation, to the population of the world. Examples of communications which qualify for the "safety" category would include communications in connection with: air traffic control aeronautical
aids to navigation (these two becoming increasingly critical as
the era of supersonic commercial aircraft approaches), highly

perishable intelligence information, warning of impending disaster, actual strike and/or retaliation orders in the nuclear age, etc. Examples of those communications which are here broadly categorized as "commercial" include those in connection with: movement and provision of goods and services, the bulk of military administrative and logistical communications, diplomatic communications, military command and control communications (except for those involving communications largely provided to allow a spreading of the responsibility for economic decisions, etc. In the category of "personal", such communications as private correspondence (including individual safety) and news and entertainment such as broadcast radio and TV, are included.

Several aspects of each communications category will now be indicated by example in an attempt to draw a better picture for the systems designer of the actual needs in each category. (In the discussion which follows, the entries on the chart of Figure 1 will be developed and discussed). Perhaps one of the most revealing aspects which can be discussed is the impact of a failure of communications in each category. The impact of a failure of "safety" communications might will be expressed in terms of the continues existence of the group of individuals involved. The impact of a failure of "commercial" communications is largely a matter of economics. The impact of a failure of "personal" communications is usually one of emotion in terms of the annoyance of the individual or individuals involved.

Speed of service is an aspect of communications that frequently relates to design requirements. Let's examine the required speed of service in each category. The speed of service requirement of "safety" communications approaches, as a limit, the length of the message to be transmitted. That is to say, these communications may be so critical that immediate delivery of message traffic through the communications systems is an actual requirement. For "commercial" communications, the required speed of service can usually be defined as, "As Soon As Possible". The feasibility of added complexity to shorten the time required is largely determined by the economic impact of delay. For "personal" communications, the requirement is best defined as "As Soon As Convenient", with a willingness to accept some delay being evident in most cases.

The system designers are invariably concerned with the kind of information to be transmitted. It is becoming more and more generally realized that for "safety" communications, digital information affords the greatest efficiency and should probably be the basis for all future planning in this area. "Commercial" communications, although tending more and more to the transmission of digital data, retains a large need for voice communications as

well, and there have been many applications of TV for commercial
communications purposes. "Personal" communications may be digital
(telegrams, etc), voice (telephone and news distribution), broader
audio (broadcast music, etc) and TV.

The bandwidth requirements vary widely across each category
but in the case of "safety" communications are clearly a function
of the actual information to be transmitted and are, in most cases, reasonably narrow except for the expansion imposed by high
speed of service requirements. The bandwidth requirement for
"commercial" communications is clearly a function of both the cost
of communications and the level of economic development and/or military defense in the Arctic region. Bandwidth requirements for
"personal" communications are directly a function of the population and the standard of living of that population.

The aspects of complexity that can be allowed in "safety"
communications are characterized by high sophistication, automation,
a large measure of redundancy (in most cases, complete redundancy)
and a maximum degree of adaptability. The allowable complexity
for "commercial" communications is largely determined by economic
factors with the goal being the maximum information transferred
per dollar invested.

On individual communications links where the volume of information to be passed is greater, a high degree of complexity is
warranted. Where volumes are lower, only simpler systems can be
justified. For "personal" communications, the prime requirement
is for simplicity of operation and dependability of terminal
equipment. That is to say, when an individual turns the equipment
knob "on", he wants it to work. The emphasis should be on a low
level of required maintenance, and simplicity of maintenance when
actually required (or perhaps individual unit replacement rather
than repair).

In terms of accuracy required in communications, the need in
the "safety" area is for absolute accuracy. Errors must be elimi-
nated through error detection and correction, through redundancy
of reception or by whatever other means can be provided, and this
absolute accuracy must be obtained within the speed of service requirement. In "commercial" communications, ultimate accuracy of
data reception is mandatory since critical economic decisions will
be affected. Some delay in attaining this ultimate accuracy can
usually be tolerated, however, telegrams with minor garbles are
usually acceptable and considerable repetition in voice communications can be tolerated.

The allowable system costs in the case of "safety" communications are dictated by policy determinations on the distribution of

the national budget. In the case of "commercial" communications, however, the profit and loss balance comes into play. That is to say, the cost of communications systems is determined by the economic impact of communications to be provided by that system and a maximization of information transferred per dollar invested is frequently, if not always, the determining criteria of system cost. The allowable costs in "personal" communications are clearly a function of the standard of living of the population to be served.

The allocation of the research budget should clearly devote the maximum effort to "safety" communications, and in fact, "safety" communications have been the justification for the majority of important research in recent years. "Commercial" communications can certainly support some portion of the research effort, but rely primarily on the results of research originally intended to support "safety" communications. Only minimal research in the area of "personal" communications can be justified in proportion to the other two categories.

Many questions have been raised by the ionospheric physicists in the past several days as to the continuing need for predictions and for forecasts services (and the need for greater accuracy in either or both). Predictions (in the sense generally agreed to in previous discussions) are certainly required by the design engineers during the design stages of all three categories of communications to aid in frequency selection, antenna design, and many other areas. The use of predictions by the operator should probably be discouraged in the future. The question of the need for forecasts and their application is determined by the degree of system adaptability provided. In the case of "safety" communications, completely automatic adaptability to changing propagation conditions should be built into the communications system with the necessary sensing devices also built in to determine when adaptation is required and to initiate the adaptation. On the other hand, the critical natur of "safety" communications may dictate complete avoidance of any need for adaptability, particularly in the Arctic. The trend in past years has certainly been to avoid the effects of the Arctic ionosphere by shifting "safety" communications wherever possible to cables and microwave relay or tropospheric scatter systems. In "commercial" communications, there will probably be developed backbone systems into the Arctic which will warrant the inclusion of automatic adaptability in response to built-in sensors. However, tributary "commercial" communications will probably warrant only a manual adaptability, for the initiation of which, forecasts will continue to be required. In the case of "personal" communications, there will probably be prime backbone and tributary routes similar in their needs to those in the "commercial" area with forecasts required on the tributary systems. The ultimate link to the final user, however,

will probably not warrant significant adaptability for many years
to come. This user is usually willing to sit and wait.

The foregoing has not tried to provide the detailed requirements of any specific Arctic communications systems, but rather,
has attempted to treat those aspects of any system which need to
be specified for the systems designer. In addition, something has
been said about each of those aspects and the manner in which they
might vary for each of three very broad categories of communications that can be expected to be required in the Arctic. What has
been said has been addressed primarily to the systems designer except for the brief comment regarding predictions and forecasts.
Even here, the need for such services will be determined by the
system that the designer produces rather then by any direct request from the user to the ionospheric physicist.

	SAFETY	COMMERCIAL		PERSONAL		
Impact of Failure	Existence	Economics		Emotions		
Speed of Service	Message Length	ASAP		ASAC		
Content	bits	bits/voice/TV		bits/voice/TV		
Band Width	f(Info)	f(Cost, Development)		f(P, SOL)		
Complexity	Sophisticated Redundant Adaptable	f(bits/$) max		Simple Equip Reliable Ease of Maintenance		
Accuracy	Absolute	Ultimate		Desirable		
Cost Determinant	Budget	Profit/Loss		SOL		
Research Effor	Maximum	Moderate		Minimum		
Reqm't for Predictions	Design	Design		Design		
Adaptability	Automate or Avoid	BACKBONE Automatic Response to Measurement	TRIBUTARY Manual Response to Forecasts	BACKBONE Automatic Response to Measurement	TRIBUTARY Manual Response to Forecasts	BRANCH Sit and Wait

Table 1.

PHYSICIST'S CONTRIBUTION TO PANEL DISCUSSION

K. Davies

Institute for Telecommunication Sciences and Aeronomy

Environmental Science Services Administration, Boulder

I believe that the answer to the communication engineer's problems must come from the engineer himself rather than the physicist. The physicist can be of great help to the engineer and, hence, the user in a consultative capacity. This interaction between the engineer and physicist is an interesting problem in (audio frequency) communications. For example, if the physicist is to be of any help he must be asked questions which he understands. Too often I feel the engineer asks a question which is couched in engineering language. Furthermore, if the engineer expects a simple answer he must be careful to ask a (physically) simple question.

The problem of education thus looms large in the dialogue between physicist and engineer (and for that matter the user). Not only is there a need to understand each other's language but it is important to appreciate the different ways of thinking. This can be illustrated by some of the discussions that have gone on at this meeting. The physicists think in terms of global patterns of particle precipitation, the engineer concentrates on correlation bandwidth, while the user is concerned with the performance of his diesel generator and/or with cooking breakfast for an airplane crew. Meetings of the sort we are just concluding are of immense value in this educational process because in them the usual trend towards specialization is reversed. Too often the physicist gives the engineer data he thinks the engineer wants whereas the engineer gives the user a piece of hardware which he thinks the user should have, instead of finding out just what the user can and will use. I think this is illustrated by the physicists' and engineers' use of the term probability: it is hard to believe that the average radio operator makes use of a statement like "the probability of an ionospheric disturbance in the next 24 hours is 35 percent". Thus there is a need for education to achieve one or both of the following ends:

(1) to enable the physicist and/or engineer to appreciate the form of information that the user can assimilate and/or

(2) to bring the user up to the level at which he can understand the sophisticated information provided by the physicists and engineers.

The former may be achieved by meetings of the present type, the latter by suitable Radio Propagation Courses.

Returning to the physicist's point of view I must confess that I am surprized to find no discussion at this meeting of the cause of ionospheric disturbances; namely the sun. To a physicist the best possibilities for successful prediction of high latitude disturbances may well come from observations and theoretical studies of solar phenomena, e.g., solar magnetic fields, plasma instabilities, etc. Thus the most useful solution to the users questions may come not from the ionospheric physicist or the radio engineer but rather from the solar astronomer.

BRIDGING THE GAP BETWEEN PHYSICIST, ENGINEER AND THE USER AND THE NEEDS FOR THE FUTURE

J. H Meek

Defence Research Telecommunications Establishment

Ottawa, Canada

During the course of this Study Institute, it has been difficult to confine the discussions strictly to Arctic problems. The
past effort in radio communications has been biased very much toward temperate and low latitude problems since most circuits cross
these regions. Very little engineering effort has been put specifically on Arctic communications systems where the main difference
is the geophysical environment.

Summing up the comments of the various speakers at this meeting, one finds a concensus of opinion that the Arctic, from a radio communications point-of-view, includes the region north of 65o
geomagnetic latitude.

In my remarks, I have the Arctic in mind although the propo-
sals for future work are not confined to the solution of Arctic
problems.

History. The present effort in ionospheric radio communications has grown out of the World War II requirement for reliable
long-distance communications. Propagation by way of the ionosphere was the most likely and most available solution but detailed knowledge of the characteristics was lacking.

Many physicists and engineers were conscripted into tasks,
analytical and observational, aimed at improving the reliability
of long distance communications circuits. For example

a)   Several countries began to prepare monthly world-wide iono-
     spheric frequency predictions. Their application to communi-

cations circuits involved some empirical formulae and some
subjective weighting factors. Results were required to serve
the immediate purpose.

b)   Solar, magnetic and ionospheric data, along with the observed
      radio circuit deterioration, were plotted regularly by some
      communications groups and were used to forecast qualitatively
      the radio communications conditions.

c)   A line of observing stations was set up from Germany through
      Norway to watch the progress of ionospheric disturbances as
      they extended southward from the auroral zone, cutting off
      radio communications.

Present Situation. Although considerable effort has been expended over the intervening years to investigate the characteristics of the upper atmosphere, the knowledge has not been applied
to produce significant improvement in forecasting and predicting
of radio circuit disturbances. With many more personnel involved
in collecting, assembling and disseminating the information, the
value of our output to the radio engineer and the operator has not
appreciably improved over the past 15 years. Nevertheless we do
have on hand enough observational information and basic knowledge
to make an improvement in our methods, which will be significant
to the operator.

In this meeting, the operators have put before us their immediate plans for coping with the problem of communications in the
Arctic. Their requirement is straightforward - 100% communications at any time and between any two specified points. You will
have noted that they have decided to use cables, land-line, troposcatter, satellites, and LF where necessary and fall back on the
ionosphere where it is the only possible means, or where they cannot afford other means.

It appears that the long path requirements are, or will be,
to a large extent covered by other communications methods. This
means that the physicist or engineer who is still convinced of the
merits of the ionosphere for Arctic communications in general,
must solve the problems to make his system competitive without ex-
pecting much support from the operator.

The requirement for HF communication will continue, for circuits with short paths, from a mobile terminal to base within the
Arctic where the mobile terminal has constraints on its equipment,
efficiency and power. One can count on continuing support from
the operator for improvements in this area.

The undisputed advantages of HF are simplicity, size of

equipment and economics. These are being offset by the complexity
which the engineer tends to build into a system to give a do-it-
all capability to the user. If he does this, he must compete with
the other large systems and the simplicity is lost.

Engineers are frequencly discouraged when their systems, worked out with considerable effort, are not accepted and implemented
by the operator. The systems engineer must take the initiative in
development of future ionospheric communications systems. To be
successful, he must have close liaison with the operating groups
in order to be sure that he produces what they want and not just
what he thinks they should have. Any system must be well developed before it will be accepted by the user.

<u>Proposed Effort</u>. Physicists should consolidate and present
the results from the large mass of data that is on hand and relate
the results of theoretical investigations to actual communications
circuit conditions.

Some things are listed that might be done:

1.   Revise synoptic ionospheric maps for the Arctic so that they
      do relate to observed communications.

2.   Investigate the types of Es, their characteristics and impor-
      tance to communications as an obstruction and as an aid. The
      typing of Es must be based on the physics of their origina-
      tion, rather than on the shape on ionograms.

3.   Produce a manual which gives the engineering data for design
      and use of Arctic LF and VLF systems (including ionospheric
      absorption and phase change effects, etc).

4.   Make a good study of the ionospheric absorption data to lead
      to information and predictions which can be used by communi-
      cators.

5.   Develop a quantitative short term ionospheric forecast which
      may be applied to circuits in various parts of the world, but
      especially polar regions. This, in my opinion, should become
      a major effort of ionospheric physicists who wish to apply
      their knowledge.

The engineer should pursue the development of real time ionosphere sampling systems with an aim of making them simple, reliable and practical for operational circuits. Some effort is required to coordinate the existing systems, e.g., CHEC and CURTS.

The physicist has been asking repeatedly "What does the operator want? The operator must not merely answer this question and

leave it to the physicist and engineer to produce a working system. There must be regular interchange of information throughout the life of a project.

It is clear that there is a need for <u>operator education</u> and I suggest the operating personnel take note of the willing attitude of the physicists and engineers at this meeting and should insist that the latter spend some time sharing their knowledge. In the process, the physicists and engineers will receive some needed education in operational matters.

<u>Human Factors</u>. One must put more attention to the human factors associated with communications systems. In the Arctic, these are principally environmental in nature, related to the climate and logistics. However, we must think more generally of the man-machine interface. It is not sufficient to ask for 100% communications reliability in the electronic and propagation parts of the system and ignore the present day glaring deficiencies in message composition, handling and delivery. All of these steps are involving humans operating at very low efficiency.

Some effort must be made to engineer systems which are designed from the start to ensure that the human and the machine work well together. The part history has required one to adapt to the other sequentially - with less than optimum results.

<u>Final Remarks</u>. It is clear that communications systems, whether in the Arctic or elsewhere, should be developed as projects involving active participation of physicists, engineers, and operators. The past habit of one group proposing a system to one of the others to take over, has not been successful and has led to many of the less-than-optimum communications systems that we put up with today.

The discussion of political and economic factors has been minimized in this meeting in order not to stifle discussion of scientific capability. One cannot limit the considerations if practical operating systems are to be conceived and developed. Perhaps we are now ready for a meeting in which such factors are purposely included. For such a meeting one might increase the number of operators attending while retaining a good representation of engineers and of basic physicists.

There has been mention of the problems of frequency management and allocation. These problems are the responsibility of I.T.U. and C.C.I.R. and I would urge that those of you who are most concerned should take an active part in your national committee of C.C.I.R.

GENERAL DISCUSSION

In a general discussion succeeding the three main contributions the following views were expressed:

(i)   It is important to learn the rules for extrapolation of ionospheric parameters to unmeasured parts of the world.

(ii)   If the physicist were aware of the sophisticated tools of the systems engineers he could use them to advantage as scientific tools.

(iii) Contours of ionospheric stability would be of considerable value for radio communications in the Arctic areas.

AUTHOR INDEX

Aarons, J., 299, 308
Akasofu, S.-I., 31, 41, 44, 185, 200
Albee, P. R., 136
Allen, K. S., 294, 299, 314
Ames, J., 221, 240
Anderson, K. A., 22, 32
Arndt, D., 68, 72
Arnold, H. R., 137
Appleton, E. V., 175, 200
Ashford, D. A., 222

Bailey, D. K., 34, 38, 43, 44
Balser, M., 229, 240, 241
Balsey, B. B., 201
Bagaryatsky, B. A., 175, 200
Barber, D., 180, 200
Barcus, J. R., 32
Baron, M. J., 204
Barrington, R. E., 68, 71
Barrow, B. B., 222
Bartels, J., 194, 200
Bartholomé, P. J., 143, 153
Bartholomew, R. R., 288
Basu, S., 314
Bates, H. F., 136, 288
Bellow, P. A., 221, 222
Belrose, J. S., 38, 44, 65, 71, 134, 136
Berg, M. R., 205
Berner, A. S., 222
Birfeld, J. G., 175, 200
Bjelland, B., 70, 72
Blackband, W. T., 129, 136
Blair, J. C., 153, 206
Blake, K. W., 272, 389
Blevis, B. C., 177, 201

Bode, L. R., 136
Booker, H. G., 162, 164, 175, 176, 201
Bourne, I. A., 72
Bowles, K. L., 196, 198, 199, 201
Boyley, D., 223
Bracewell, R. N., 127, 136
Bramley, E. N., 240
Brennan, D. G., 389
Brice, N. M., 10, 12, 15, 31
Brice, P. J., 222
Briggs, B. H., 296-299, 310, 313, 314
Brunette, G., 220
Budden, K., 107, 128, 240
Bullough, K., 195
Buneman, O., 198, 201
Burgess, B., 91, 107, 127, 137
Burrows, J. R., 22, 23, 32

Chamberlain, J. W., 175, 201, 204
Chapman, A. C., 221
Chilton, C. J., 137
Chipp, R. D., 223
Cohen, R., 196, 201
Cole, A. E., 65, 71
Coll, D. C., 341, 358
Collins, C., 26, 32, 34, 45, 153, 155, 158, 164, 177, 201, 206
Coon, R. M., 380
Corke, R. L., 137
Cosgrove, F., 223
Croisdale, A. C., 223
Crombie, D. D., 98, 107, 108, 114, 116, 137
Crompton, R. W., 66, 72
Currie, B. W., 199, 201

455